CONTROL AND DYNAMIC SYSTEMS

Advances in Theory and Applications

Volume 54

CONTRIBUTORS TO THIS VOLUME

UWE L. BERKES

JUHN-HORNG CHEN

HUBERT H. CHIN

ROY R. CRAIG, JR.

AUGUSTINE R. DOVI

WODEK GAWRONSKI

EDWIN E. HENKEL

RENÉ HEWLETT

DANIEL C. KAMMER

MORDECHAY KARPEL

AN-CHEN LEE

YI LU

RAYMOND MAR

MICHEL MARITON

T. SREEKANTA MURTHY

V. R. MURTHY

TZU-JENG SU

DAVID D. SWORDER

GREGORY A. WRENN

CONTROL AND DYNAMIC SYSTEMS

ADVANCES IN THEORY
AND APPLICATIONS

Edited by
C. T. LEONDES

School of Engineering and Applied Science
University of California, Los Angeles
Los Angeles, California
and
Department of Electrical Engineering
and Computer Science
University of California, San Diego
La Jolla, California

VOLUME 54: SYSTEM PERFORMANCE IMPROVEMENT
AND OPTIMIZATION TECHNIQUES AND
THEIR APPLICATIONS IN AEROSPACE
SYSTEMS

ACADEMIC PRESS, INC.
Harcourt Brace Jovanovich, Publishers
San Diego   New York   Boston
London   Sydney   Tokyo   Toronto

Academic Press, Inc.
1250 Sixth Avenue, San Diego, California 92101-4311

United Kingdom Edition published by
Academic Press Limited
24–28 Oval Road, London NW1 7DX

Library of Congress Catalog Number: 64-8027

International Standard Book Number: 0-12-012754-7

PRINTED IN THE UNITED STATES OF AMERICA
92 93 94 95 96 97   QW   9 8 7 6 5 4 3 2 1

CONTENTS

CONTRIBUTORS

Numbers in parentheses indicate the pages on which the authors' contributions begin.

Uwe L. Berkes, (23), *European Space Agency, 31055 Toulouse, France*

Juhn-Horng Chen, (417), *Department of Mechanical Engineering, National Chiao Tung University, Hsinchu, Taiwan, People's Republic of China*

Hubert H. Chin, (69), *Grumman Aircraft Systems Division, Grumman Corporation, Bethpage, New York 11714*

Roy R. Craig, Jr., (449), *Aerospace Engineering and Engineering Mechanics, The University of Texas at Austin, Austin, Texas 78712*

Augustine R. Dovi, (1), *Lockheed Engineering and Sciences Company, Hampton, Virginia 23666*

Wodek Gawronski, (373), *Jet Propulsion Laboratory, California Institute of Technology, Pasadena, California 91109*

Edwin E. Henkel, (341), *Space Systems Division, Rockwell International Corporation, Downey, California 90241*

René Hewlett, (341), *Space Systems Division, Rockwell International Corporation, Downey, California 90241*

Daniel C. Kammer, (175), *Department of Engineering Mechanics, University of Wisconsin-Madison, Madison, Wisconsin 53706*

Mordechay Karpel, (263), *Faculty of Aerospace Engineering Technion, Israel Institute of Technology, Haifa 32000, Israel*

An-Chen Lee, (417), *Department of Mechanical Engineering, National Chiao Tung University, Hsinchu, Taiwan, People's Republic of China*

Yi Lu, (297), *Department of Mechanical and Aerospace Engineering, Syracuse University, Syracuse, New York 13244*

Raymond Mar, (341), *Space Systems Division, Rockwell International Corporation, Downey, California 90241*

Michel Mariton, (483), *Signal and Image Processing Laboratory, MATRA, 78052 St Quentin en Yvelines Cedex, France*

T. Sreekanta Murthy, (225), *Lockheed Engineering and Sciences Company, Hampton, Virginia 23666*

V. R. Murthy, (297), *Department of Mechanical and Aerospace Engineering, Syracuse University, Syracuse, New York 13244*

Tzu-Jeng Su, (449), *Spacecraft Dynamics Branch, NASA Langley Research Center, Hampton, Virginia 23665*

David D. Sworder, (483), *Department of AMES, University of California, San Diego, La Jolla, California 92093*

Gregory A. Wrenn, (1), *Lockheed Engineering and Sciences Company, Hampton, Virginia 23666*

PREFACE

The modern era of aviation systems began with the end of World War II and accelerated further with the introduction of jet propulsion in military and commercial aircraft. The launch of the *Sputnik* satellite marked the beginning of the space age and all the many remarkable achievements that followed thereafter. As a result, the term *aerospace systems*, to include both aeronautical and space systems, was introduced.

The past decade has seen the increasingly strong rise of another trend, namely, the expanding and increasingly sophisticated utilization of system performance improvement and optimization techniques in the design of aerospace systems. These techniques include multiobjective optimization methods in the design of aerospace systems, the optimization of aerospace structures, knowledge-based system techniques for pilot aiding, optimal sensor placement techniques in aerospace systems, balanced systems and structural design, and many others. Many of these techniques are presented in this volume. These system performance improvement and optimization techniques will continue to grow and expand in their development and increasingly effective utilization in the design of future aerospace systems. The positive implications are many, including increased systems capabilities and performance, improved system reliability, safety efficiency, and cost effectiveness. As a result, this is a particularly appropriate time to treat the issue of aerospace system performance and optimization techniques in aerospace systems in this international series. Thus, this volume is devoted to the most timely theme of "System Performance Improvement and Optimization Techniques and Their Application in Aerospace Systems."

The first contribution to this volume is "Techniques for Aircraft Conceptual Design for Mission Performance Comparing Nonlinear Multiobjective Optimization Methods," by Augustine R. Dovi and Gregory A. Wrenn. In the future, virtually every new aircraft system design will utilize the techniques presented in this contribution, which, perhaps, constitutes the only such treatment available in any book.

The next contribution is "Optimization of Aerospace Structures Using Mathematical Functions for Variable Reduction," by Uwe L. Berkes. In the balanced and optimized design of aircraft and spacecraft structures by the use of finite-element procedures, a reiterative design process must be used and it should be controlled by using numerical optimization procedures. However, due to the complex search algorithms needed for structural optimization and the computer time required for the finite-element analysis, this procedure becomes very time consuming if large structures have to be optimized. In this contribution, effective means for reduction of the number of variables are presented and their effectiveness is illustrated by examples. This is, perhaps, a unique reference in the book literature on these effective and powerful design techniques.

The next contribution is "Knowledge-Based System Techniques for Pilot Aiding," by Hubert H. Chin. The need for a knowledge-based system stems from the heavy workload and demanding decision-making requirements that modern aircraft can place on a pilot. A pilot aiding system can reduce a pilot's workload in a stressful environment by taking care of low-level decisions, enabling the pilot to concentrate on the high-level decision-making processes. This contribution is an in-depth treatment of this area of potentially major significance in the development of future aerospace systems.

The next contribution is "Techniques for Optimal Sensor Placement for On-Orbit Modal Identification and Correlation of Large Aerospace System Structures," by Daniel C. Kammer. One of the key problems that will be faced by the designers of on-orbit modal identification experiments is the placement of the sensors that will gather the data. This contribution presents techniques for doing this optimally in the case that the actuator locations are already designated.

The next contribution is "Investigation on the Use of Optimization Techniques for Helicopter Airframe Vibrations Design Studies," by T. Sreekanta Murthy. Even though excessive vibrations have plagued virtually all new helicopter development, until recently there has been little reliance on the use of vibration analyses during design to limit vibration. As a result, optimization techniques have also only recently come into play and this contribution is an in-depth treatment of this major new area.

The next contribution is "Size Reduction Techniques for the Determination of Efficient Aeroservoelastic Models," by Mordechay Karpel. Aeroservoelasticity deals with stability and dynamic response of control augmented aeroelastic systems. Flight vehicles are subjected to dynamic loads, such as those caused by atmospheric gusts, which excite the aeroelastic system and cause structural vibrations. The application of various modern control design techniques, simulation, and optimization procedures require the aeroservoelastic equations of motion to be transformed into state space form. A key question the analyst faces is how many structural modes should be taken into account, and in what manner. This contribution is an in-depth treatment of these issues, and is, perhaps, a unique treatment in the book literature of this area of major significance.

The next contribution is "Sensitivity Analysis of Eigendata of Aeroelastic Systems," by V. R. Murthy and Yi Lu. In the design of engineering systems, sensitivity analyses play a key role in arriving at optimum solutions. Sensitivity analyses call for the sensitivity derivatives, which are defined as the ratios of the variations in the system characteristics to the variations in the design parameters. The sensitivity analyses of systems with stability constraints often involve the calculation of derivatives of eigenvalues and eigenvectors with respect to the design parameters. This contribution is an in-depth treatment of methods to calculate the derivatives of eigendata and, as such, will represent an important reference source for this significant issue in system design.

The next contribution is "A Simplified General Solution Methodology for Transient Structural Dynamic Problems with Local Nonlinearities," by Edwin E. Henkel, René Hewlett, and Raymond Mar. The linear transient solutions for large structural problems are usually performed in modal coordinates that are obtained from an eigensolution. There are a number of aerospace system problems that involve local nonlinear phenomena. Any practical solution technique for transient solutions for large structural problems would need to use an invariant set of eigenvalues and eigenvectors, while at the same time accurately accounting for such local nonlinear phenomena effects in the structural system's modal characteristics. Techniques for doing this are presented in this contribution.

The next contribution is "Balanced Systems and Structures: Reduction, Assignment, and Perturbations," by Wodek Gawronski. Balanced representation is defined for systems with poles at an imaginary axis or at the origin. The system grammians do not exist in this case, but introduced antigrammians exist, making balanced reduction possible. In a manner similar to grammians, antigrammians reflect controllability and observability properties of a system, but, unlike grammians, they do exist for systems with integrators, making their reduction possible. In this contribution, a reduction algorithm for systems with integrators is presented. Properties and procedures for sensor/actuator configuration are presented such that a required degree of controllability and observability is acquired. Because the issues and techniques presented in this contribution are of fundamental importance in a number of aerospace system problems, it will constitute a most valuable reference source for workers in the field.

The next contribution is "Response Only Measurement Techniques for the Determination of Aerospace System Structural Characteristics," by An-Chen Lee and Juhn-Horng Chen. Experimental modal analysis has become an increasingly important engineering tool during the past 40 years in the aerospace, automotive, and machine tool industries. This contribution presents techniques for overcoming the difficulty of nonuniqueness of mode shape in modal analysis when random input data are not or cannot be measured. Once again, because of the importance of this area, this contribution is an important element of this volume.

The next contribution is "Krylov Vector Methods for Model Reduction and Control of Flexible Structure," by Tsu-Jeng Su and Roy R. Craig, Jr. Although all real structures are, in fact, distributed-parameter systems, it is usually inevitable for the structures to be modeled as discrete systems for the purpose of design and analysis. The finite element method is the discretization approach that is most frequently used. The finite element model of a large or geometrically complicated structure may attain tens of thousands of degrees, and, therefore, model order reduction plays an indispensable role in the dynamic analysis and control design of large structures. In this contribution, Krylov vectors and the concept of parameter-matching are combined to develop model-reduction algorithms for structural dynamic systems. The significant effectiveness of these techniques is illustrated by several examples.

The final contribution is "Maneuvering Target Tracking: Imaging and Non-Imaging Sensors," by Michel Mariton and David D. Sworder. As the term is commonly interpreted, tracking refers to estimating the current state of a target from spatiotemporal observations. In this contribution, techniques are presented toward the development of new tracking algorithms that would incorporate explicit models of the maneuvering/nonmaneuvering phases of target encounter. The significant degree of effectiveness of these techniques are illustrated by a number of examples.

This volume rather clearly manifests the significance and power of the techniques that are available and under continuing development for system performance improvement and optimization in aerospace systems. The coauthors are all to be commended for their splendid contributions to this volume that will provide a significant reference source for students, research workers, practicing engineers, and others on the international scene for years to come.

Techniques for Aircraft Conceptual Design for Mission Performance Comparing Nonlinear Multiobjective Optimization Methods

Augustine R. Dovi
Gregory A. Wrenn

Lockheed Engineering and Sciences Co.
Hampton, Virginia 23666

I. INTRODUCTION

In this chapter, a recently developed technique which converts a constrained optimization problem to an unconstrained one where conflicting figures of merit may be simultaneously considered has been combined with a complex mission analysis system. The method is compared with existing single and multiobjective optimization methods. A primary benefit from this new method for multiobjective optimization is the elimination of separate optimizations for each objective, which is required by some optimization methods. A typical wide body transport aircraft is used for the comparative studies.

Aircraft conceptual design is the process of determining an aircraft configuration which satisfies a set of mission requirements. Engineers within several diverse disciplines including but not limited to mass properties, aerodynamics, propulsion, structures and economics perform iterative parametric evaluations until a design is developed. Convention limits each discipline to a subset of configuration parameters, subject to a subset of design constraints, and typically, each discipline has a different figure of merit.

Advanced design methods have been built into synthesis systems such that communication between disciplines is automated to decrease design time [1,2] The design systems of the type described in [1] and summarized in [2] are automated systems of interdisciplinary computer analysis programs. The system of [1] generates structural weights data which includes the effect of aeroelasticity for the design of advanced aircraft configurations. Some of the disciplines included within this synergistic system are loads, finite element modeling/analysis, computational fluid dynamics and flutter analysis. The initial computed weights information is typically provided to a mission performance analysis system which has the task of sizing the aircraft based on weight and performance estimates. Systems and computational codes of this nature are available at major aircraft design and manufacturing companies or research organizations.

Portions of this chapter have been reprinted from the authors' paper in the Journal of Aircraft, Volume 27, Number 12, December 1990, Copyrights © AIAA, used with permission.

Each discipline may select its own set of design goals and constraints resulting in a set of thumbprint and/or carpet plots from which a best design may be selected within constraint boundaries and mission requirements. In addition, the conceptual design problem has been demonstrated to be very amenable to the use of formal mathematical programming methods, and these algorithms have been implemented to quickly identify feasible designs [3,4,5].

The purpose of this chapter is to investigate the use of multiobjective optimization methods for conceptual aircraft design where conflicting figures of merit are considered simultaneously. Three multiobjective methods [6,7,8] have been combined with a complex mission analysis system [5]. These methods convert a constrained optimization problem to an unconstrained one. Design trade-offs using these methods are compared with single objective results. In addition, the methods are compared against each other. Classical parametric results of the design space are presented showing the optimum region bounded by the most influencial constraints. The aircraft chosen for this investigation is a typical wide body transport, for which copious data is readily available from the manufacturer and in the open literature.

II. GENERAL MULTIOBJECTIVE OPTIMIZATION

The constrained multiobjective optimization problem stated in conventional formulation is to minimize

$$F_k(X), \qquad k = 1 \text{ to number of objectives} \qquad (1)$$

such that,

$$g_j(X) \le 0, \qquad j = 1 \text{ to number of constraints}$$

and

$$x^l_i \le x_i \le x^u_i, \qquad i = 1 \text{ to number of design variables}$$

where,

$$X = \{x_1, x_2, x_3, ..., x_n\}^T, \qquad n = \text{number of design variables}$$

The fundamental problem is to formulate a definition of $F_k(X)$, the objective vector, when its components have different units of measure thereby reducing the problem to a single objective. Several techniques have been devised to approach this problem [7]. The methods selected for study in this chapter transform the vector of objectives into a scalar function of the design variables. The constrained minimum for this function has the property that one or more constraints will be active and that any deviation from it will cause at least one of the components of the objective function vector to depart from its minimum, the classic Pareto-minimal solution [9,10]. One should add that multiobjective optimization results are expected to vary depending on the method of choice since the conversion method to a single scalar objective is not unique.

III. FORMULATION OF THE MISSION/PERFORMANCE OPTIMIZATION PROBLEM

The purpose of the optimization is to rapidly identify a feasible design to perform specific mission requirements, where several conflicting objectives and constraints are considered. The use of a formal optimizer takes into account the synergism of all disciplines considered in the design process. A primary benefit from multiobjective conceptual design optimization is the ability to obtain designs which consider the influence of all disciplines simultaneously yet each discipline may formulate its own objective or figure of merit and governing constraints . When using optimization methods the design function is transferred from the analysis, that task is given over to the optimization. An example of this is the fuel volume limit of the fuselage. The analysis computes a required fuel volume to complete the mission, this is termed demand , this is compared with the available fuel volume limit, this is termed capacity. The ratio of demand versus capacity is supplied to the optimizer as a computable function which acts as a constraint to the design. Other constraints on the mission may be formulated by the analysis and passed to the optimizer in a similar manner as either equality or inequality constraints. This subject will be discussed later in this chapter in more detail. The point made here is that sizing and performance limitations are treated as constraints, thus placing the design effort on the optimizer and relieving the analysis of this computational effort.

The aircraft type selected for this study is a typical wide body transport, Fig. 1, in the 22680 kg weight class [11].

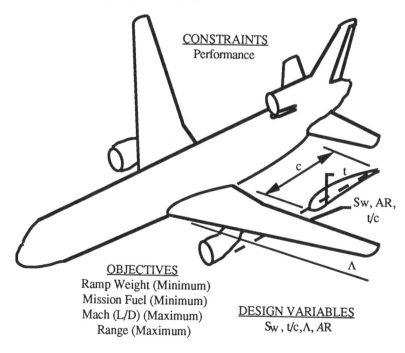

CONSTRAINTS
Performance

OBJECTIVES
Ramp Weight (Minimum)
Mission Fuel (Minimum)
Mach (L/D) (Maximum)
Range (Maximum)

DESIGN VARIABLES
S_W , t/c, Λ, AR

Fig. 1 Wide-body transport.

The aircraft has three high-bypass ratio turbofan engines, with 6915 newtons thrust each. The mission requirements are design range = 7413.0 km; cruise Mach number = 0.83; cruise altitude = 11.9 km; payload = 42185.0 kg; number of passengers and crew = 256. The primary and reserve mission profiles are shown in Fig. 2.

The design variables considered, Fig 1, are aspect ratio (AR), area (S_w),

quarter chord sweep (Λ) and thickness to chord ratio (t/c) of the wing, where the initial values chosen for all cases are

$$X_o = \left\{ \begin{array}{c} AR \\ S_w \\ \Lambda \\ t/c \end{array} \right\} = \left\{ \begin{array}{c} 11.0 \\ 361.0 \ m^2 \\ 35.0 \ deg \\ 0.11 \end{array} \right\}$$

Using the classical parametric design approach the design variables used for sizing a wing would be wing loading W/S_w, defined as the ratio of the gross weight to the wing reference area and the wing thickness to chord ratio t/c. This creates a non-convex design space where local minima exist and a range of designs provide the same gross weight. Using the wing area S_w creates a convex feasible design space which has a distinct optimum occurring at a vertex in the design space. Thus, the concept of formulating the optimization problem such that the design space does not contain local minima is a consideration to be evaluated by the designer, but may not always be possible [12].

The objectives to be minimized or maximized for this investigation include

$F_1(X)$ = ramp weight (minimize)

$F_2(X)$ = mission fuel (minimize)

$F_3(X)$ = lift-to-drag ratio at constant cruise Mach number (maximize)

$F_4(X)$ = range with fixed ramp weight (maximize)

Minimum ramp weight is typically a figure of merit for structures and performance. Minimum mission fuel is a factor in the direct operating cost. Maximum lift to drag ratio is a figure of merit for aerodynamic efficiency. Maximum range at a fixed ramp weight is an objective influenced by aerodynamic efficiency and fuel requirements. For this chapter the terms performance function, figure of merit and objective function have the same definition and are used interchangeably.

The functions to be maximized were formulated as negative values so that they could be used with a minimization algorithm. These objectives are first optimized for feasible single objective designs. The objectives are then considered simultaneously for multiobjective designs. Tables Ia and Ib list fourteen cases, six multiobjective and eight single objective, along with the unconstrained objective function formulation used for each. Each of the three formulations use the Davidon-Fletcher-Powell variable metric optimization method to compute the search direction S for finding a local unconstrained minimum of a function of many variables $F(X + \alpha S)$ where X is the design variable vector and α is a magnitude indicating the amount of design change [13]. This first order method replaces the local hessian $H^{-1}q$ by an approximate metric Mq. This eliminates the need for second derivatives and matrix inversions. The metric Mq is improved with each iteration and converges to $H^{-1}min$. This optimization method when used with the penalty function or KSOPT formulations described later tends to smooth out sharp peaks and valleys observed in constrained problems. Unconstrained minima are therefore less prone to converge at local minima due to the exisitence of non-convex constraints as discussed earlier. This allows the solution a better chance to converge at the global optimum.

PRIMARY MISSION PROFILE

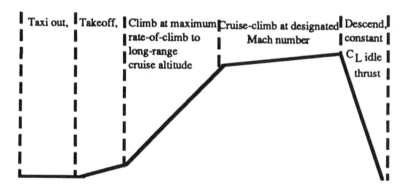

RESERVE PROFILE DOMESTIC OPERATIONS

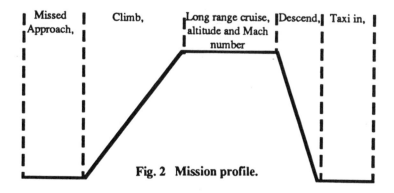

Fig. 2 Mission profile.

The inequality behavioral constraints used in each case are

$g_1(X)$ = lower limit on range, (1853.2 km)

$g_2(X)$ = upper limit on approach speed, (280.0 km/h)

$g_3(X)$ = upper limit on takeoff field length, (2700.0 m)

$g_4(X)$ = upper limit on landing field length, (2700.0 m)

$g_5(X)$ = lower limit on missed approach climb gradient thrust, (3458.0 newtons)

$g_6(X)$ = lower limit on second segment climb gradient thrust, (3458.0 newtons)

$g_7(X)$ = upper limit on mission fuel capacity, (fuel capacity of wing plus fuselage)

Table Ia.   Multiobjective cases.

Case number	KSOPT multiobjectives
1	$F_1(X); F_2(X)$
2	$F_1(X); F_2(X); F_3(X)$
	Penalty method weighted composite multiobjectives
3	$F_1(X) + F_2(X)$
4	$F_1(X) + F_2(X) + 10{,}000\, F_3(X)$
	Global criterion method target objectives
5	$F_1^T(X) = 201{,}629.0$ kg; $F_2^T(X) = 60{,}954.0$ kg
6	$F_1^T(X) = 201{,}629.0$ kg; $F_2^T(X) = 60{,}954.0$ kg; $F_3^T(X) = M\,(28.1)$

Table Ib.   Single Objective Cases.

Case number	KSOPT single objectives
7	$F_1 (X)$
8	$F_2 (X)$
9	$F_3 (X)$
10	$F_4 (X)$
	Penalty method single objectives
11	$F_1 (X)$
12	$F_2 (X)$
13	$F_3 (X)$
14	$F_4 (X)$

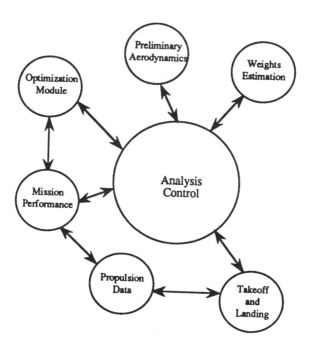

Fig. 3   FLOPS primary modules.

Table II.   Single Objective Design Results.

	X_0 initial conditions	Final Values Misson Fuel (minimize)		Final Values Ramp Weight (minimize)	
		Case 8 KSOPT	Case 12 PF	Case 7 KSOPT	Case 11 PF
Design variables					
AR, x_1	11.00	18.20	18.94	11.35	11.10
Sw, x_2, m^2	361.0	304.0	295.0	281.1	281.4
Sweep, x_3, deg	35.00	26.16	27.62	22.00	22.22
t/c, x_4	0.11	0.091	0.0913	0.0996	0.0989
Objective functions					
Ramp Weight, $F_1(X)$, kg	207729.0	219248.0	220155.0	201629.0	201763.0
Mission Fuel, $F_2(X)$, kg	67136.0	60954.0	60728.0	66891.0	66981.0
M (L/D), $F_3(X)$	0.83 (19.34)	0.83 (24.50)	0.83 (24.76)	0.83 (18.92)	0.83 (18.78)
Range, $F_4(X)$, km	7413.0	7413.0	7413.0	7413.0	7413.0
Constraints					
g_1	-	- 1.0	-1.0	-1.0	-1.0
g_2	-	- 0307	-.0140	-.0327	-.0332
g_3	-	- 0601	-.0181	-.114	-.129
g_4	-	- 0227	-.00166	-.0326	-.444
g_5	-	- 568	-.554	-.00227	-.532
g_6	-	- 459	-.451	-.217	-.211
g_7	-	- 0249	-.000754	-.0293	-.0266
Other quantities					
Span, (b), m	63.0	74.3	70.6	56.5	56.5
L/D	19.34	24.50	24.76	18.92	18.78
W/S	117.80	147.60	152.70	147.00	146.80
T/W	0.327	0.318	0.309	0.337	0.337
Function evaluations	-	483	255	158	146

Table II. (Concluded) Single Objective Design Results.

	X_o initial conditions	Final Values Mach L/D (maximize)		Final Values Range (maximize)	
		Case 9 KSOPT	Case 13 PF	Case 10 KSOPT	Case 14 PF
Design variables					
AR, x_1	11.00	22.13	22.14	10.68	10.39
Sw, x_2, m^2	361.0	381.0	361.0	331.0	361.0
Sweep, x_3, deg	35.00	30.21	36.39	22.00	22.22
t/c, x_4	0.11	0.087	0.107	0.099	0.098
Objective functions					
Ramp Weight, F_1 (X), kg	207729.0	256156.0	239332.0	219248.0	219248.0
Mission Fuel, F_2 (X), kg	67136.0	62791.0	62882.0	79492.0	79040.0
M (L/D), F_3 (X)	0.83 (19.34)	0.83 (28.09)	0.83 (22.86)	0.83 (19.25)	0.83 (19.31)
Range, F_4 (X), km	7413.0	7413.0	7413.0	8974.0	8922.0
Constraints					
g_1	-	- 1.0	-1.0	-0.210	-0.203
g_2	-	- 0.0636	-0.0699	-0.0701	-0.0910
g_3	-	-0.410	-0.0781	-0.112	-0.165
g_4	-	- 0.0479	-0.0628	-0.0626	-0.0870
g_5	-	-0.565	-0.601	-0.391	-0.403
g_6	-	-0.475	-0.509	-0.110	-0.112
g_7	-	-0.102	-0.114	-0.126	-0.0635
Other quantities					
Span, (b), m	63.0	91.8	89.36	59.40	61.2
L/D	19.34	28.09	22.86	19.25	19.31
W/S	117.80	137.70	135.9	136.00	129.80
T/W	0.327	0.280	0.284	0.310	0.310
Function evaluations	-	213	242	190	180

where the constraint functions g_j are written in terms of computable functions stated as demand(X) and capacity. These functions provide the measure of what a design can sustain versus what it is asked to carry:

$$g_j(X) = \text{demand}(X)/\text{capacity} - 1 \qquad (2)$$

In addition, side constraints were imposed on wing sweep and wing area in the form of upper and lower bounds.

IV. DESCRIPTION OF THE ANALYSIS SYSTEM FOR MISSION PERFORMANCE

The Flight Optimization System (FLOPS) is an aircraft configuration optimization system developed for use in conceptual design of new transport and fighter aircraft and the assessment of advanced technology [5]. The system is a computer program consisting of four primary modules shown in Fig.3: weights, aerodynamics, mission performance, and takeoff and landing. The weights module uses statistical data from existing aircraft which were curve fit to form empirical wing weight equations using an optimization program. The transport data base includes aircraft from the small business jet to the jumbo jet class. Aerodynamic drag polars are generated using the empirical drag estimation technique [14] in the aerodynamics module. The mission analysis module uses weight, aerodynamic data, and an engine deck to calculate performance. Based on energy considerations, an optimum climb profile is flown to the start of the cruise condition. The cruise segment may be flown for maximum range with ramp weight requirements specified; optimum Mach number for maximum endurance; minimum mission fuel requirements; and minimum ramp weight requirements. Takeoff and landing analyses include ground effects, while computing takeoff and landing field lengths to meet Federal Air Regulation (FAR) obstacle clearance requirements.

V. DESCRIPTION OF OBJECTIVE FUNCTION FORMULATION METHODS

A. ENVELOPE FUNCTION FORMULATION (KSOPT)

This algorithm is a recently developed technique initially used in structural optimization for converting a constrained optimization problem to an unconstrained one [6] and is easily adaptable for multiobjective optimization [15]. The conversion technique replaces the constraint and objective function boundaries in n-dimensional space with a single smooth surface. The method is based on a continually differentiable function [16],

$$KS(X) = \frac{1}{\rho} \log_e \sum_{k=1}^{K} e^{f_k(X)} \qquad (3)$$

where $f_k(X)$ is a set of K objective and constraint functions and ρ controls the distance of the KS function surface from the maximum value of this set of functions evaluated at X. Typical values ρ range from 5 to 200. The KS function defines an envelope surface in n-dimensional space representing the influence of all constraints and objectives of the mission analysis problem. The initial design may begin from a feasible or infeasible region.

B. GLOBAL CRITERION FORMULATION

The optimum design is found by minimizing the normalized sum of the squares of the relative difference of the objectives and are referred to as fixed target objectives. Computed values then attempt to match the fixed target objectives. Written in the generalized form

$$F^*(X) = \sum_{k=1}^{K} \left[\frac{F_k^T(X) - F_k(X)}{F_k^T(X)} \right]^2 \qquad (4)$$

where F^T_k is the target value of the kth objective and F_k is the computed value. F^* is the Global Criterion performance function [7]. The performance function F^* was then minimized using the KSOPT formulation described earlier.

C. UTILITY FUNCTION FORMULATION USING A PENALTY FUNCTION METHOD

The optimum design is found by minimizing a utility function stated as

$$F^*(X) = \sum_{k=1}^{K} w_k F_k(X) \qquad (5)$$

where w_k is a designer's choice weighting factor for the kth objective function, F_k, to be minimized. This composite objective function is included in a quadratic extended interior penalty function [17]. This function is stated in generalized form as

$$\tilde{F}(X, r_p) = F^*(X) - r_p \sum_{j=1}^{m} G_j(X) \qquad (6)$$

and

$$G_j(X) = \begin{cases} \dfrac{1}{g_j(X)} & \text{for } g_j(X) \geq \varepsilon \\ 2\varepsilon - g_j(X) & \text{for } g_j(X) < \varepsilon \end{cases}$$

where the $r_p \sum\limits_{j=1}^{m} G_j(X)$ term penalizes $\widetilde{F}(X, r_p)$, the performance function in proportion to the amount by which the constraints are violated and ε is the designer's choice transition parameter. The value of the penalty multiplier, r_p, is initially estimated based on the type of problem to be solved and is varied during the optimization process. The penalty multiplier, r_p, is made successively smaller (referred to as the draw down) to arrive at a constrained minimum.

VI. RESULTS AND DISCUSSION

A. SINGLE OBJECTIVE FUNCTION OPTIMIZATION

Single objective results for two of the methods are presented, the envelope function KSOPT and the classic penalty function PF methods. Single objective cases were run to establish a base line for comparison of multiobjective performance. In addition, target objectives are obtained for the Global Criterion Method. Final optimization values are presented in Table II for both methods. Both techniques converged to very similar designs for all cases listed in Table Ib. Greatest modifications from the initial design are seen in lift-to-drag ratio (L/D), cases 9 and 13 and range, cases 10 and 14.

Lift-to-drag was modified by increasing the aspect ratio and wing area thus minimizing the wing loading (W/S). Thrust-to-weight requirements (T/W) increased due to the larger ramp weight. In addition, the wing was made thinner and the wing sweep angle reduced. The KSOPT method converged to a 23% higher L/D versus the PF method. This is typically due to the way constraint boundaries are followed.

Range improvements, cases 10 and 14, were accomplished by reducing the sweep angle of the wing to the lower limit allowed and wing volume was adjusted to carry the maximum fuel load with reserves at the penalty of increased ramp weight. In addition, both optimizers reduced wing thickness, area and aspect ratio from there initial values. Wing loading was kept at a minimum. KSOPT again produced a slightly better design compared with the PF method.

To minimize mission fuel requirements, cases 8 and 12, the aspect ratio was increased, and the wing area was decreased. In addition, the wing sweep angle was reduced and the wing made thinner. This design improved aerodynamic performance by over 20% from the initial value while ramp weight increased slightly. The PF method converged to a slightly better design for this case.

Ramp weight, cases 7 and 11, was decreased by reducing the wing sweep angle to the lower limit of 22.0 degrees. Aspect ratio is essentially unchanged from the initial condition design point. The wing thickness was decreased, along with a decrease in area. Aerodynamic performance was not penalized significantly from the initial design value. KSOPT produced a slightly lower ramp weight.

The chart in Fig. 4 summarizes the final design objective's percent change from their initial design point for the KSOPT and PF methods.

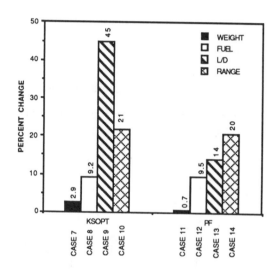

Fig.4. Single objective optimization change from initial conditions.

B. PARAMETRIC RESULTS OF THE DESIGN SPACE

Point designs, obtained parametrically, for minimum ramp weight, minimum mission fuel and maximum Mach (L/D) are shown in Fig. 5 through Fig. 7. Wing aspect ratio and thickness-to-chord ratio were varied, while other design variables were set to optimum values given in Table II, Case 8, Case 7 and Case 9, respectively. The design space is shown with the most critical constraints or criteria governing the design. To arrive at the optimum point designs shown by traditional parametric trade studies, over 256 evaluations would have been required. The parametric design is typically done by varying the independent design variables throughout their specified range and plotting the results obtained from a sizing program. This is highly time consuming and labor intensive.

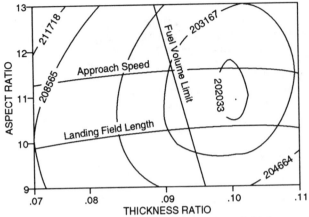

Fig. 5 Ramp weight as a function of aspect ratio and thickness ratio.

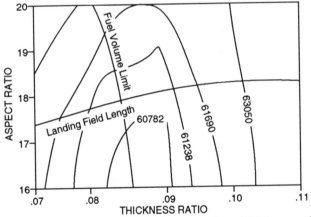

Fig. 6 Mission fuel as a function of aspect ratio and thickness ratio.

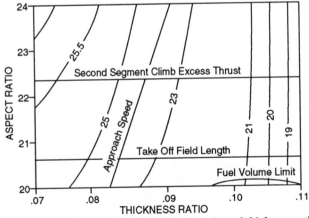

Fig. 7 Mach (L/D) as a function of aspect ratio and thickness ratio.

C. MULTIOBJECTIVE OPTIMIZATION

Multiobjective optimization considers all conflicting design objectives and constraints simultaneously to meet mission specifications. Three methods are compared, the envelope function KSOPT, the Penalty Function (PF) method and Global Criterion (GC) method. Feasible designs were obtained for two objectives, Table III, and three objectives, Table IV, satisfying all constraints.

Table III. Two Objective Design Results.

	Ramp Weight and Mission Fuel (minimize)		
	Case 1 KSOPT	Case 3 PF	Case 5 GC
	Design Variables		
AR, x_1	14.51	10.28	12.31
Sw, x_2, m^2	289.0	369.0	282.0
Sweep, x_3, deg	24.50	22.17	22.00
t/c, x_4	0.0948	0.0946	0.0958
	Objective functions		
Ramp Weight, F_1 (X), kg	206268.0	205499.0	202360.0
Mission Fuel, F_2 (X), kg	62353.0	65803.0	64647.0
M (L/D), F_3 (X)	.83 (21.77)	.83 (19.84)	.83 (19.97)
Range, F_4 (X), km	7413.0	7413.0	7413.0
	Constraints		
g_1	-1.0	-1.0	-1.0
g_2	-0.0352	-0.148	-0.0334
g_3	-0.116	-0.349	-0.120
g_4	-0.0334	-0.161	-0.0334
g_5	-0.537	-0.529	-0.481
g_6	-0.385	-0.263	-0.281
g_7	-0.0341	-0.254	-0.0243
	Other quantities		
Span, (b), m	64.7	58.9	55.5
L/D	21.77	19.84	19.97
W/S	146.20	114.0	146.80
T/W	0.329	0.331	0.336
Function Evaluations	325	121	98

Table IV.   Three Objective Design Results.

	Ramp Weight and Mission Fuel (minimize) and M (L/D) (maximize)		
	Case 2 KSOPT	Case 4 PF	Case 6 Global Criteria
	Design variables		
AR, x_1	16.87	15.49	11.643
Sw, x_2, m^2	365.0	291.0	286.0
Sweep, x_3, deg	26.39	22.12	24.20
t/c, x_4	0.083	0.089	0.099
	Objective functions		
Ramp Weight, F_1 (X), kg	228716.0	210065.0	202162.0
Mission Fuel, F_2 (X), kg	62041.0	61564.0	65980.0
M (L/D), F_3 (X)	0.83 (25.09)	0.83 (22.81)	0.83 (19.29)
Range, F_4 (X), km	7413.0	7413.0	7413.0
	Constraints		
g_1	-1.0	-1.0	-1.0
g_2	-0.0956	-0.0305	-0.0405
g_3	-0.171	-0.0921	-0.134
g_4	-0.0961	-0.0267	-0.0416
g_5	-0.599	-0.543	-0.466
g_6	-0.465	-0.402	-0.249
g_7	-0.118	-0.0839	-0.0485
	Other quantities		
Span, (b), m	78.4	63.4	54.4
L/D	25.09	22.08	19.29
W/S	128.50	147.60	144.60
T/W	0.297	0.323	0.336
Function Evaluations	62	174	73

Table V.   Best Single Objective Results.

Case	Objective	Method	Final value
12	Fuel	PF	60,728.0 kg
7	Weight	KSOPT	201,629.0 kg
9	L/D	KSOPT	28.09

D. COMPARISON OF TWO OBJECTIVE WITH SINGLE OBJECTIVE DESIGN

Fig. 8 shows the percent deviation or compromise from each method's single objective design. KSOPT treated ramp weight and mission fuel equally where the PF and GC methods favored ramp weight, preferring to pay a larger penalty for mission fuel. This behavior is expected with the PF and GC methods since the ramp weight is larger in magnitude giving this objective greater influence. This effect could have been eliminated by judicious normalization or weighting.

E. COMPARISON OF THREE OBJECTIVE WITH SINGLE OBJECTIVE DESIGN

Fig. 9 shows the percent deviation or compromise from each method's single objective design. KSOPT traded aerodynamic efficiency L/D and ramp weight to keep fuel requirements down. The PF method weighted L/D to a greater extent since the weighting coefficient w_k was 10,000, with small penalties in ramp weight and mission fuel. The GC penalty behavior is similar to the two objective results in that the ramp weight was weighted more over mission fuel and aerodynamic efficiency. The overall compromise is lowest for the PF method.

F. COMPARISON WITH OVERALL BEST SINGLE OBJECTIVE DESIGNS

The best single objective design results are listed in Table V along with objectives and methods. Since L/D was not part of the objective function set, Fig. 10, two objective compromised results behaved very similar to Fig. 8. Three objectives, Fig. 11, caused the design space to be more constrained. KSOPT again traded ramp weight and L/D to keep mission fuel requirements down. The PF method traded in a similar way but compromised L/D to a greater extent. The GC method gave more priority to ramp weight because of its magnitude. The overall compromise of KSOPT and PF were about the same at 26.3 and 24.4 percent respectively and the GC method 40.4 percent.

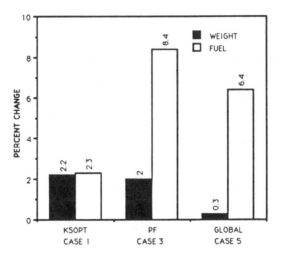

Fig. 8 Two objective optimization compromise from single objective cases.

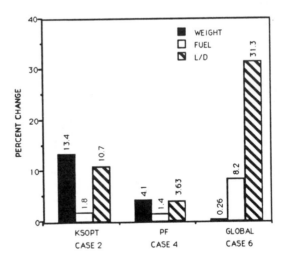

Fig. 9 Three objective optimization compromise from single objective cases.

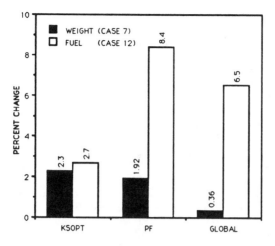

Fig. 10 Two objective optimization compromise from best single objective cases.

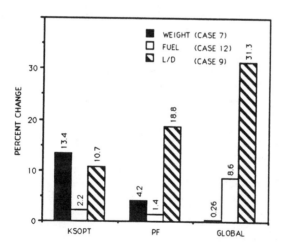

Fig. 11 Three objective optimization compromise from best single objective cases.

VII. CONCLUSIONS

A typical wide body subsonic transport aircraft configuration was used to investigate the use of three multiobjective optimization methods, 1) an envelope of constraints and objectives, KSOPT, 2) a Penalty Function and 3) the Global Criterion. The methods were coupled with a complex mission performance analysis system. The optimizer used with all three methods was the Davidon-Fletcher-Powell variable metric method for unconstrained optimization. Multiobjective compromised solutions were obtained for two and three objective functions. Feasible designs for each objective were also obtained using single objective optimization as well. The initial value design variable vector Xo and the constraints g1 through g7 were the same for all cases in this comparative study.

The KSOPT method was able to follow constraint boundaries closely and considered the influence of all constraints and objectives in a single continuously differentiable envelope function. KSOPT defines the optimum such that the function component with the greatest relative slope dominates the solution. The PF method also produced feasible designs similar to the KSOPT final designs for single objective optimization. This method, however, weights the individual objective functions in the multiobjective cases. The GC method is usually applied to multiobjective problems but may be used in the single objective problem if a target objective is supplied. This would be equivalent to imposing an upper or lower bound on the performance function. The GC method has a disadvantage in resource requirements, requiring separate single objective optimizations to provide target objectives.

Computational effort has been measured in functional evaluations, shown in the tables of results. They are defined as the number of calls to the analysis procedures from the optimization procedures. Function evaluations are very similar for single objective cases except for mission fuel using KSOPT. This deviation is due to the methods implementation, convergence criteria and the way constraint boundaries are followed. The multiobjective tables show the GC method with the least functional evaluations, however single objective function evaluations must be included with these values thereby making it the most costly in terms of number of analyses.

All of the methods produced feasible solutions within the design space. Attributes of the methods, such as ease of use, data requirements and programming should also be considered when evaluating their performance along with computational efficiency. Many cases have been compared, too numerous to report herein, where initial design variables were changed up to 40 percent above and below the initial values given in this chapter. KSOPT, a recently developed technique, continued to perform in a robust manner compared to the penalty function method, producing similar final designs within 1 percent of the mean. Based on the results of this study and the above considerations, KSOPT is thus concluded to be a viable general method for multiobjective optimization. Finally, one should add that multiobjective optimization results are expected to vary depending on the method of choice.

VIII. REFERENCES

1. N. A. Radovcich, "Some Experiences in Aircraft Aeroelastic Design Using Preliminary Aeroelastic Design of Structures [PADS]", Recent Experiences in Multidisciplinary Analysis and Optimization, Part 1, NASA CP-2327, (1984), pp. 455-503.

2. F. K. Ladner, A.J. Roch, "A Summary of the Design Synthesis Process", SAWE paper No. 907 presented at the 31st Annual Conference of the Society of Aeronautical Weight Engineers; Atlanta, Georgia, (1972).

3. B. A. M. Piggott, and B. E. Taylor, "Appication of Numerical Optimization Techniques to the Preliminary Design of a Transport Aircraft", British Royal Aircraft Establishment, Fanborough, England, UK, TR-71074, (1971).

4. S. M. Sliwa, and P. D. Arbuckle, "OPDOT: A Computer Program for the Optimum Preliminary Design of a Transport Airplane", NASA TM-81857, (1980).

5. L. A. McCullers, "FLOPS - Flight Optimization System", Recent Experiences in Multidisciplinary Analysis and Optimization, Part 1, NASA CP-2327, (1984), pp. 395-412.

6. G.A. Wrenn, "An Indirect Method for Numerical Optimization Using the Kreisselmeier-Steinhauser Function", NASA CR-4220, (1989).

7. S.S. Rao, "Multiobjective Optimization in Structural Design with Uncertain Parameters and Stochastic Processes", AIAA Journal, Vol. 22, No. 11, (1984), pp. 1670-1678.

8. R. L. Fox, Optimization Methods for Engineering Design, Addison-Wesley Publishing Company, Inc., Menlo Park, CA, (1971), pp. 124-149.

9. L. A. Zadeh, "Optimality and Non-Scalar-valued Performance Criteria", IEEE Transactions on Automatic Control Vol. AC-8, No. 1, (1963), pp. 59, 60.

10. V. Pareto, "Cours d'Economie Politiques Rouge", Lausanne, Switzerland, (1896).

11. L. K. Loftin, Jr., "Quest for Performance the Evolution of Modern Aircraft", NASA SP-468, (1985), pp. 437-452.

12. K. Svanberg, "On Local and Global Minima in Structural Optimization", "New Directions in Optimal Structural Design", Edited by Atrek, E., et. al., John Wiley & Sons, New York, New York, (1984), pp. 327-341.

13. W. C. Davidon, "Variable Metric Method for Minimization", Argonne National Laboratory, University of Chicago, ANL-5990 Rev., (1959).

14. R. C. Feagin, and W. D. Morrison, Jr., "Delta Method An Empirical Drag Buildup Technique", NASA CR-15171, (1978).

15. J. Sobieski-Sobieszczanski, A. R. Dovi, G.A. Wrenn, "A New Algorithm for General Multiobjective Optimization", NASA TM-100536, (1988).

16. G. Kreisselmeier, R. Steinhauser, " Systematic Control Design by Optimizing a Vector Performance Index", International Federation of Active Controls Symposium on Computer Aided Design of Control Systems, Zurich, Switzerland, (1979).

17. J. H. Cassis, and L. A. Schmit, "On Implementation of the Extended Interior Penalty Function", International Journal for Numerical Methods in Engineering, Vol. 10, (1976), pp.. 3-23.

Optimization of Aerospace Structures using Mathematical Functions for Variable Reduction

Uwe L. Berkes

European Space Agency
Toulouse, France

I. Introduction

A lightweight design, low operation costs, and advanced mission capabilities are major drivers when developing aircraft and spacecraft structures. To enable reaching these objectives, the development of those vehicles is based upon sophisticated analysis tools which allow results to be obtained with the necessary accuracy even in pre-design phases. Within structural design in particular, the stresses and strains of the structures are analyzed using finite-element methods. Since those finite-element procedures do not size structures, but analyze them based on boundary conditions, such as loads, the design process is an iterative one. Initial conditions have to be assumed, the performance has to be analyzed and, depending on the result, re-iterations have to be performed. This process may be controlled using numerical optimization procedures. But due to the complex search algorithms needed for structural optimization and the required computer time needed for the finite-element analysis, this procedure becomes very time consuming if large structures have to be optimized.

Besides utilization of powerful computers and implementation of sophisticated optimization algorithms, reduction of the variable number within the design process yields best results when reduction of the computer time is required. Figure 1 shows the effect of variable reduction on the computer time needed for thickness optimization of a wing-box structure [1,2,3]. A coarse finite-element model using 430 elements, with each element thickness being a design variable, yields 111 minutes computer time. If the variables are reduced to 57, the computer time needed is only 27 minutes. Thus, a reduction of a factor of 4.2 can be stated. With a large finite-element representation of the wing-box using 4300 elements, the reduction of computer time is even larger, a factor of 7.4 results when variables are reduced from 4300 to only 57. Coupling of the variables is a frequently used method for variable reduction. A set of variables is coupled to a master variable and the condensed variable set is then treated as one variable in the optimization process. This method is known as 'variable slaving' or 'variable coupling' and it may be useful in cases where manufacturing constraints have to be respected, such as a constant thickness in a certain structural area. However, reduction of the variable number by variable slaving reduces the variational freedom, thus, this procedure cuts feasible parts of the design space that are useful for optimization.

Current procedures for structure shape optimization make use of mathematical functions for coupling the node-point locations [4,5]. The mathematical function is controlled by a set of control points, or its coefficients, which are treated in the optimizer instead of the node-point locations themselves. The locations of node-points are then extrapolated from the mathematical function. Thus, the number of variables is reduced, the optimization process is smoothed, but the variational freedom remains virtually unchanged, if the chosen mathematical function allows the global optimum to be represented.

A similar method is applied to thickness optimization (sizing) of large aerospace structures and is described in this contribution. The expected advantage is to reduce the computer time needed for optimization of the thickness distribution, without reducing significantly the variational freedom when mathematical functions are applied which allow optimum thicknesses to be represented.

First, implementation of mathematical functions into the analyzer-optimizer loop of structural optimization is discussed in chapter II. Suitable mathematical functions are then described in chapter III and it is shown how to implement them for one-dimension and multiple-dimensions. Optimization of an aircraft wing-box structure using this method of variable reduction is presented in chapter IV. Several functions are introduced within the optimization process and the results obtained are compared. The range of examples is enlarged in chapter V for delta shaped wing-boxes and form optimization. Chapter VI gives recommendations and summarizes 'Cookbook' methods for the procedure of variable reduction using mathematical functions for optimization of large structural systems.

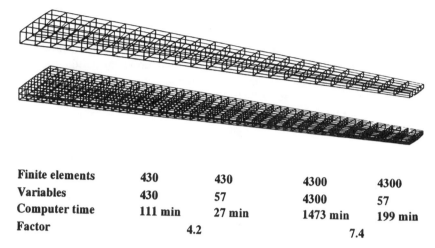

Finite elements	430	430	4300	4300
Variables	430	57	4300	57
Computer time	111 min	27 min	1473 min	199 min
Factor		4.2		7.4

Times are given for 10 iteration cycles on CDC 180-860

Figure 1: Comparison of computer time using variable reduction for optimization

Figures 1 through 4 adapted / reprinted from the author's paper in the Journal of Aircraft, Volume 27, Number 12, December 1990, Copyright © by the American Institute of Aeronautics & Astronautics. Used with permission.

II. Structural Optimization and Variable Reduction

Structural optimization itself may be categorized into two major classes, sizing of a structure and shaping of a structure. Within the first class of optimization the geometry remains fixed and the structure is sized, whereas in the second class the geometry is optimized and the size, e.g. thickness, remains fixed. A typical problem for the first class of aerospace structural optimization is mass-minimization of a wing-box structure by optimizing the thickness distribution considering a maximum stress level in the structure and a minimum flutter frequency (sizing problem). For the second class a typical problem is the optimization of the shape of an access door with objective minimization of the stress differences around the hole by variation of the shape considering a minimum required open surface.

A. Definition of the Optimization Problem

As can be seen from the description of the two general structural optimization problems, the optimization process consists of three elements, the objective function, the boundary conditions and the variables. The related mathematical formulation for optimization reflects these three elements.

An objective function $F(\mathbf{x})$ is minimized with respect to constraints $g(\mathbf{x})$ and $h(\mathbf{x})$ for a variable set $\mathbf{x}$, with nv being the number of variables:

$$\min F(\mathbf{x}) \tag{1}$$

$$g(\mathbf{x}) < 0 \tag{2}$$

$$h(\mathbf{x}) = 0 \tag{3}$$

$$\mathbf{x} = [x_1, \dots, x_{nv}]^T \tag{4}$$

In the sizing process a set of finite-element thicknesses $t_1, \dots, t_{ne}$ (ne = number of elements) forms the variable vector such that

$$\mathbf{x} = [t_1, \dots, t_{ne}]^T \tag{5}$$

The node point locations $x_1, y_1, z_1, \dots, x_{nn}, y_{nn}, z_{nn}$ (nn = number of node points) form the variable vector for shape optimization such that

$$\mathbf{x} = [x_1, y_1, z_1, \dots, x_{nn}, y_{nn}, z_{nn}]^T \tag{6}$$

A related numerical optimization process which enables the computing of these structural optimization problems, consists of a preprocessor-analyzer-postprocessor-optimizer loop, fig. 2, with

- The preprocessor, generating the required element grids, and the stiffness, mass and load matrices,
- The finite-element analyzer, calculating the stresses, strains and modal-modes,
- The postprocessor, calculating the constraints from the analyzer's set of information, and,
- The optimizer, consisting of a complex control procedure, being able to search for extreme values (minimum or maximum) by variation of the design variables.

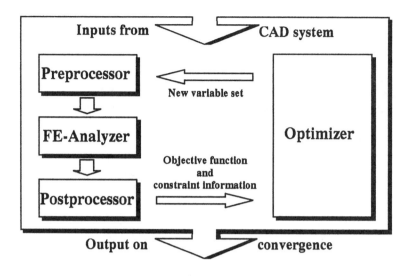

Figure 2:   The preprocessor-analyzer-postprocessor-optimizer loop of
structural optimization

The pre- and postprocessor modules are adapted to the specific optimization
type intended to be performed and they are specifically adapted to the
finite-element procedure used.

B. Numerical Optimization Methods

Optimization methods used for computer aided structural optimization may be
categorized into three groups:

- Recursive methods, optimizing with respect to an optimality criteria, such
  as the stress-ratio method,
- Indirect searching methods, searching for the unconstrainted optimum,
  utilizing a modified objective function for considering the constraints,
  such as the Penalty-Functions method, and,
- Direct searching methods, searching for the optimum by considering the
  constraint information directly, such as the Feasible-Directions method.

Table I summarizes the numerical efficiency of the methods mentioned. Recursive methods are efficient when comparing the computer time needed to carry out the optimization process, but their utilization is strongly limited to the optimality criterion used. Possibilities to implement additional constraint information are limited [6,7].

The well known indirect searching Penalty-Functions method [10,11] has been successfully used for structural optimization and was implemented for structure optimization carried out in this contribution. This method is second order dependant on the variables and first order dependant on the constraints. For the nonlinear multivariate search algorithm with the Penalty-Functions method, the Conjugate Gradient algorithm was implemented [12].

Direct searching methods are very efficient, especially when used with approximative algorithms and dual formulation [8,9]. But the direct searching methods are second order dependant on the variable number and on the constraint number, thus, they yield long computer times when optimizing using large finite-element sets. If these more efficient direct searching algorithms are intended to be used, it is recommended to reduce both the variable number and the constraint set information to gain additional computer time, chapter II, D.

Table I: Comparison of the efficiency of the major numerical optimization methods suitable for structural optimization

++ good + acceptable − poor	Variable dependancy	Boundary condition dependancy	Convergence behaviour	Required computer time	Gradient of objective function necessary ?	Gradient of boundary conditions necessary ?
Stress-Ratio Method	n_v	0	+	++	no	no
Penalty-Function Method	n_v^2	n_v	+	+	possible	possible
Feasible-Direction Method	n_v^2	n_v^2	++	−	yes	yes

n_v = number of variables

C. Implementing Variable Reduction

Since the computer time needed for the optimization process depends to a large extent on the number of variables treated in the optimizer, it is very inefficient to size a structure with all finite-elements being variables. For this reason the variable set has to be reduced. Moreover, the finite-element model and its variable set (element thicknesses or node point locations), is in general not best suited to be identical to the variable model used within the optimizer. An example may better describe this. High accuracy required on output from the analyzer leads to large finite-element sets which allow local effects to be assessed. However, a low mass obtained by thickness optimization is not obtained primarily by local optimization, but more by global low thicknesses. Thus, the local high number of variables available from the analyzer is not needed for global low mass optimization in the optimizer.

The recommended decoupling of the analyzer from the optimizer is automatically carried out by implementing mathematical functions into the preprocessor-analyzer-postprocessor-optimizer loop. Figure 3 shows how to implement the functions. Instead of treating the finite-elements or node point

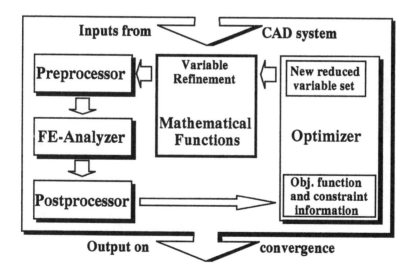

Figure 3: Implementation of the variable reduction procedure using mathematical functions.

locations directly, the control point values $y_1, \dots, y_{nc}$ (nc = number of control points) of the mathematical function $G(x_1, \dots, x_{nc}, y_1, \dots, y_{nc})$, with $x_1, \dots, x_{nc}$ being the control point locations, are handled by the optimizer. The variable vector in the optimization process is then reduced to the control point values such as follows:

$$\mathbf{x} = [y_1, \dots, y_{nc}]^T \tag{7}$$

Within the optimization process the function, or more generally, a set of functions, is then optimized with respect to the full constraint set. The variable values needed in the preprocessor are interpolated from the mathematical functions. This allows a coarse finite-element set with some hundred elements to be handled the same way as a refined finite-element set with several thousand elements, with an identical function in both cases controlled by a very low number of variables.

Figure 4 shows schematically how to derive the data set for the preprocessor from the mathematical function. The function itself is controlled by its control points. From this function the data set of finite-elements, or node point locations, is refined by simply interpolating the function values from a given set of locations.

Advantage of this procedure is not only strong reduction of the variable number but also smoothing of the optimization process. In particular at the beginning of optimization the variable set may be such that constraints are partly feasible, partly unfeasible and a large difference between the search gradient values occurs. This induces oscillating effects within most optimization procedures and reduces efficiency, requiring additional iteration steps before convergence is reached. Due to the continuous behaviour of the mathematical functions the variable set is virtually coupled. This reduces the tendency to oscillate and convergence of the optimization process is improved.

D. Reducing the Constraint Set Information

Most efficient optimization methods, which are direct searching methods, are square order dependant on the constraint information. Reducing the variable number alone would not utilize the full potential for reducing the computer time, as can be achieved by using variable reduction methods together with these optimization algorithms. The constraint set has to be reduced as well, using the same technique as was shown for the variables. Figure 5 shows how to implement a function for reduction of the constraint set information. The

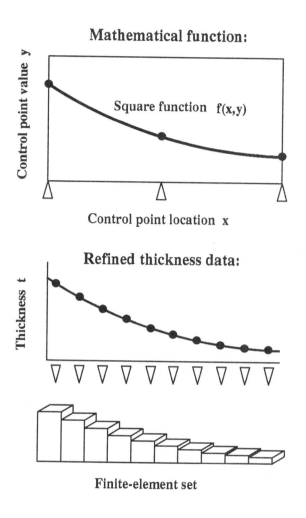

Figure 4: Deriving the variable set from the mathematical function

constraints g_j and h_k with $j = 1, \dots, n_{ng}$ and $k = 1, \dots, n_{nh}$ (n_{ng} being the number of boundary conditions and n_{nh} being the number of equality criterions) is reduced to a reduced set g_{jr} and h_{kr} with $jr = 1, \dots, n_{gr}$ and $kr = 1, \dots, n_{hr}$ (n_{gr} the number of reduced boundary conditions and n_{hr} the number of reduced equality criterions). This may easily be carried out using mathematical functions which are controlled and calculated from the large set

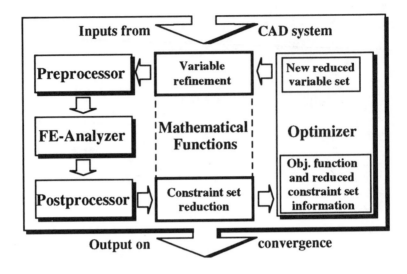

Figure 5:   Reduction of constraint set information within the structural optimization process

of boundary point values. A strongly reduced set of constraints is then calculated from the mathematical functions.

Another method frequently used is to consider only those finite-elements which are expected to yield critical values when carrying out the optimization process. For example the elements in areas of high load concentrations. But although this method works fine it is recommended to use it with great care, since elements which are not considered in the constraint set and which become unfeasible during optimization cannot influence the optimizer, yielding non acceptable optimization results.

E. Methods of Variable Reduction

As was justified before, variable number of the optimization process needs to be reduced, to enable large structures to be optimized with acceptable computer time. Several methods are known which may be utilized for reduction of the variables in the structural optimization process. Figure 6

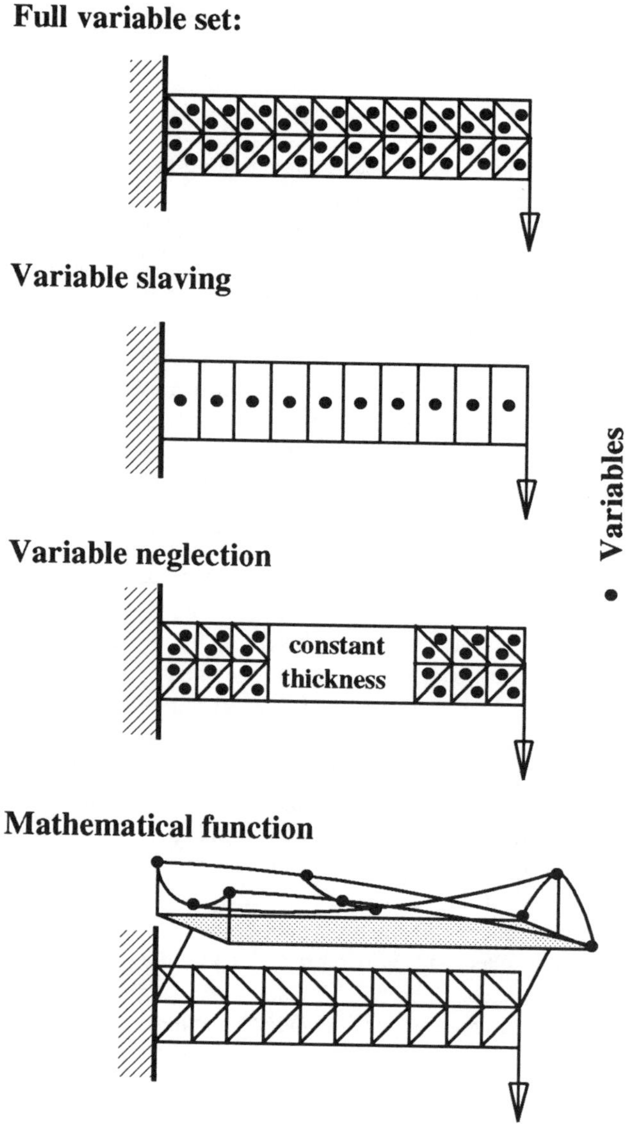

Full variable set:

Variable slaving

Variable neglection

constant thickness

Mathematical function

• **Variables**

Figure 6:   Description of different procedures for variable reduction

shows the working methods of the following procedures:

- Variable slaving or variable coupling
- Variable neglection
- Mathematical functions

The first method, variable slaving or variable coupling is used in cases were a set of variables has to change its values identically, for example identical thickness of a shell or plate structure in a certain geometry range due to manufacturing reasons, or, the geometry of a node-point set when constant shape is required in a certain range. This method is efficient, but it cuts feasible parts of the design space which may be useful for the design process.

Variable neglection is another possibility for variable reduction. It is utilized when a certain range of the finite-element thicknesses (sizing process), or node point locations (shape optimization), does not need to be modified during the optimization process. This may be the case when using standardized structural elements for example which cannot be modified. In such a case the variable neglection process may be utilized. But this method strongly cuts feasible parts of the design space.

The third method makes use of interpolating mathematical functions which allow a large number of variables to be calculated from a small set of control points which control the function simply by interpolating as many values as needed from the function itself. If the function has been properly chosen, the optimization result is virtually identical to the full variable set optimization, thus the feasible region of the design space remains valid and the global optimum may be reached by the optimizer. It is obvious, that if, for example, a square mathematical function type has been chosen for a cubic thickness distribution, the optimum cannot be approached and the optimization process will not converge to the global optimum. If in this case a cubic mathematical function is used the optimum can be reached. Mathematical functions implemented with a higher order than needed are not advantageous, since the optimum is already reached with a lower order function with smaller variable number. Furthermore, the higher order function may reduce stability of the optimization process, due to stronger tendancy to oscillate.

III. Mathematical Functions for Variable Reduction

For variable reduction the most appropriate mathematical functions are those which enable interpolation of an arbitrary number of points and which are themselves controlled by a small set of control points. The following features are important for structural optimization using mathematical functions:

-   First, the mathematical function used must permit the optimum condition to be represented. Thus, if in a sizing process the thickness distribution is a cubic one, a square function would not permit the optimum to be represented, thus yield unfavourable results.
-   A stable behaviour is required. At the beginning of the optimization process, the variable vector values might differ to a large extent one to another and in between the optimization's iteration steps. If in this case the mathematical function utilized yields a strong tendency to oscillate, the optimization process may in the best case run to a local optimum value, or, which is generally the case, may not converge.
-   The aim of implementing mathematical functions is to reduce the computer time when carrying out large structural optimizations, thus, only a little additional computer time is acceptable for evaluation of the variable set from the control point set. Although time needed for the interpolating process is small compared to the time needed for the optimization process itself, functions should be implemented, which yield low computer times.

The following gives a short description of three different types of mathematical functions, which are investigated and applied to the optimization process:

-   Polynomial functions, the explicit function type,
-   Spline functions, based on a piecewise polynomial representation,
-   Bezier functions, based on an implicit representation of a function.

All of these functions are standard numerical mathematical functions [13]. It should be mentioned, that these three mathematical functions do not represent the limit of what can be used in the process of variable reduction for optimization. More complex functions such as sine- and cosine functions together with Fourier transformation, or higher order surface functions used in CAD applications, have not been utilized, the reader may refer to the large amount of documentation available in related fields and should feel free to utilize those mathematical functions which are best known to him. The following gives an overview of possible function types suitable for optimization, and describes advantages and disadvantages of several easy to implement functions.

Definition of the nomenclature, of numbering the function's control points and of the interpolated values is shown in figure 7. A function $f(x)$ is determined by a set of control points $[x_1, y_1, ... , x_i, y_i, ... , x_{nc}, y_{nc}]$ with $i = 1, ... , nc$ (nc =

UWE L. BERKES

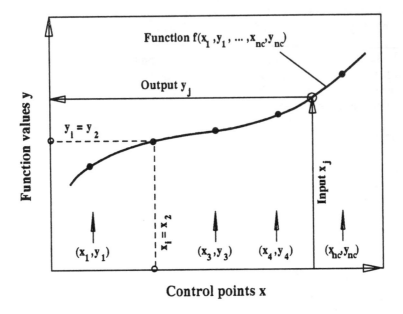

Figure 7: Definition of the mathematical function, controlled by a set of control points

number of control points, x_i the control point locations and y_i the function values at these control point locations). From this function, an arbitrary set of function values $[y_1, \ldots, y_j, \ldots, y_{np}]$, depending on $[x_1, \ldots, x_j, \ldots, x_{np}]$ with $j = 1$, $\ldots, np$ (np = number of points to be represented) is then calculated, with x_j the location of points to be calculated and y_j the values at these locations, derived from the function itself. Based on this, the three functions are described below.

A. Polynomial Mathematical Function

Polynomial functions are an universal type of function and easy to implement. The general form of a polynomial function is the following (cubic order):

$$f(x) = c_1 + c_2 x + c_3 x^2 + c_4 x^3 \tag{8}$$

Two possibilities exist to control the function, either by its constants $c_1, ... ,c_4$, or via a control point set x_i, y_i, with $i = 1, ... ,nc$ (nc = number of control points), the order of the polynomial being nc-1. It is not recommended to use the constants of the polynomial within the optimization process, since small changes of their values do deform the function very much, which yields oscillating and non-converging optimization processes. Using the control point values x_i, y_i to determine the function is the preferred way to generate the parameters $y_j = f(x_j)$ from a given set x_j, with $j = 1, ... ,np$ (np = number of points to be represented). The well known Lagrange algorithm allows this process to be performed without calculating the function coefficients (eq. 9):

$$f(x_j) = Sum_{(i=1..nc)} ((Prod_{(k=1..nc,k \text{ not } i)} (x_j-x_k))/(Prod_{(k=1..nc,k \text{ not } i)} (x_i-x_k)) *y_i)$$

Written for a square function:

$$f(x_j) = ((x_j-x_2)(x_j-x_3))/((x_1-x_2)(x_1-x_3))*y_1 + \qquad (10)$$
$$((x_j-x_1)(x_j-x_3))/((x_2-x_1)(x_2-x_3))*y_2 +$$
$$((x_j-x_1)(x_j-x_2))/((x_3-x_1)(x_3-x_2))*y_3$$

The procedure for higher order polynomials is straightforward.
Utilizing this algorithm, the process of interpolating a large set of parameters from a small set of control points may easily be performed without knowing the coefficients. Furthermore, additional control points are easy to implement without changing the overall process. But is has to be remarked that this function tends to generate large variations even with moderate changes of the control point set, possibly leading to oscillations when being implemented within the optimization process.

Advantages of the polynomial mathematical function:

- Very simple to implement, yields lowest additional computer time,
- Easy implementation of additional control points.

Disadvantages of the polynomial mathematical function:

- Tends to oscillate if control point values are very different and if control point locations are near to one another,
- Additional control points change the order of the function.

B. Spline Mathematical Function

The behaviour of a flexible ruler when fixed to a certain set of points is simulated by the spline function based on a piecewise description of the curve by polynomials. The continuous behaviour of the function over the control point range is assured by identical control point values and identical second order derivatives at the boundaries between one function and another.

For the polynomial, implemented in the range between two control points, a cubic form is used:

$$f_i(x_j) = a_i + b_i(x_j - x_i) + c_i(x_j - x_i)^2 + d_i(x_j - x_i)^3 \tag{11}$$

with x_i the locations of the control points ($i = 1, \dots, nc-1$), x_j the input for computation of the values $f_i(x_j)$, to be calculated in each interval ($j = 1, \dots, np$ with np = number of points in each interval). Constants are $a_i = y_i$ the control point values at the beginning of each interval i, constants b_i, c_i and d_i have to be computed and c_i and d_i are the second order derivatives. Several methods exist to calculate the constants of the polynomial [13, 14]. Most processes use a two-step algorithm. In the first step the second order derivatives are computed for the control point set. In the second step, this information is used to compute the constants of the set of polynomials. With this information an arbitrary number of function values may be computed with $nc-1$ polynomials, each valid for an interval between two control points.

A method has been chosen in the following (which calculates the necessary information using a tri-diagonal scheme of the control point values), based on a set of eight auxiliary points aa to ah [15].

As was already mentioned, the first constant a_i is equal to the control point value y_i at the beginning of an interval ($i = 1, \dots, nc-1$):

$$a_i = y_i \tag{12}$$

First, the differences between the control point locations have to be computed ($i = 1, \dots, nc-1$):

$$aa_i = x_{i+1} - x_i \tag{13}$$

Second, derivatives of the curve at the beginning and the end are set to zero (natural spline):

$$c_1 = 0 \tag{14}$$

$$c_{nc-1} = 0 \tag{15}$$

Another auxiliary point computes the slope between the control points ($i = 3$, ... ,nc):

$$ab_{(i-1)} = 3((y_i - y_{i-1})/aa_{i-1} - (y_{i-1} - y_{i-2})/aa_{i-2}) \tag{16}$$

A further three auxiliary points ($i = 2$, ... ,nc-1) form the tri-diagonal scheme:

$$ac_i = aa_{i-1} \tag{17}$$
$$ad_i = 2(aa_{i-1} + aa_i) \tag{18}$$
$$ae_i = aa_i \tag{19}$$

And then:

$$af_2 = ad_2 \tag{20}$$
$$ah_2 = ab_2/af_2 \tag{21}$$

Last three auxiliary points ($i = 3$, ... ,nc-1):

$$ag_i = ae_{i-1}/af_{i-1} \tag{22}$$
$$afi = ad_i - ac_i * ag_i \tag{23}$$
$$ah_i = (ab_i - ac_i * ah_{i-1})/af_i \tag{24}$$

Thus, the constants c_i can be calculated ($i = nc-2, nc-1$, ... ,1):

$$c_{nc-1} = ah_{nc-1} \tag{25}$$
$$c_i = ah_i - ag_{i+1} * c_{i+1} \tag{26}$$

The constants b_i and d_i are ($i = 1$, ... ,nc-1):

$$b_i = (y_{i+1} - y_i)/aa_i - aa_i/3 * (2c_i + c_{i+1}) \tag{27}$$
$$d_i = (c_{i+1} - c_i)/(3 * aa_i) \tag{28}$$

This calculation of the auxiliary values and the coefficients has to be carried out from a set of ordered control point values. Once this procedure has been performed for the given control point set, computation of the function values is straightforward, using the set of polynomials each of which is valid in a range between two control points.
Compared to the polynomial function a more rigid behaviour will result,

although this function type is based on polynomials as well.

Advantages of the spline mathematical function:

-   The number of control points may easily be changed.
-   Once the auxiliary values are computed and the coefficients are
    determined, any number of points may be rapidly computed.

Disadvantages of the spline mathematical function:

-   Although the spline functions behaviour is much better than that of the
    polynomial function, it is based on the latter one, thus, instabilities may
    occur.

C. Bezier Mathematical Function

The previously described polynomial and spline functions are explicit
functions. It is therefore possible to calculate the function itself and to
interpolate function points directly from the formulation used. The Bezier
functions belongs to the implicit function type. These functions are
determined by a parameter representation from which interpolated points have
to be calculated indirectly.
A two-dimensional curve $c(u)$ is represented in a parametric way:

$$x = f(u), y = f(u) \qquad (29)$$
$$c(u) = (x(u), y(u)) \qquad (30)$$

with the parameter $u = 0...1$.

This curve $c(u)$ shall be controlled by a set of points p_1, ... ,p_{nc} and a
fit-function $B_{i,nc}(u)$ ($i = 0$, ... ,$nc-1$), with nc the number of control points and
$nc-1$ the order of the curve, such that:

$$c(u) = Sum_{i=0...(nc-1)} \ (p_i * B_{i,nc}(u)) \qquad (31)$$

Instead of $c(u)$, the coordinates x and y are implemented (definitions
according to fig. 7), thus the mathematical formulation may also be written:

$$x(u) = Sum_{i=0...(nc-1)} \ (x_{i+1} * B_{i,nc}(u)) \qquad (32)$$

$$y(u) = \text{Sum}_{i=0...(nc-1)} \ (y_{i+1}*B_{i,nc}(u)) \tag{33}$$

with x_i, y_i are the nc control point values $(x_1, y_1, \ ... \ , x_{nc}, y_{nc})$.

Bezier [17] used this type of parametric representation of a curve and defined the fit-function as follows:

$$B_{i,nc}(u) = (nc-1)!/[i!*((nc-1)-i)!]*u^i*(1-u)^{(nc-1)-i} \tag{34}$$

Due to the implicit parametric definition of the function, the points y_j for a set of x_j cannot be computed directly. First, the u-values $(u_1, \ ... \ , u_{np})$, corresponding to the x_j-values $(x_1, \ ... \ , x_{np})$, with np the number of points being interpolated, have to be computed from equation (32). Then, taking the set of known u-values, the necessary y_j-values may be computed from equation (33). Thus, an additional step is necessary to get the interpolated points. Since the mathematics of the Bezier formula is quite simple, this additional process does not take very long.

The Bezier functions curve fits exactly the first and last control points x_1, y_1 and x_{nc}, y_{nc} and is controlled by intermediate points, which make up a polygon, figure 8. Thus, the control point values in between the first and last point do not correspond to the value of the curve in this area, which is important when using this function type for variable reduction, since control point values derived from the physical model, such as thicknesses for example, do not match the curve.

Advantages of the Bezier mathematical function:

- Virtually no instabilities occur even with tight control points with very different values.
- Several curves may be continuously coupled, since the slope of the controlling polygon at the first and last point is equal to the functions slope at those points, thus, a continuous curve may easily be generated.

Disadvantages of the Bezier mathematical function:

- The degree of the curve and its rigidity is directly dependant on the number of control points, a change of only one control point value changes the curve globally, a local change is not possible
- Control point values of the intermediate control points do not correspond to the actual curve values.

D. Comparison of the Efficiency of the Mathematical Functions

Three mathematical function types have been presented in the previous section, the polynomial function, the spline function and the Bezier function. Comparisons between them have to be based on the following characteristics, when considering the use of these functions in the optimization process for variable reduction:

- Ease of control (by a reduced set of control points),
- Rigid behaviour even for strong variations of the control point values or control point locations,
- Low additional computation time necessary for the interpolating process.

All of the discussed mathematical functions are controlled by a small number of control points. For the polynomial function it is not possible to change the number of control points without changing the order of the curve. The Bezier mathematical function is also dependant on the number of the control points. Important for the efficiency of the optimization process is a stable behaviour of the function itself when large variations of the control point values occur. An unstable behaviour might lead to non-convergent optimization or non-optimal solutions. Considering this point the Bezier function is best. The curve remains stable even for tightly located or strongly different control points. By contrast, the polynomial function tends to oscillate particularily when the control point values are located close to one another. The spline function offers a convenient compromise, which yields good stiffness. Figure 8 compares the behaviour of the three described functions.

Of further importance is the additional computer time needed to reduce the variable set by mathematical functions. Table II compares the computer time needed for interpolation, without considering the optimization process itself, [2]. The polynomial function results in the lowest computer time. By comparison, Bezier yields computer times which are longer by a factor of 3 and spline by a factor of 4. Although Bezier needs to compute on an implicit basis, this function type is quite fast due to the simple mathematics involved.

Finally, a recommendation should be given. If other function types are known to the reader or are already installed on the computer, do not hesitate to use them. It is the experience gained with a certain type of function which is most likely to avoid unreasonable results, when using mathematical functions to reduce the variable number in structural optimization processes.

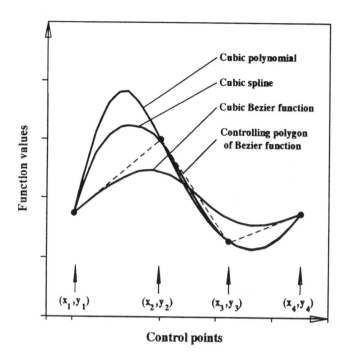

Figure 8: Behaviour of the polynomial, spline and Bezier functions

Table II: Comparison of the computer time needed for interpolation

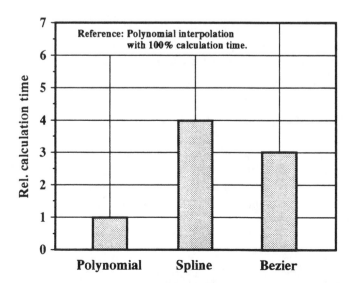

E. Two-Dimensional Mathematical Functions

Mathematical functions have been presented for the one-dimensional case. Since numerous structural optimization problems are two-dimensional, for example optimization of thickness of a wing-box plate, the functions already mentioned have to be extended in order to be applied to two-dimensions.

A two-dimensional description exists for each of the described functions, the reader should feel free to consult specific literature on this, especially that literature dealing with CAD mathematics (curve and surface fitting). The following paragraph decribes a simple way to achieve the required two-dimensional formulation from the one-dimensional ones.

A two-dimensional control point set $(x_1,y_1,z_{1,1}, \dots ,x_{ncx},y_{ncy},z_{ncx,ncy})$ with x_{ix} and y_{iy} the control point locations and $z_{ix,iy}$ the control point value (n_{cx} the number of control points in x-direction and n_{cy} the number of control points in y-direction), is first interpolated in one direction, for example in x-direction, utilizing a set of n_{cy} one-dimensional functions with the information $(x_1,z_{1,ncy}, \dots ,x_{ncx},z_{ncx,ncy})_{ncy}$. From this set of one-dimensional functions, an arbitrary set of two-dimensional function points may be computed, using an arbitrary set of one-dimensional functions in y-direction for interpolation. Thus, using the one-dimensional functions, the two-dimensional problem is readily computed. Figure 9 shows this simple method. This works perfectly and is recommended if one-dimensional mathematical functions are already implemented on the computer.

F. Choosing the Stepsize

When utilizing mathematical functions for reduction of variable number, a large set of parameters is controlled by a small set of control points of the function. The previously outlined mathematical functions are based on two values for each control point of the one-dimensional function (x_i,y_i) and for each parameter to be calculated (x_j,y_j). The values treated in the optimization process are the y_i's, and the finite-element program's preprocessor needs the y_j's from the mathematical functions. How are the necessary x_i's and x_j's derived?

These are the control point locations, which may be chosen arbitrarily. The geometrical model of the structure being optimized may be chosen for locating the control point values. Or, a fictious control point set is used, derived from a transformation of the physical points of the finite-element

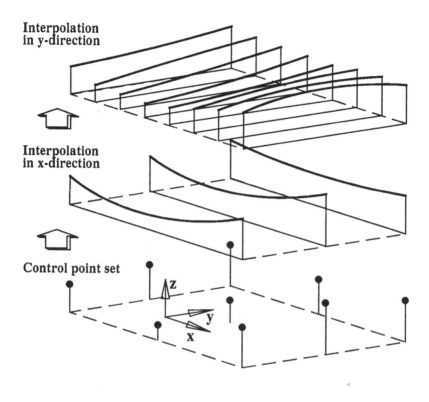

Figure 9:   Interpolation in two dimensions with one-dimensional functions

method to the range between 0 and 1. Figure 10 shows the two possibilities. An advantage when using the physical values is that the function also represents the physical value, for example the thicknesses of the finite-elements. From this function the parameters (thicknesses, node point locations, etc.) may be calculated and used without further treatment. This works quite well, as long as the range of variable values x_i is not too large.

Problems may occur if such a curve with very different x and y ranges is controlled in the optimizer, particularily when the optimization process is under convergence. Variations become small and remain right in the error band (oscillation) of the functions themselves.

Transformation of the parameter values x_i and x_j, as well as the y_i's and y_j's into the range 0 to 1 independantly has been performed with considerable

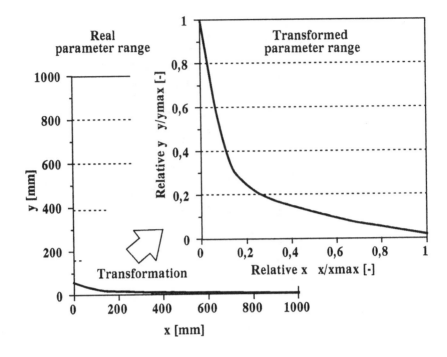

Figure 10:   Transformation of the parameter range

success, figure 10. The optimization process when utilized with a finite-element structural problem was well controlled in that case and we had virtually no convergence problems.

G. Multifunction and Patch-Technique

If within a large and complex structure it is necessary to introduce local forces, or, if the structure does change its geometry, the physical properties such as the thickness or the stiffness change as well. Such local changes of the variables cannot efficiently be represented by only one mathematical function. It is recommended to use multiple mathematical functions, each in an undisturbed region in between those inconsistencies. This technique is called 'patch technique' for two-dimensional functions, since the functions are used only in certain parts of the structure, which gives an impression of being a patchwork [2]. Figure 11 shows an example for a delta shaped wing-box

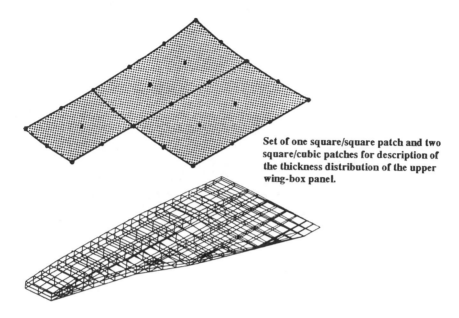

Set of one square/square patch and two square/cubic patches for description of the thickness distribution of the upper wing-box panel.

Figure 11:   Patch technique applied to an aircraft wing-box panel

structure of a hypersonic aircraft. Due to the very large structure and locally complex loads, three two-dimensional functions have been implemented for description of the thickness distribution of the upper wing-box panel. The order of the functions in the two directions have been chosen so that, depending on the complexity of the thickness distribution expected, the optimum may be reached.

If a continuous curve is required, first and second order derivatives have to be used at common control points. If these derivatives are set equal for either curve at a specific control point, a continuous curve is obtained.

IV. Optimization of a Wing-Box Structure

The method of variable reduction using mathematical functions has been applied to thickness optimization of a twin-engine aircraft's wing-box structure. For reference purpose, the wing-box structure is first optimized using a full variable set consisting of all finite-elements.

The three described mathematical functions:

- Polynomial function,
- Spline function, and
- Bezier function

are then implemented into the optimization process and results are compared to the reference optimization. A similar control point set has been chosen for the various functions, consisting of 21 variables for each wing-box panel and 57 variables in total for the wing-box.

The spline function was used in a second investigation on the function behaviour with change of variable number and variable location. The variable number of the wing-box panels was modified to:

- 21 variables,
- 27 variables, and
- 30 variables

for each of the upper and lower wing-box panels.

To further enhance the knowledge of using mathematical functions for variable reduction in structural optimization, this method is then used in combination with the variable slaving method. Two formulations are investigated:

- A variable set reduced using the variable slaving method, and
- A variable set using a combination of the variable reduction method by mathematical functions and the variable slaving method.

In order to make a clearer comparison, only the upper wing-box thicknesses or stress distributions are shown in the following discussion. However, the conclusions also hold for the lower wing-box panels, the spars and ribs.

A. Description of the Wingbox Structure used for Optimization

Figure 12 shows the aircraft and the wing-box structure's finite-element model. The model consists of 125 elements for each upper and lower wing-box panel, 25 elements for each forward spar and rear spar, and 5 elements for each of the 25 ribs. The wing-box is clamped at its fuselage junction point. Loads such as the aerodynamic lift and drag forces, the fuel mass, point loads from the landing gear and the engine are considered within this model. Loads from the landing gear are introduced at the rearend of rib

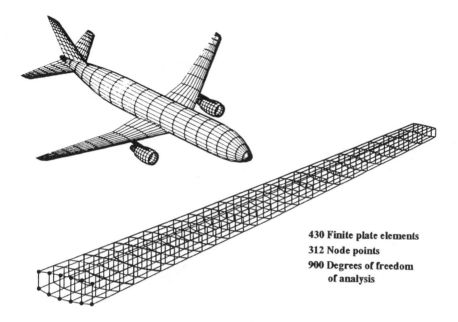

430 Finite plate elements
312 Node points
900 Degrees of freedom
of analysis

Figure 12:   Twin-jet aircraft and finite-element model of the wing-box

number 5, counting from the clamped end, and engine loads are introduced at
the forward and aft section of rib number 7. A sample of different loadcases
such as the touchdown shock, the maximum g-loads and a full-thrust loadcase
have been considered in a load envelope for the wingbox, which allows design
for the most severe combination of the mentioned loads.

The wing-box is considered as being built from aluminium material AlCuMg
with a limit stress level of $300N/mm^2$. Skins are integrally stiffened and are
considered as free from buckling due to local optimization of stiffener
dimensions. A minimum thickness constraint has been implemented for
manufacturing reason. This lower thickness boundary is 1mm and becomes
active in the wing-tip zones of the wing-box. In this area of the wing the loads
are relatively small and sizing would lead to very thin skin thicknesses.

B. Reference Computation

First, the wing-box finite-element model was used for reference optimization
with all finite-element thicknesses treated as variables. Corresponding stresses
of the finite-elements are treated as constraints. The variable set yield 430

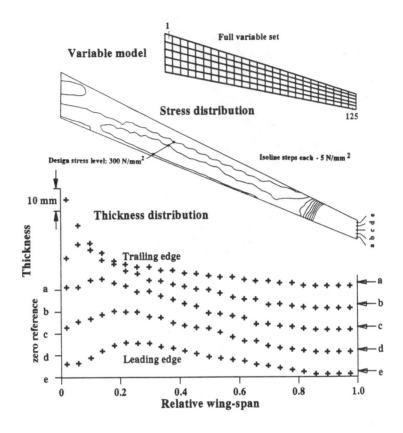

Figure 13:   Stress- and thickness distributions of the reference computation

variables for this full variable set formulation. Figure 13 shows the optimum thickness and stress distribution of the upper wing-box panel. The panel is virtually fully stressed to the allowable stress level of 300N/mm², except for the already indicated wing-tip zone, were the minimum thickness constraint becomes active. The optimized mass of the wing-box is 2295kg in total.

Optimization itself has been carried out utilizing the already mentioned Penalty Function method with a Conjugate Direction optimizer (chapter II). Computer time of the full variable reference optimization was 111 minutes on a CDC 180-860 mainframe.

C. Optimization using Mathematical Functions

The three mathematical functions have been implemented in the optimization procedure for variable reduction. An identical set of control points has been chosen for all three functions for better comparison. It became necessary to use a formulation with two functions in spanwise direction for the plates, spars and ribs, due to the wing-kink and introduction of concentrated loads in this area, yielding a change of the thickness distribution. One function was implemented from the wing/fuselage junction to the kink and another from the kink to the wingtip in order to better handle this expected change. Two square/cubic two-dimensional functions are used on each of the panels (upper and lower), yielding 21 variables for each plate. Two square one-dimensional functions are implemented for each of the spars (forward and rearward), yielding 5 variables for each spar. The ribs have been treated differently since the reference computation yields virtually constant thicknesses for the five finite-elements which make up a rib. They have therefore been coupled to one variable, such as it is done when using variable slaving. The resulting 25 variables of the 25 ribs are controlled by two one-dimensional square functions, which are controlled by 5 variables. With this formulation, the overall variable number of the complete wing-box is 57. Figure 14 shows the structural elements and the location of the control points.

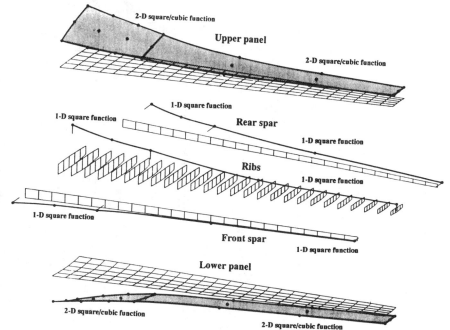

Figure 14:   Finite-element model and mathematical functions

Comparison of the results of the optimization procedures using mathematical functions for variable reduction is based on three important objectives:

- The result (mass) gained with the procedure,
- The computer time needed until convergence of the optimization process, and,
- The behaviour of the optimization process itself.

Compared to the reference optimization with 111 minutes computer time, the time needed with variable reduced optimization was 27 minutes, thus a gain of a factor of 4.2 can be stated. A difference between the three function types implemented is virtually non existant, since computing the mathematical

Table III: Comparison of mass, stress-level and computer time

	Relative weight of upper wing-box panel	Maximum difference of stress level on upper wing-box panel	Number of variables of upper wing-box panel	Computer time
Full variable reference	1.0	5 N/mm^2	125	111 min
Polynomial function	1.038	10 N/mm^2	21	27 min
Spline function 21 variables	1.025	8 N/mm^2	21	27 min
Spline function 27 variables	1.043	14 N/mm^2	27	30 min
Spline function 30 variables	1.022	8 N/mm^2	30	32 min
Bezier function	1.079	18 N/mm^2	21	27 min
Variable slaving	1.11	30 N/mm^2	25	41 min
Variable slaving with functions	1.11	33 N/mm^2	5	19 min

functions is much faster than the optimization procedure with its gradient search algorithms and the time consuming finite-element analyzer. The ratio between the interpolation time and the optimization time, including analysis, was approximately 1/400. Thus, the computer time gained is influenced little by the mathematical function chosen.

Table III compares the masses of the three calculations to the reference full variable set. The polynomial mathematical function yields 3.8% higher mass, the spline function was best with only 2.5% higher mass and the Bezier function is worst with 7.9% higher mass, comparing the upper wing-box panels. Comparing the thickness distribution of the three formulations, figures 15 to 17, show the differences to the full variable set optimization.

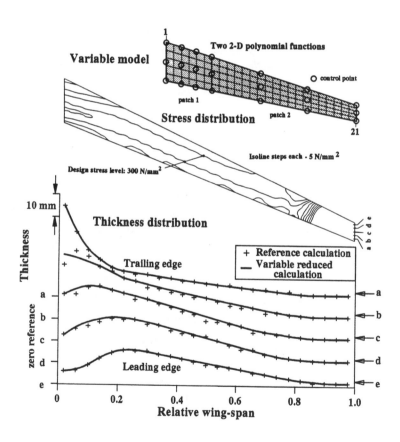

Figure 15: Optimization applying the polynomial function

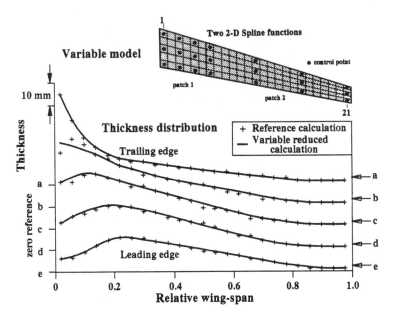

Figure 16:   Optimization applying the spline function

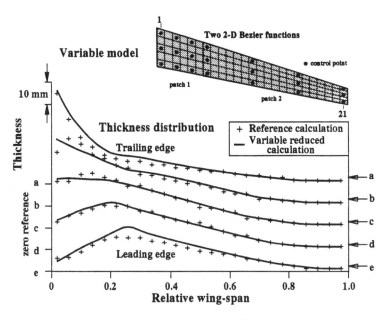

Figure 17:   Optimization applying the Bezier function

The functions very closely represent the optimum thickness distribution, except in the area close to the fuselage, where the order of the function in chord direction should have been higher due to a more complex thickness distribution in this direction. Difficulties observed with the Bezier function can be extracted from this figure as well. The behaviour of the Bezier function in spanwise direction is rigid and therefore hardly follows the required thicknesses in the chordwise direction. In addition, the initial variable vector had to be modified since the control point values of the Bezier function are those of the controlling polygon, not those of the thickness distribution itself. Thus, convergence behaviour is reduced and the result is less accurate when compared with the first two methods.

The optimization behaviour was smoothed, due to coupling of the variables by the function. Optimum values have been approached within 3 to 5 iterations and convergence has been reached within 8 to 10 iterations.

Summarizing the results: Variable reduction by using mathematical functions proved to be efficient. The advantage of the polynomial and the spline functions is evident, control points are also function points for these functions. The Bezier function yields less accurate results due to the different control point and function values and its rigid behaviour.

D. Variation of the Number of Control Points

Within the previous investigation a common variable set has been used. As was described in chapter III, the behaviour of the functions change with variation of the control point location and the number of variables used. The following investigation compares two additional formulations for the spline function [2]:

-   One with additional control points in spanwise direction, yielding 27 variables for each wingbox panel,
-   Another using only one two-dimensional function for the wing-box panel but with increased variable number in chord and spanwise direction, for better control of the complex thickness behaviour.

Figures 18 and 19 show the control point sets. The functions of the spars and ribs have not been changed with respect to previous computations.

Table III compares the masses. Increase of the variable number for the two 2-D function representations did not work out. The functions became more flexible but were not able to follow the optimum thickness distribution, which seemed to be a problem resulting from the chordwise thickness representation, such as was also the problem for the other formulations. The resulting mass

UWE L. BERKES

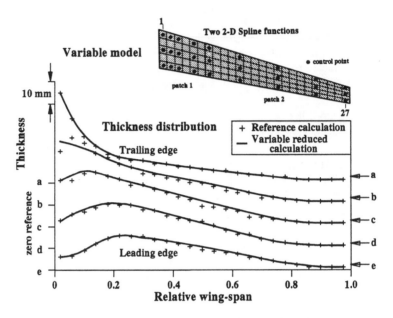

Figure 18: Optimization with variation of the spline function

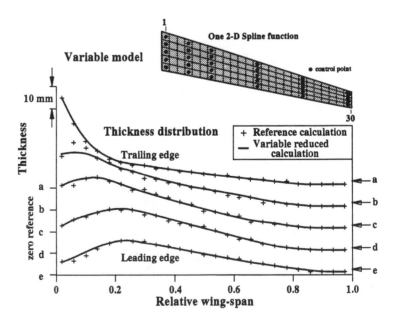

Figure 19: Optimization with one spline function for the wing-box panel

was 4.3% larger than with the global optimum. By contrast, the spline mathematical function with a formulation, using only one two-dimensional function for the complete wingbox panel and increased control point set in chordwise direction, yields best results with only 2.2% higher mass compared to the reference computation. The thickness distribution of this formulation, figure 19, shows that the function better follows the optimum thickness distribution in chordwise direction and that it smooths the thicknesses in spanwise direction. Thus, it yields a lower mass.

Computer time was higher for these formulations due to larger variable numbers, but a significant reduction compared to the full variable set optimization is apparent. The relative gain was 3.6 with the 27 variables for each panel, yielding 69 variables for the complete wing-box model. The gain was 3.2 with the 30 variables for each panel, yielding 75 variables in total.

Using a larger variable number does not always yield good results, if the areas which are most critical are not well controlled. Thus, if the expected optimum requires exact local representation, it might be preferable to use only one function of high order in the critical direction, which is able to better represent the critical local optimum distribution.

E. Combination of Mathematical Functions and Variable Slaving

Two other formulations have been investigated to assess a variable reduction method using a combination of the mathematical functions method with the variable slaving method. First, a formulation is presented, which uses the variable slaving method for optimization of the wing-box. Another formulation combines this method with the mathematical function procedure [1].

A variable slaving procedure has been implemented for the wing-box plates only. Element thicknesses of all finite-elements of the wing-box plates in between two ribs have been coupled to one value. Other elements such as the spars and ribs have been treated without variable reduction. Thus, the overall optimization model yields 225 variables. Figure 20 shows the variable sets of the wing-box plates.

Combination of the variable slaving method and the variable reduction using mathematical functions used two one-dimensional square order functions for each plate, in order to describe the values of the coupled thicknesses of the finite-elements in between two ribs. The two one-dimensional functions already successfully used for the spars and for the ribs have again been implemented, the ribs finite-elements have been coupled such as is done for the plates finite-elements in between two ribs. The model consists of only 25 variables, 5 variables for each wing-box plate, 5 variables for each spar and 5 more for the ribs.

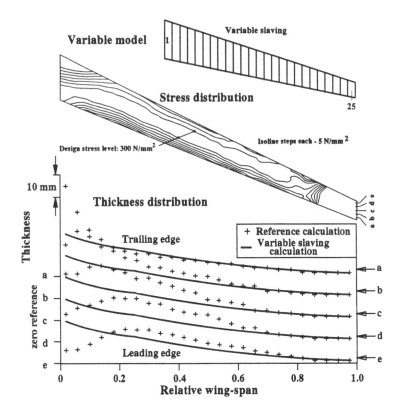

Figure 20: Optimization using the variable slaving method

Figure 21 gives an overview on this variable model used for the upper wing-box plate.

Table III compares the masses of the optimized wing-boxes. Due to the reduced design space, masses are 11% higher than for the reference full variable set computation, showing virtually no difference for both formulations. Comparing the thickness distributions with that of the reference shows the coupled thicknesses of the upper panel. Comparable results can be stated for the lower panel. Slaving of the variables does not allow the thicknesses to be adapted in chordwise direction, thus, a higher mass is the result. Comparing the stress distributions of the upper wing-box panel, figure 20 and 21, yields large differences compared to the reference computation, figure 13. The panel is only stressed to the limit value in the center area,

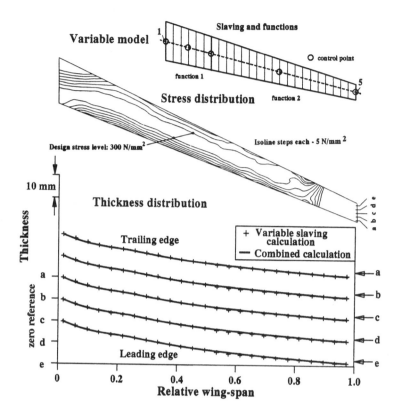

Figure 21: Optimization using a combination of mathematical functions and variable slaving

whereas in the forward and aft region stresses of about 30N/mm^2 lower than the allowed design value resulted for the slaving formulation and up to 33N/mm^2 lower stress level for the slaving and mathematical function combination. Comparison of the computer times needed for these optimizations shows additional reduction of computer time depending on reduction of the variables. The formulation using the combination with only 25 variables yields lowest computer time of only 19 minutes, instead of 111 minutes for the reference full variable set computation.

As could have been extracted from the investigated examples on optimization of a wing-box structure, the variable reduction method using mathematical functions yields virtually the same masses compared to a full variable set optimization, but strongly reduces the computer time. If results achieved from a variable slaving method are acceptable, the computer time is further decreased by combination of the variable slaving method with the variable reduction method using mathematical functions.

V. Further Applications using Mathematical Functions

Two further examples enlarge the field of structural optimization problems which may be suitable for variable reduction using mathematical functions. First, a delta shaped wing-box model is presented, and a set of mathematical functions is proposed for reduction of the variable number of this complex structure [1,2]. As was already mentioned in the introduction to this contribution, variable reduction by using mathematical functions is frequently used for form optimization. The second example reflects this and shows shape optimization of the edge of a hole of a plate structure which is carried out by a strongly reduced variable set yielding only three variables [3].

A. Delta-Wing Structure

High-speed aircraft and spacecraft are designed having delta shaped wings because of their advantageous aerodynamic and aerothermodynamic behaviour. This wing type is much more complicated to analyze and optimize from a structural point of view due to structural hyperstaticity induced by a multiple beam and rib design. Optimality criterion methods, such as the fully-stressed design method, do not compute the global mass optimum when sizing due to hyperstaticity. Direct or indirect search methods have to be used, especially when implementing constraints such as flutter behaviour or thermal conductivity. This latter constraint becomes effective for spaceplanes.

Moreover, due to their complexity, these wing-box structures yield large finite-element models, which makes variable reduction a 'must' when optimizing them. The variable reduction method using mathematical functions has been implemented to a delta shaped wing-box structure, figure 22. The structure yields 982 finite-elements, which are equivalent to the number of variables in a sizing process where the objective is a minimum mass. Extrapolation of preliminary calculations showed this model would require 57 hours computer time on a mainframe (CDC-CYBER 180/860) with full variable set optimization and 10 iterations needed to achieve convergence.

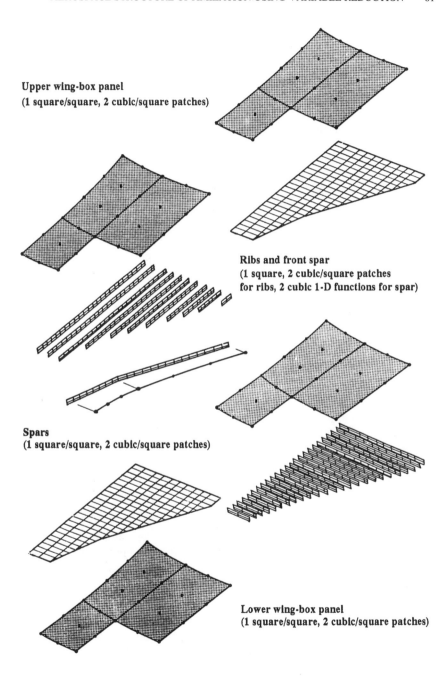

Upper wing-box panel
(1 square/square, 2 cubic/square patches)

Ribs and front spar
(1 square, 2 cubic/square patches
for ribs, 2 cubic 1-D functions for spar)

Spars
(1 square/square, 2 cubic/square patches)

Lower wing-box panel
(1 square/square, 2 cubic/square patches)

Figure 22: Finite-element model of a delta shaped wing-box and related mathematical functions for optimization

Figure 22 shows a possible set of 2-D mathematical functions used to describe the thickness distributions of upper- and lower wing-box panels, spars and ribs. In addition two one-dimensional functions are used to describe the thickness distribution of the leading edge spar. In order to better describe the resulting complex thickness behaviour of the delta shaped wing-box, each two dimensional function for the plates consists of three patches, double square in the front part, and square/cubic for the two rear surfaces, yielding 26 variables for each set of three patches. The one-dimensional function set consists of only two cubic functions. This set of mathematical functions and patches yields 111 variables in total, compared to which the unreduced optimization model yields 982 variables. Optimization utilizing this variable reduced model reduces computer time by a factor of 5.8, thus less then 10 hours computer time is required. Or, if the computer time of 57 hours remains the same, it would be possible to optimize a finite-element set consisting of approximately 9000 finite-elements with the same variable model, allowing a much more detailed analysis.

B. Form Optimization of a Hole

As was already mentioned in the introduction to this contribution, the method of variable reduction using mathematical functions is utilized as a standard procedure for form optimization when continuous shapes need to be achieved. The following example implements this method for the form optimization of a hole in a plate structure, figure 23. The finite-element model consists of 130 finite-elements. A Bezier function controlled by only three control points is sufficient to allow a wide variation of the shape, including variations to the global optimum which it is possible to achieve for the given boundary conditions. Three load conditions have been applied, a ratio of $n_x/n_y = 5$, $n_x/n_y = 1$ and $n_x/n_y = 0.1$. The objective was to achieve an optimum shape for all three load conditions together, minimizing the stress peak at the edge of the hole. This requires a multi-purpose optimization [3]. Figure 23 shows the three single-purpose solutions, their shapes and related stress distributions. As can be seen, an elliptic shape is the optimum result for the first and third load-cases, whereas for equal load conditions ($n_x/n_y = 1$), the optimium is the circle. The multi purpose optimum is shown in figure 23, which is a curve, connecting the most interior points of the single purpose solutions and taking into account the degree of freedom of the mathematical function.

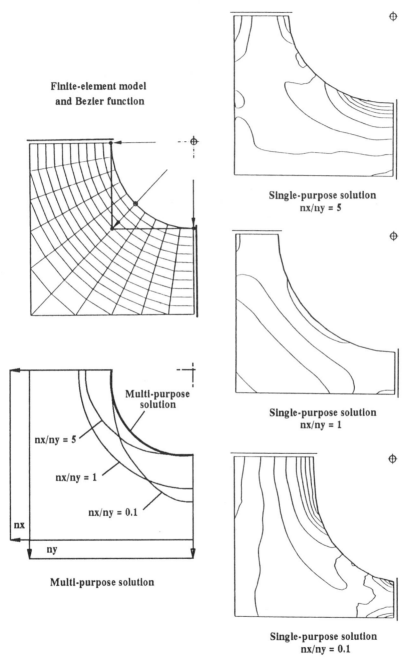

Figure 23:   Multi-purpose form optimization of a hole in a plate structure

VI. Recommendations and 'Cook-Book' Methods

Optimization results achieved by implementing variable reduction using mathematical functions, yield a very efficient behaviour if functions are correctly chosen and control points are carefully implemented. During development of this method, a set of recommendations have been developed, which are useful guidelines when working with this method [1].

- Avoid using functions of higher order than necessary.

    As was already mentioned when presenting different function types (chapter III), a high order function tends to oscillate, since small changes to control points which are located close together, may result in large local changes of the function itself. Therefore, convergence of the optimization process is reduced. To avoid this, the user should always have an idea of the probable final characteristics of the optimized thickness distribution, or optimized shape, before chosing a function type and its order.

- Use several functions when inconsistent thickness, or shape distributions are expected.

    If single loads are introduced to the structure, or if the structure changes its geometry, this will have an influence on the thickness distribution or the shape. Mathematical functions are in general uniform, thus they cannot follow local gradient changes. In such a case it is recommended to utilize several functions, each representing the variable set of a region in between the loaded points or changed geometries. But, as could have been observed from the examples on optimization of the wing-box, it may still be worthwile to try single functions, if the local load conditions or geometry changes are not extreme.

- Compute the function from a set of control points instead of controlling the function by its coefficients.

    For most function types both possibilities exist for control. It is strongly recommended to always use the control points as a variable set. In this case small changes of the control point values induce only small changes in the functions behaviour. If the function is controlled by its coefficients, a small change of one coefficient may considerably alter the functions behaviour, thus leading to unstable optimization.

- The first iteration of the optimization process should lead to a feasible initial variable vector.

As was already discussed, the optimization behaviour and the convergence rate is improved by implementing the mathematical functions. However, if functions are used which tend to oscillate with larger changes to control point values (such as is the case at the beginning of the optimization process with feasible and unfeasible variable vector values), the optimization process may oscillate as well. In such a case it is recommended to start with an all feasible variable vector, which causes the optimization process to search for all variables in the same gradient direction avoiding the oscillations.

VII. Conclusions

Variable sets within large structural optimization problems have to be reduced in order to allow the objectives to be reached within an acceptable amount of computer time. Efficient optimizers are second order dependant on the variable number, thus when reduction of the computer time is required reduction of the variables yields best results. The method investigated implements mathematical functions into the preprocessor-analyzer-post-processor-optimizer loop. These functions are controlled from a small variable set of the optimizer. From these functions an arbitrarily large finite-element set may be controlled. If functions are used which allow the optimum to be represented, this method significantly reduces the variable number virtually without reducing the variational freedom, thus, it allows the global optimum to be calculated.

The method has been used for thickness optimization of a wing-box structure. Three function types, the polynomial function, the spline function and the Bezier function have been implemented. Spline functions yield best results, comparing the optimized mass of the structure, the computer time needed and the optimization behaviour. Variations of control point locations and of the number of functions on the wing-box panels have been investigated in addition, using one 2-D function with 30 control points instead of two 2-D functions with 21, or 27 variables. Increase of the control point number and an increase of the order of the function does not always yield better results. If the control points of a single function are located so that a complex thickness distribution is better reproduced, a single function with a larger control point set may yield better results than a multiple function set.

Using mathematical functions for variable reduction is extremely efficient. Compared to the full variable set optimization, a reduction of computer time of a factor of 4.2 for thickness optimization of the wing-box structure was recorded, using 57 variables instead of the full variable set with 430 variables. The mass when compared to the full variable set optimization was only 2.5% higher on the upper panel when comparing the spline function results.

Further examples for implementing this variable reduction method were given for a delta shaped wing-box of a hypersonic aircraft, yielding a significant reduction of the expected computer time by a factor of 5.8, when reducing the variables from 982 to only 111, using sets of 2-D functions representing the thickness behaviour of the panels, spars and ribs.

Application of this method is extended to shape optimization of a hole, for which the contour has to be optimized respecting a multi-purpose load configuration. The shape of the hole is represented by a Bezier function, controlled by only three variables.

A further advantage of this method is that it provides the possibility to chose the finite-element representation with varying degrees of refinement. Pre-design optimization may be carried out using a coarse finite-element set, detailed design optimization is then carried out with a refined element set, yielding several thousand elements. Both stages of optimization use the same variable set in the optimizer.

Further potential exists for the use of this method and these mathematical functions throughout the design and manufacturing process. Starting from the CAD design, where mathematical functions are utilized for representation of surfaces, thicknesses and geometries, and where these functions may therefore also be used in the optimizer-analyzer loop for optimization of size and shape. This process may be applied with varying degrees of refinement and detailing, depending on the development stage, without change of the functions used for variable reduction.

Finally, these functions may be applied to the workshop's milling machines, allowing the structural parts to be manufactured. The complete process is carried out without change of the function, chosen at the beginning. Thus, information handled throughout this process is strongly reduced, since the function's control point values are the only pieces of information which need to be transferred in addition to the geometry of the structural parts. A significant reduction of computer resources may be envisaged when large structures are designed, optimized and machined.

VIII. Acknowledgement

Results of investigations utilized in this contribution have been obtained during a 1987 research program on structural optimization, carried out by the author at the Institute of Aeronautics and Astronautics of the Technical University of Berlin, Germany. The author wishes to express his thank's to Mr. Bernd Ziegler, who carried out major parts of the research work on this topic. Further thank's are to Professor Johannes Wiedemann, for fruitful discussions, expertise and motivation on the research project.

IX. References

1. U.L. Berkes, "Efficient Optimization of Aircraft Structures with a Large Number of Design Variables", AIAA Journal of Aircraft, Vol. 27, No. 12, December 1990, pp. 1073-1078.

2. B. Ziegler, "Untersuchungen zur Parameterreduktion bei der Optimierung von Flugzeug-Tragflügelstrukturen" ("Investigations on the Reduction of Variables when Optimizing Aircraft Wing-Structures"), Publication of the Institute of Aeronautics and Astronautics of the Technical University of Berlin, Berlin, Germany, ILR Mitteilung 203 (1988).

3. U.L. Berkes, "Zur numerischen Multi-Purpose Dimensions- und Formoptimierung von Scheibentragwerken" ("On the Numeric Multi-Purpose Size- and Shape-Optimization of Plate Structures"), Doctor Thesis, Technical University of Berlin, Berlin, Germany, D 83 (1988).

4. M.E. Botkin, "Shape Optimization of Plate and Shell Structures", Technical Papers of the 22nd Structures Structural Dynamics and Materials Conference, Atlanta, GA, April 1981, pp. 242-249.

5. J.A. Bennet, and M.E. Botkin, "Structural Shape Optimization with Geometric Description and Adaptive Mesh Refinement", AIAA Journal, Vol. 23, March 1985, pp. 458-464.

6. M.A. Save, "Remarks on the Minimum Volume Design of a Three Bar Truss", in: H. Eschenauer, and N. Olhoff, "Optimization Methods in Structural Design", Proceedings of the Euromech Colloquium 164, University of Siegen, 1982, Bibliographisches Institut, Mannheim Wien Zürich (1983).

7. L.A. Schmit, "Structural Design by Systematic Synthesis", Proceedings of the 2nd Conference on Electronic Computation, American Society of Civil Engineers, New York (1960).

8. C. Fleury, and L.A. Schmit, "Dual Methods and Approximation Concepts in Structural Synthesis", NASA Contractor Report 3226 (1980).

9. L.A. Schmit, and C. Fleury, "Structural Synthesis by Combining Approximation Concepts and Dual Methods", AIAA Journal, Vol. 18, No. 10, 79-0721R (1980).

10. J.H. Cassis, and L.A. Schmit, "On Implementation of the Extended Interior Penalty Function", International Journal of Numerical Methods in Engineering, Vol. 10, No. 1 (1976).

11. R.T. Haftka, and J.H. Starnes, "Application of a Quadratic Extended Interior Penalty Function for Structural Optimization", AIAA Journal, Vol. 14, No. 6 (1976).

12. M.J.D. Powell, "An Efficient Method for Finding the Minimum of a Function of Several Variables without Calculating Derivatives", Computer Journal, Vol. 4, No. 4 (1964).

13. W.H. Press, B.P. Flannery, S.A. Teukolsky, and W.T. Vetterling, "Numerical Recipes", Cambridge University Press, Cambridge, New York (1987).

14. H. Akima, "A New Method of Interpolation and Smooth Curve Fitting Based on Local Procedures", Journal of the Association for Computing Machinery, Vol. 17, No. 4 (1970).

15. J. Becker, H.-J. Dreyer, and W. Haacke, "Numerische Mathematik für Ingenieure" ("Numerical Mathematics for Engineers"), Teubner Verlag, Stuttgart, Germany (1977).

16. W.M. Newman, and R.F. Sproull, "Principles of Interactive Computer Graphics", McGraw-Hill International Book Company, Hamburg London Sydney (1981).

17. P. Bezier, "Mathematical and Practical Possibilities of UNISURF", in: R.E. Barnhill, and R.F. Riesenfeld, "Computer Aided Geometric Design", Academic Press (1974).

Knowledge-Based System Techniques For Pilot Aiding

Hubert H. Chin

Grumman Aircraft Systems Division
Grumman Corporation
Bethpage, New York 11714-3582

I. INTRODUCTION

The purpose of this article is to introduce the state-of-the-art in "reasonable-to-assist" applications of knowledge-based system techniques as applied to pilot aiding. The need for a knowledge-based system stems from the heavy workload and demanding decision-making requirements that modern aircraft place on a pilot. A pilot aiding system can reduce a pilot's workload in a stressful environment by taking care of low-level decisions, enabling the pilot to concentrate on the high-level decision making processes [1]. One of the design goals is the pilot aiding system that can operate at varying degrees of autonomy in accordance with NASA's "Autonomous Aircraft Initiative Study" issued in February 1990. Therefore, the pilot can still control the system with a high level of decision-making autonomy and simple overrides [2].

The design of a pilot aiding system requires the merging of real-time operations, embedded computer technology, software engineering, information processing, and knowledge-based techniques into a single focused effort. The Pilot's Associate Program [3], sponsored by the Defense Advanced Research Projects Agency (DARPA) and administered through the Air Force's Wright Aeronautical Laboratories is an important software development program related to pilot aiding. The purpose of the Pilot's Associate Program is to provide an application environment for evaluating the feasibility and potential benefits of knowledge-based system techniques in enhancing mission effectiveness. The program has demonstrated the feasibility of some significant capabilities that could be incorporated into military applications as soon as 1995.

The major difference between the pilot aiding system and a stand-alone conventional knowledge-based system is that in the pilot aiding system the control flow and data flow are in real time [4] while in a conventional knowledge-based system, response time is not so critical to the system. Straightforward implementations of real-time systems are extremely expensive in terms of memory and processing requirements [5]. This poses particular problems in an embedded system environment, such as the pilot aiding system, where these resources are limited. Therefore, more sophisticated partitioning and sharing schemes are required to significantly improve efficiency. The pilot aiding system must depend on efficient control algorithms, scheduling algorithms, and software solutions rather than looking to the hardware to solve these problems [6].

In discussing the pilot aiding system a broad approach is taken, general design principles are discussed, and the knowledge gained through many hard-learned lessons is offered. However, prior to discussing the design details and criteria, the next section (Section II) presents the background that has led to the need for a pilot aiding system. The section also outlines the technological developments that have made the system feasible.

This article covers the design phases, software development techniques [7], system architecture, and knowledge-based construction techniques required to develop, test, and deploy a pilot aiding system. The design phases and software development techniques are discussed in detail in Section III. The seven design phases are the feasibility study, requirements specification, prototype system design [8,9], prototype software development, evaluation, implementation, and maintenance.

The proposed system architecture consists of an executive system, an intelligent component system, and a knowledge base management system. The executive system maintains policies and is responsible for the entire system operation. The intelligent component system incorporates heterogeneous inference engines that make decisions based on stored knowledge and real-time data inputs. The knowledge base management system is designed to maintain and coordinate a knowledge source sharing paradigm. The system architecture is discussed in detail in Section IV.

The heart of the pilot aiding system is the knowledge base. Constructing the knowledge base is a most formidable task because of the difficulty in gathering and representing the knowledge required to implement an autonomous system. For the system to be truly autonomous, the pilots' skills and decision making processes must be represented in a knowledge library. This library combined with real-time inputs forms the basis for decisions made by the intelligent component system. The knowledge base construction methodologies are outlined in Section V. Finally, Section VI provides a summary and conclusions.

II. BACKGROUND

Currently the pilot must perform a variety of tasks within a complex and stressful environment. Knowledge-based system techniques that have been developed for many diverse applications, offer ways to enhance the pilot's situation awareness and improve overall performance [10, 11, 12]. These techniques provide a way of capturing, replicating, and distributing expertise; a way of fusing the knowledge of many experts; a way of managing complex problems; and a way of amplifying and managing knowledge [13,14,15]. Cooperative inference mechanisms work on a global knowledge base to offer advice to the pilot in given situations [16]. The next generations of avionic systems [17] are likely to make use of a pilot aiding system to address the problems of situation assessment, system status, mission planning, tactical planning, and pilot vehicle interface. The knowledge-based system techniques [18], applications, implementations, and design philosophy are introduced in the following sections.

A. Techniques

Humans have a good ability to cope with ambiguous, vague, or uncertain environments. They can make inductive decisions in novel situations and generalize from analogous experience. Knowledge-based techniques are the key towards incorporating these capabilities into a system. The term "knowledge-based system" refers to a knowledge base and inference engine [19]. The knowledge base contains the mission goals, heuristic rules, value judgments, and procedural parts. The inference engine constitutes a reasoning mechanism to postulate assertions, to identify targets, and to select actions. The knowledge-based system technique includes knowledge acquisition methods [20,21], knowledge representations [22, 23], reasoning mechanisms [24], explanation facilities, knowledge management [25], and knowledge validation and verification [26, 27].

The reasoning mechanisms are chosen from the discriminate nets, model-based reasoning [28], rule-based reasoning [29], procedural reasoning [30], probability reasoning [31], fuzzy logic reasoning [32], and evidential reasoning paradigms [33]. Several possible control strategies are developed to formulate a cooperative expert problem solver [34]. One possible approach is based on the "blackboard" paradigm [35]. The blackboard is made available in a shared memory scheme. Applicable rules from each knowledge base are brought into the blackboard, scheduled, and executed.

However, the shared blackboard is highly complex and cannot be easily implemented in a time critical system. The suggested approach is based on a cooperative control algorithm to identify the dependency of tasks, generate a time chart, match the reasoning capability of intelligent components to tasks, and select an appropriate cooperative plan. This plan consists of a set of actions to achieve a system-wide goal.

An important feature of the pilot aiding system is the flexibility of a knowledge base combined with the efficiency of a large data base [36]. This combination allows dynamic, real-time, checking and modification of decisions and actions with a system supporting strong data types and knowledge base compilation. Knowledge and data representations are integrated using object-oriented technology [37, 38, 39]. Within the object-oriented technology, problems of synchronization [40] and security as well as knowledge inconsistency and incompleteness have to be solved [41].

Heuristic and analytic algorithms have to be incorporated into the pilot aiding system. An algorithm is a recipe and if the recipe is followed a certain result will be achieved. On the other hand, a heuristic is a rule of thumb. A result is not guaranteed for a heuristic. The fact is that one algorithm is often worth a thousand rules, but it suffers when the real world does not conform to rigid assumptions. The heuristic provides a means for dealing with a complex situation by transferring expert experiences in response to unanticipated circumstances and operating conditions.

Most of the time a pilot aiding system depends more on access to a knowledge base than on ingenious analytic techniques. This is particularly true when the problem cannot be solved with an algorithm. This occurs because the experts usually have trouble translating their solutions into computer language representations. Knowledge engineers must ferret out solutions and represent them in computer programs.

B. Applications

Modern civilian aircraft systems need to improve in six functional areas: flight-path (autopilots) management; aircraft system management; mission (flight) management; information management; air traffic and environment management; and automatic landing management. The potential benefits of using a pilot aiding system are reduced operating costs, better safety, reliability, maintainability, work-load reduction, increased navigation accuracy, better control precision, and economy of cockpit space. Reduced fuel consumption may prove to be the most important. Performance improvements are typical requirements for future systems, though reduction of work load is often an explicit design goal that guides the design of a pilot aiding system.

Future high performance military aircraft will be required to exhibit substantial enhancements in aerodynamics, propulsion, and avionics. They will need greater maneuvering performance coupled with greater sustained capability. They will also need advanced avionics that are capable of gathering, correlating, and operating on vast amounts of data for many critical mission functions. The Pilot's Associate Program suggested the basic principles of knowledge-based techniques. Emphasis is placed on specific functional areas including situation assessment, system status, mission planning, tactical planning, and pilot-vehicle interface. These functions allow the pilot to maintain situation awareness, plan tactical maneuvers and navigation, monitor internal systems, and perform other tasks required to achieve mission success. The pilot aiding system will demonstrate competence and proficiency in terms of intellectual monitoring, controlling, and decision making ability.

C. Implementations

Most of the automation that can be found in aircraft systems today simply replaces human controlling functions with computerized machine functions. They are mechanical and algorithmic designs based upon well-understood principles of modern structure and control theories. The Air Force Studies Board, in a 1982 study, defined automation as "Those processes by which essential functions can be performed with partial, intermittent, or no intervention by the pilot." Such systems are impressive in their precision and reliability, but in no way challenge the human intellect. The new approach to aircraft automation is autonomy. Autonomy is a subjective property which refers to the degree of independence a system tends to elicit. An autonomous system provides reasoning mechanisms, planning, and explanation facilities.

The major differences between automation and autonomy are the decision making capability and the level of functional responsibility in flight. The electronic auto-pilot assumes responsibility for all control activities required during a flight. The human pilots first enter their initial command and then merely monitor the progress of the flight and the performance of the systems. Even though the auto-pilot acts automatically, the system still requires the human pilot to provide sufficient information about his decisions and alternative actions. Although the Pilot's Associate Program is designed to perform many of the control tasks, it acts more as a co-pilot and interacts with the human pilot. In working with it, the pilot remains very much "in the loop" procedurally. The pilot aiding system should be capable of intellectual function and decision making ability, even to the point of monitoring the performance of a human pilot and recommending that corrective actions be taken to compensate for a human mistake or system malfunction.

D. Design Philosophy

The pilot aiding design philosophy is based on the principles of adaptive aiding and information fusion. Adaptive aiding optimizes information transfer to match pilot ability and task loading, and provides physical assistance if appropriate. Information fusion exploits expert perceptual abilities and computerized information processing strengths to enhance pilot's situation awareness. The pilot aiding system is designed to provide capabilities complementary to the pilot. Many pilot's limitations are pilot aiding system strengths. The strengths are used to compensate in attention, allocation, and planning by providing a comparative reference.

Delegation can be accomplished through pilot consent before system responses or through non-interference with system responses. The exact mechanism should be defined at the system design level. What actions should the pilot delegate to the pilot aiding system? There are two approaches to this problem. One approach is for the aiding system to control the intensive time-critical operations. Another approach is to share tasks between pilot and aiding system based on resource availability. The latter approach allows the pilot to maintain proficiency with manual skills. In practice, a combination of these two approaches allows the pilot to exploit the system's capability but remain prepared to operate the aircraft manually if the autonomous system fails.

The task allocation scheme should optimize the pilot aiding system effectiveness with no performance degradation. Future pilot aiding systems are moving toward situations involving inordinate amounts of data and information covering multiple and diverse fields of interest. Compounding the problem is the need to make decisions in less time than it takes for most human beings to review the incoming information. The pilot's fundamental role has undergone a significant shift. The pilot becomes more of a manager than an operator and will remain a critical part of the aviation system for the future. He will spend an increasing proportion of his time monitoring the pilot aiding system that is taking over a growing share of his flight tasks. The role of the pilot aiding system will continue to change dramatically as knowledge-based systems become more sophisticated and complex.

III. KNOWLEDGE ENGINEERING DESIGN

The design of a pilot aiding system presents a special problem to knowledge engineers. The Defense Science Board Task Force Report on Military Software highlighted the concern that the traditional software process models were discouraging. The challenge is not just to *"design the system right"* but to *"design the right system right."* Designing the right system means building products and services that meet real customer needs. Currently, the requirements for pilot aiding systems are not well defined at the beginning of the development life cycle. These requirements are constantly in a state of flux, and once deployed, the systems will continue to undergo constant change. Proven and effective approaches, such as prototype methodology and reusable software components presented in this article, appear to offer a great deal of promise on cost reduction, schedule maintenance, and user satisfaction.

Prototype design is a refinement procedure between pilots, customers, and designers. Furthermore, experience with the evolving prototype leads to a better understanding of system functionality, dynamics, and performance so that the resulting products and services are improved to match actual requirements. Software reuse is the process of incorporating the iterative prototype methods into a productive system without reinventing the same logic repeatedly. The advantages are greatly reduced development time and cost, and increased reliability because the reused components have already been shaken down and debugged.

The prototype design refinement process consists of the design of three different systems, each with a different degree of sophistication: rapid prototype, evolutionary prototype, and operational prototype. Initially, a rapid prototype is created in a quick fashion for providing the system interfaces and simply deploying the major functions, while not necessarily being bound by the same hardware speed, size, or production constraints. Demonstration and iterative refinement produce a variety of design alternatives for improving the system performance. The evolutionary prototype is constructed using an assumed set of requirements to gain experience, to uncover new requirements, and to validate the design approach. The selected version becomes an operational prototype that is modified and redeployed whenever new information is learned. The operational prototype incorporates those features that are agreed on and can be deployed in an operational environment. After the operational prototype has been verified and validated, it becomes a production system. The eight phases of the prototype development life cycle are shown in the Figure 1. The second, third, fourth and fifth phases iterate until they produce a production system.

A. THE FIRST PHASE: Feasibility Study

Before any actual development, heavy emphasis is placed on conducting an initial investigation of the application domain and feasibility of the proposed project. The objective is to ensure the pilot aiding system will satisfy a real need and minimize all possible risks. The feasibility study concentrates on four major feasibility areas: human factors feasibility, technical feasibility, cost feasibility, and alternatives. Feasibility and risk studies are related

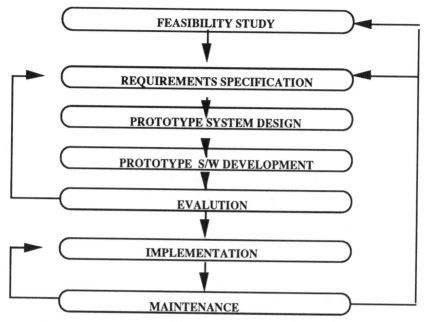

FIGURE 1. PILOT AIDING SYSTEM DEVELOPMENT LIFECYCLE

in many ways, but, in general, if the risk is great, the feasibility is reduced. The risk study consists of identification, estimation, assessment, monitoring, and controlling of the risks.

1. Human Factors Feasibility

The human factors study examines the pilots' perceptions and feelings about the information processing tasks that they currently perform. The feasibility study is to identify the specific activities within an information processing problem area where the potential systems are best regarded as "tools," "advisers," "surrogates," "replacements," "supervisors," or "subordinates." The specific functions of the pilot aiding system are designed according to the purported strengths of each. These functions are carried out to enhance overall system performance, reduce risk, and encourage the pilots' acceptance.

2. Technical Feasibility

Technical feasibility begins with an assessment of knowledge-based system techniques. The considerations include development risk, resource availability, and technology mutuality. Development risk concerns the necessary function and performance that will be achieved within the constraints. Resources include skilled staff who is competent and available to develop the system, and facilities to build the expected system. The relevant technology is required to progress to a state that will support the system. Evaluation metrics can be used as direct quantitative measurements, but may not be the best way. It is difficult to establish quality.

The best way is to establish a mockup of how the system can be produced. Building a product mockup means constructing a user interface scenario and generating a videotape of the prototype functions. The videotape is used in the technical feasibility study for two reasons. First, it is shown to the group participants who are encouraged to vocalize their reactions to the system. Second, it is also shown to potential users who are asked to fill out a questionnaire designed to assess their willingness to learn the use of the system being suggested. The survey probes the perceived needs of the user population and asks if system depicted in the videotape will respond to these needs. The future users' reactions to the mockup will generate information as to whether the pilot aiding system, as currently envisioned, will be acceptable. This information is usually acquired through the group studies and user surveys.

3. Cost Feasibility

The cost feasibility study is complicated by criteria that vary with the operational functions of the system to be developed. The cost estimation is categorized by set-up costs, start-up costs, project-related costs, ongoing costs to determine a return on investment, a break-even point, and a pay-back period. The cost-effect analysis considers incremental costs associated with improvement of the functional capability and system performance.

4. Alternatives

This study examines the effects of individual technologies. Alternatives always exist and it is necessary to evaluate their solutions based upon a sample selection, optimization techniques, and statistical techniques. The process consists of evaluation metrics, tradeoffs, and risk assessment. The selection of comparative parameters includes performance, effectiveness, reliability, maintainability, life cycle costs, etc.

The feasibility study phase is not warranted for a system in which cost justification is accurate, technical risk is low, pilot's acceptance is expected, and no reasonable alternatives exist. The major accomplishments are statements of system architecture, reasoning mechanisms, knowledge base, data base, operational environment, and test scenarios. The statements may not be very detailed at this point and they will be used in the next phase.

B. THE SECOND PHASE: Requirements Specification

After the feasibility study, the second phase must represent the study of both the information and functional domains of the system, and development of a reasonable requirements specification. The requirement analysis must focus on top level architecture, interface, and knowledge-based design rather than detail procedural design. The users' requirements will most likely change during each cycle of the prototype development lifecycle. These changes could be faster processing requirements, new information needs, or reformatting. A final version of the requirements specification is produced at the end cycle of the operational prototype. In many cases the final version is accompanied by an executable operational prototype and preliminary user's manual. The manual can serve as a valuable tool for uncovering problems at the pilot interface. The knowledge and inference mechanisms allocated to each task are refined to an acceptable level of the task allocation scheme. The requirements specification will establish a complete information description, a detailed functional description, an indication of performance requirements and design constraints, appropriate validation criteria, and other knowledge pertinent to the requirements. The following several baseline requirements are described.

1. Cooperative System

The cooperative system requirement comprises of cohesion within each inference engine and the coupling of inference engines into a system. The pilot aiding system usually requires a set of inference engines for problem solving on a global knowledge base or distributed knowledge base. Breaking up the problem domain into manageable sub-domains that are distributed to inference engines is essential to the requirements. The cooperative system requirement involves not only mapping of the global knowledge base to locate relevant inference engines, but also selecting an appropriate scheduling plan along with conflict resolution strategies and communication methodologies. The cooperative scheduling plan is proposed at the system architecture stage.

2. Open Architecture

An open architecture approach fulfills the requirements of legibility, device independence, structure, and self-containment along with portability to a different but similar computer and easy-to-use foreign system interface. In general, a large scale system cannot be introduced into an avionic system, it must be subtly and incrementally embedded into the overall system. It is important to adhere to global standards and to rely on highly modular approaches to architecture design and implementation. When building prototype systems, the incremental approach that has been utilized is acceptable to keep up with the complexities of a deliverable system.

3. Time Critical Issue

The time critical issue is important to a pilot aiding system that must operate under extremely rigid performance and reliability constraints:

- Data acquisition must be conducted at per-defined intervals
- Analysis must be performed within specified execution time constraints
- The database must be updated at defined intervals
- Controller interaction should not impede any other system functions.

For these reasons, emphasis is placed on design requirements for real-time control and response-time analysis. To provide a timely answer to the task at hand, the knowledge-based system must know how intelligent it can afford to be, versus how intelligent it has to be. The response time is dependent on the real time design and computational efficiency. Interrupt processing, data transfer rate, distributed data base, operating systems, specialized programming languages, and synchronization methods are concerns of a system with time critical issues.

The prototype methodology follows a dictum: "Get it to work, then make it fast." Sometimes, this is acceptable in the prototype of a small scale knowledge-based system, but it can be a fatal blow when expanding the prototype to a large scale system. Thus, time critical issues have to be addressed aggressively and explicitly in requirements as follows:

- It is not necessary to demonstrate and refine the system prototype to consider the time critical issue

- During operational design, selection of modules must be done in a manner that achieves time critical optimization
- Redesign and recording in a machine-dependent language are required to improve time efficiency.

4. Pilot Interface

The functions of the physical pilot interface should be separated from the underlying functionality in a pilot aiding system. The interface changes dramatically because many of the design decisions are based on the cockpit design. These include the Hands-on-Throttle-and-Stick, the Helmet-Mounted Sight, the Voice Interactive System, and manual inputs to the touch screen or any switches/levelers/valves in the cockpit. This separation will emerge later on as part of the system.

The requirements of the pilot interface are tuned to match the pilot's thinking processes about tasks and his methods of mental data manipulation. During operation, pilots can easily interact directly with the system, if required. In the real-time, most of the input comes from the mission computer data bus, data links, and sensor management system.

For simplicity, the pilot interface is defined as "the things the pilot sees or hears, the way information is entered, and the way interactions are structured." In demonstration, the interfaces are messages telling the pilot what the system is doing and the selection of function keys to control the system. Commercial user interface tools can be used for offering menu selection techniques and focusing on services for the exploration and specialization of interface characteristics. Such tools would allow the designer to build a system I/O manager and to capitalize on distinctions between the physical pilot interface and the underlying functions as they emerge.

5. Task Allocation

Task allocation requires consideration of the pilot aiding system capabilities and the pilots' expertise. The allocation of a task must be dynamic flexible to adapt to such factors as task complexity, processing load, and the problem solving strategies. The requirements describe the circumstances under which a task should be allocated. The following steps will be helpful in allocating a task for the pilot aiding system.

a. Task invention

Task invention requires analysis of how pilots perform assigned tasks to apply them to the pilot aiding system. A principle objective of dynamic allocation is to equalize cognitive effort relative to capacity and diverse pilot and computer devices.

b. Task identification

Task identification is based on problem solving behavior to define functions, verbal protocols, rules, procedures, and principles of each task.

c. Performance evaluation

The measurements of "difficulty" and "quality" in analytic methods, concurrent performance, or direct subjective judgment are assessed. These measures are used to determine whether the task should be performed by the pilot alone, the system alone, or some combination of the two.

d. Task dependence

In the effort to share tasks between pilot and the pilot aiding system, it is important to investigate the relationship among all tasks. Task dependence is measured by two qualitative criteria: cohesion and coupling. Cohesion is a measure of the relative functional strength of a task. Cohesion can be sequential, logically, or temporal. Coupling is a measure of the relative dependence among tasks. Tasks can be classified as functional control coupled or data content coupled.

C. THE THIRD PHASE: Prototype System Design

The prototype design focuses on the system architecture and incorporates application-specific requirements into the design phase. The prototype design is a process consisting of three cycles. The first cycle uses a small team of highly skilled designers to build a prototype in a quick-and-dirty way and demonstrate its functional capability. In the second cycle, the designers collect the customer's new requirements and system improvements. These changes are incorporated into a new prototype in an evolutionary manner. Once the evolutionary prototype is built, the third cycle starts. The third cycle improves the prototype to meet operational environment requirements. The prototype is integrated with the pilot interface and operated in real-time. Finally, as these features are bundled into new versions of software, new releases are baseline and redeployed.

The prototype strategy is based on design changes in each cycle that are derived from a more thorough understanding of system performance gained by observing the effects of previous modifications. A more detailed description of the three design cycles is as follows.

1. Rapid Prototypes

The major objective of the rapid prototype stage is to demonstrate the technical and economic feasibility of the intended system. An early demonstration has advantages from both the management and technical perspectives. Management may be reluctant to fund a major activity in a new technology without a demonstration of likely success. Technical designers value an early prototype as a way of stating system requirements precisely and providing a testbed for future system development. The rapid prototype is itself a statement of the requirements specification in terms of interfaces, goals, and principal reasoning paradigms. It also provides experience solving system problems and helps develop methods for organizing the knowledge base.

There are numerous approaches available to construct a rapid prototype, including the following:

- Use the quick-and-dirty manner to create a prototype to enhance the communication with users
- Reuse software components from a software repository to reduce development costs
- Use expert shells available to assist in rapid creation to reduce cost risks.

Where possible, rapid prototype should assemble rather than build. The assembling method uses a set of existing software components. A software component may be a data base, knowledge base, software program, or procedural component. The software component must be designed in a manner that enables it to be reused without detailed knowledge of its internal workings. In order to maintain the program component's reusability, a system library must be established so that the existing components can be catalogued and retrieved. It should be noted that an existing software product can be used as a rapid prototype for the system. The rapid prototype is concerned typically with only a subset of the problem and may not be satisfying all the requirements.

2. Evolutionary Prototypes

Following the rapid prototype and project approval, the objective of the evolutionary prototype is to develop the range of functions needed to deal with the full complexity of the problem. Improvements from the rapid prototype typically include expansion of the knowledge base in terms of both coverage and sophistication. It also includes expansion of the representation and links to other systems and data bases such as transaction flows and data storage devices.

The evolutionary prototype is an expansion of the rapid prototype and is tested with the same verification and validation procedures, but in a more detailed form. The differences between rapid and evolutionary prototypes are as follows.

a. Degree of Requirement

In rapid prototype, only those parts of the system that are not well understood are built. In evolutionary prototype, those parts of the system that are well understood are built, so that there can be continuous development on a solid foundation. Each increment is slightly more risky than the preceding one, but the experience with preceding versions has provided enough insight into the problem to make a risky endeavor much less so.

b. Difference in Quality

Rapid prototypes are built with little or no robustness while evolutionary prototypes must have all the quality built in up front or they will not be able to withstand the necessary levels of use and modification. This stage has to verify features, retrofit quality, and maintain a schedule of planned enhancements.

There are numerous design problems associated with evolutionary prototype.

a. The user-friendly development features of powerful expert system shells are left behind and only those features important to the ultimate users are included. This may involve a rewrite in an object-oriented or object-based programming language that will run on mini-computer or mainframe.

b. Setting standards and guidelines for appropriate levels of quality and documentation is very important. If any new increment results in a dead end, it has to start over. The level of quality should not be insufficient or unnecessarily high.

c. Comments from users about the use of the evolutionary prototype must be obtained expeditiously, so that they can be incorporated into the next design cycle.

d. As experience is gained and more requirements are incorporated with the addition of each new feature, the new changes become ever more risky. These changes must be incorporated with as low risk as possible.

e. To build an evolutionary prototype to accommodate large numbers of major changes is easier if based on an open architecture software. Some techniques are very useful for changes, such as information hiding, low inter-module coupling, high module cohesion, object-oriented development methods, and sound documentation practices.

If the evolutionary prototype offers adequate performance, development of an operational prototype may not be necessary and it may be possible to go directly from the evolutionary prototype to the deliverable system.

3. Operational Prototype

The operational prototype combines the best of rapid and evolutionary prototype features and optimizes critical system parameters such as performance or memory usage. The features incorporated are well known, understood, and agreed on. This design cycle is appropriated for changing and testing the design requirements, for exploring alternative approaches, and for evaluating improvement in an operational environment. These design requirements are incorporated into the prototype for immediate verification and validation.

The verification and validation are performed on a flight simulator. Pilots are commissioned to test the prototype operations. The pilots' experience will help uncover problems and suggest improvements that can be made. Desired changes are made on top of the working baseline.

When the new features are incorporated into the new operational baseline, the pilots use the modified prototype to ensure that the requested feature really is what is needed. If the operational prototype stages have been conducted and documented clearly, a set of requirements will be available for modifying the deliverable system.

D. THE FOURTH PHASE: Prototype Software Development

The objective of the software development phase is to specify completely the software architecture required for the prototype being designed, and to translate the design descriptions into code modules in a development environment. For developing software components, Object-Oriented Design (OOD) is a promising technology and Ada is a suitable programming language. Ada is not truly an object-oriented programming language, because it does not support dynamic binding and inheritance. Dynamic binding can be an alternative to using types and accessing types. Inheritance can be accomplished by a variant record for the class type. The different variants of the record define the alternate sets of attributes for the associated subclasses. However, Ada does support the concepts of encapsulation (package), information hiding (private types and package body), and concurrent processing (tasks).

OOD enables builders to indicate the object classes, to specify their structures, and to use them. It is important to note that the use of OOD can lead to extremely effective implementation of top-down, progressive refinement, and reusable techniques. The top-down technique is based on two main ideas. One involves representing the control structure and its interaction with data. The other involves establishing a guideline for grouping functions into object classes (or simply classes). The classes that are specified and implemented for the current prototype can be catalogued in a library.

For functional refinement, the builder can specify the prototype using existing classes contained in the reusable library rather than inventing new ones. By using existing classes, the development time will be reduced substantially. Even more important, the software will be more reliable.

During program development, all components in the system are represented as a collection of individual class packages. There is no coherent system by which the classes can interact. The first step in building the system is to couple all the classes together by making declarations in a class package that allows any other class package to call a method in the class package where the declaration is made. The main subprogram allocates the necessary data values of the basic classes and starts them executing. The goals of implementing classes are:

- Keep the classes independent of the application
- Facilitate the interface support packages in coupling classes
- Make the number of data values of a class variable dependent solely on the application.

The stages of prototype software development are initial development, functional refinement, and program development.

1. Initial Development

This stage starts with a high-level design description of the prototype. The objective of initial development is to define all important classes, attributes, methods, messages, and interfaces in the prototype design. During the initial development, several questions often trouble designers, such as, "Which elements of the system are to be classes, and which are not?", "What is the best way to decompose the knowledge structure into a set of classes and their relationships?", "Which classes provide methods that can perform the required algorithms or inferences?" and "What kinds of messages are required?".

The answers to these questions are concerned in the areas of data abstraction, knowledge abstraction, and communication discussed below:

a. Data Abstraction

The data abstraction process is performed in a top-down analysis method to generate classes and their relationships. Often the classes correspond to system functions and elements based upon the prototype being modeled. At the top level, attributes, constraints, and methods are combined into a new class. As the top-down analysis proceeds, this process is repeated at each new level of abstraction. New levels of abstraction should provide some attributes or methods that can not be expressed at the previous level. If the new abstraction inherits from another class, inspect the methods of that class to see if any need to be overridden by the new class. It is important to note that a new class may or may not have meaning as a subsystem by itself. The purpose is to collect common data variables and methods called an abstract class. It may be viewed as a link between data access and data processing.

The most difficult step is to determine the aggregation, generalization, and association structures of classes in addition to single or multiple inheritance. These structures do not use often in object-oriented programming, but in database they are crucial. Aggregation structure is the "is-part-of" relationship of an owner class and its subclasses. For example, electronic warfare(EW) is organized into electronic warfare support measures(ESM), electronic counter-measures(ECM), and electronic counter-counter-measures(ECCM). The EW is an aggregate of ESM, ECM, and ECCM. These three subclasses inherit different attributes and methods of the EW. Subclasses are depends on the existence of owner class. Each subclass exists independently of other subclasses unless existential constraints are specified. Aggregation relationships are validated when the owner class is modified.

Generalization structure is the "is-class-of" relationship of an owner class and is specified by the AND/OR constrain between subclasses. For example, airborne electronic warfare(AEW) is a subclass of EW. The AEW inherits the EW and also inherits its aggregation structure. Each subclass is allowed multiple inheritance from different owner classes. Subclasses have the AND/OR option of coexistence with the other subclasses under the owner class. Therefore, relationships specify the conditions of "is-class-of" to be a generalization structure.

Association structure is the "is-set-of" relationship among a group of related classes of different types. For example, a current EW design method is through mission requirements. Each mission requirement is formed a subclass. The EW design method is an association structure of subclasses of mission requirements. Subclasses associated together do not form a new class and do not refer to a combined data structure. They can be a set of declared constraints against forward chaining and backward chaining to achieve a goal. A method of an association, referred to as a forward/backward chaining, is defined on the constraint-satisfaction of these multiple subclasses. The advantage of this is to reduce the search space for a method.

b. Knowledge Abstraction

Knowledge abstraction is the knowledge representation structure embodied by a program that maintains a control mechanism over the use of the knowledge. Knowledge representation deals with the structures used to represent the knowledge provided by domain experts. There is no single global structure to represent knowledge in the most effective manner. A number of different approaches are suitable based on the problem domain, the potential set of solutions, and the level of knowledge stored. These difficulties helped motivate the development of domain-level knowledge on frames, and semantic networks. They also spurred the representation of knowledge through an object-oriented representation at the programming level.

A frame provides a structured representation of knowledge. The knowledge consists of concepts and situations, attributes of concepts, relationships between concepts, and procedures to handle relationships as well as values of attributes. Each frame could be represented as a separate class. In a semantic network scheme, knowledge is represented in term of objects and associations between objects. Objects are denoted as nodes of a graph and associations as edges connecting nodes. A semantic network can be viewed as a set of classes and their relationships.

Knowledge abstraction is a combination of stepwise refinement and program verification. Stepwise refinement starts with a knowledge structure (facts) included in a class description and refines that description into smaller and smaller subclasses. Program verification specifies the desired structure behavior type to conform to a desired result. The knowledge structure becomes a knowledge base when the data variables of each class are instantiated. The essential counterparts to knowledge structures are control mechanisms referred to as inference engines. The inference engines and algorithms are combined with the use of tasks that depend on a library package. Some provision must be made for keeping the main subprogram running until all library tasks are completed.

c. Communication

The communication step is to design all messages. Classes are polymorphic, that is, the same message can be sent to both a class and its subclasses. A message is the description of methods to be executed in a class. Classes consist of three parts: private, public, and interface. Private sections are encapsulated and can only be accessed or modified with the activation of their methods within a class. Public sections can be accessed by accepting messages through an interface section. Upon receiving a message, a class may process it and take a number of actions. These include modifying its states, sending one or more messages to other classes, and creating new classes.

The query messages to classes are used to get the data value of attributes. In querying the class, it is often required to use a retrieving method which can access the values of a class and its subclasses. Updating methods include inserting, deleting, adding, and modifying the attribute value of classes. For carrying out necessary inference procedures, calling messages are passed to a specific association. The returning message is true when appropriate conditions are satisfied, otherwise it stops evaluation and generates a "failure" message. Ada provides a rich set of options for task communication through the rendezvous mechanism.

2. Functional Refinement

The functional refinement is a process of elaboration achieved by successively refining the prototype development cycle to provide increasing detail in structures, relationships, and functionalities. Each refinement cycle ends when all the classes and their methods are checked. The functional refinement consists of class refinement, method refinement, and relationship refinement.

a. Class Refinement

The top-down refinement strategy is utilized for refining classes. The attributes of a new class are required to specify. The constraint part has to be expanded to include any additional constraints that were identified. The class refinement is an outcome of the attempt to enhance the characteristics of the prototype. Therefore, these new classes may be regarded as nested within the old class or they are the parent classes of the old one.

b. Method Refinement

During the "method refinement" step, new methods are created to satisfy the description of functional refinement. Parameters of a new method are required to match the data type of attributes. For example, when the return argument of a method is specified as a constant, all data values of the class have the new values. Method refinement on the classes may be very complex or very simple depending on the requirements.

c. Relationship Refinement

The relationship refinement defines the relationships based on the control flows of the knowledge base modification . Relationships describe semantics in the classes for a particular application domain. To construct new relationships between classes, the semantic description is a combination of structure and control. Control is developed and driven by the relationships on classes and the message passing, respectively. Structures ensure that the relationships that carry out the class are semantically significant.

3. Program Development

In the program development stage, the data/knowledge base design comes first and the application program comes next. However, these two steps are coupled together closely since the algorithms and procedures that manipulate parameters must be included in the class definitions. The class hierarchy is analogous to the conceptual scheme of the conventional data structure except that the class definition contains more behavior constraints for each class. The application program can be viewed as collections of messages in which only the name of the class and the methods are known. Object-oriented programming can be viewed as a computer program development tool. When programs are grown, classes become numerous and relationships among classes become more complex. A system reusable library of classes will be important to support the development process.

Object-oriented programming as used here is concerned with the organization of the messages to allocate the data values and the manipulation of methods. In Ada programming language, each package defines a single class within the library package. The run-time attributes are referred to as data values. A conventional approach is to declare the task as a method of a class. The task itself must represent a method of the class and the body of the task must contain the class. In order for the method of a class to access the attributes, the task must have an entry call for each class. A second approach (polymorphism) is to declare the task as a message. The benefit of this approach is that a particular data value is available to the methods without making entry calls to the class. The tasks support concurrence in a real-time implementation. One of the most valuable features of Ada is its capability to represent and manipulate data (value of attributes), meta-knowledge (structure of classes), and knowledge (relationships among classes) in a uniformed manner based on object-oriented techniques.

E. THE FIFTH PHASE: Evaluation

This phase is the key to successfully building a pilot aiding system. Evaluation consists of two steps: validation and verification, and demonstration and testing.

1. Validation and Verification

Validation is the process of ensuring the system satisfies its customers' needs. Verification of a prototype is to check the coding accuracy of the programs developed and eliminate all technical errors. The Verification and Validation (V&V) are composed of a number of activities that occur between stages in a development process and apply to the various modules produced. These activities are categorized by the reviewing requirements, checking programs, certifying inference engines, validating knowledge base, and validating performance.

a. Reviewing requirements

Prototypes serve as the basis for writing requirements. Even where they are incomplete or go slightly astray, a V&V review enables the requirements to be defined at a detailed internal level.

b. Checking Programs

After the "high-level" design is completed, the verification may proceed to a lower, more detailed design, or go directly to the code. A desk check or walk-through can result in a level of confidence in the program's correctness. Even the code-level examination leads to confidence in the program's correctness. This verification can be done with the code, while validation could consists of observing the execution of program.

c. Certifying Inference Engines

Certification means looking at the knowledge base and verifying the facts, rules, classes, and other components to determine whether an inference engine carries out the knowledge processing paradigms specified in the requirements specification correcting. Certification of an inference engine involves:

- A formal definition of an inference engine
- Specification of how a given inference engine operates on a given knowledge base
- Development of scenarios for testing the inference engine
- Performance of certification tests on the inference engine
- Validation of all explanatory information provided by the inference engine.

d. Validating Knowledge Base

The knowledge base consists of knowledge sources and structure. The purpose of validation is to go over the knowledge structure to see if it matches the high-level design. Validation also checks the knowledge source of each association/method (rule/reasoning), and goals for correctness. The following are specific items to check:

- Confirm that the knowledge base conforms to the requirements at each level of functionality
- Verify that the knowledge structure includes classes, relationships, and methods to merge additional informations
- Confirm that the knowledge source is correct and reasonable within the content of a particular scenario and within mission requirements
- Identify the portions of the knowledge base that are independent of the inference engine (Independence provides the changes of knowledge base that will not affect the inference engine.)

e. Validating Performance

Even if each step along the way has been checked out, it is necessary to validate the behavior of the integrated system separately. This is required to discover errors that only appear at this point and to check against the customers' real needs. The followings are specific items to check:

- Determine the performance criteria such as accuracy, adaptability, adequacy, appeal, availability, precision, realism, reliability, resolution robustness, sensitivity, technical and operational validity, usefulness, plain validity, and wholeness
- Develop scenarios specific to the problem and demonstrate the system's proper response

- Validate the system performance by an independent panel of experts who are not connected with the system development effort
- Maintain detailed information on system performance as the knowledge base is elaborated.

2. Demonstration and Testing

To assure conformance of each prototype cycle to the requirements, demonstration and evaluation are a necessary part of the design process. This is particularly true regarding consistency and cost considerations. The customer must provide concrete evidence of the system usefulness that can only be obtained by observing the pilots in realistic situations interacting with the system.

The primary purpose of the prototype system tests is to collect measurements of the pilot performance and workload while working with the pilot aiding system. Performance measurements are used to evaluate how well the pilot completed tasks related to the mission goal. Videotape review and verbal protocol procedures are suggested to assist the pilot in evaluating workload levels using six criteria of ergonomics. They are mental demand, physical demand, temporal demand, amount of effort required, workload related to satisfactory performance, and frustration level.

The procedure is to have pilots perform the same generic mission using both the baseline and pilot aiding system. They are asked to rank the "level of help" in situation awareness, decision aiding, workload reduction, and performance enhancement. Some caution is needed when interpreting the demonstration results. For example, the evaluations are based on simulation, not on a real-world mission environment, and pilots may need considerable training before optimal performance with the system can be attained.

Evaluation also relies ultimately on pilot skills to identify test cases and produce meaningful results. The results can often have serious consequences on the reliability assessment of the system. It is well known, however, that testing can only establish the presence of errors but cannot assure their absence. If the system does not satisfy what the customer needs, the requirement definition stage is re-opened and new requirements are drawn up to meet the new challenges. The following evaluation facilities can be used separately for testing a rapid prototype, evolutionary prototype, operational prototype, and delivery system.

a. Review Demonstration

Design reviews prove the concept feasibility by demonstrating a rapid prototype that can be run on personal computers or work stations. Recommendations often depend on thoroughness with which the review is conducted and the experience and attention of those conducting the reviews. The reviews are part of an overall management strategy to meet the system goals and to assure conformance with the specifications of the pilot aiding system.

b. Computer Simulation

Computer simulation involves the demonstration of the system capabilities in some simplified manner before implementation. The computer facilities utilized are either mini-computers, mainframes, or super-computers. The software is executed on a system in a simulated environment. Since the evolutionary prototype is not required to operate in real-time, a slow response is expected. The results can be used to study the capabilities of operations, system resource utilization, program design, and the adequacy of inferences.

c. Laboratory Simulator

Flight simulation is a real-time interactive simulation that produces detailed mission scenarios for testing operational prototypes. The major testing emphasis is on the embedded hardware and real-time software. Scenarios are used to identify the types of activities for which the prototype must provide output information to meet the requirements. Therefore, the evaluation will determine a system performance, system configuration, and availability characteristics.

d. Flight Test

The primary purpose of flight test is to collect measurements of pilot performance, situation awareness and workload while aided by the deliverable system. Results are compared with the performance and workload of pilots, flying the same missions in the same type of aircraft, without the pilot aiding system. A secondary objective is to obtain pilot opinion on the system implementation and usefulness. The goal is to determine the advantages and benefits resulting from use of the knowledge-based pilot aiding system.

F. THE SIXTH PHASE: Implementation

The implementation phase consists of acceptance test, user training, and conversion. The pilot aiding software is installed on an appropriate embedded system. The final pilot aiding system may not be completely satisfactory, but the essential behavior of the system is defined and complete. Compromises may be made in areas that have dependencies with the embedded hardware, functional areas, and real-time responses. The implementation involves system integration, software implementation, and real-time control that are described as follows.

1. System Integration

Any changes that are replacements for unfitted modules should be replaced with new modules. If the changes go beyond the original specifications or cannot perform the new requirements, they should be modified. The upgrades and modification of the system are through step-by-step additional changes and monitoring of the system performance after the changes are made. New or modified components that disrupt the system's operation are removed or restored to their earlier version until their implementation flaws are corrected. When the operation and performance of the system meet the requirements specification, the final system can be transmitted to the field for release.

The knowledge engineering environment should be capable of generating the system front end. This modification and synergy become even more important when one considers the life cycle of an application. It is very desirable to avoid having any change in either the stable storage architecture or the in-memory knowledge base which impact the user interface. Indeed, in maintaining a large scale system it becomes critical to be able to make a change in one location and to have the knowledge engineering environment reflect this throughout the infrastructure of the entire application. This ability would greatly reduce the development time as well as the associated maintenance of any full scale system.

2. Software Implementation

In Ada, the tasking model has been pulled out of the realm of the operating system and made a part of the high-level programming language itself. The real-time multi-tasking kernel for a 16-bit microprocessor or 32-bit RISC system is specifically designed to answer the needs of real-time embedded applications.

The Ada compiler vendors are presently engaged in the process of designing run-time compilers for cross-development efforts in embedded system software design. Cross-development is required any time that a target computer cannot conceivably host its own development tools. In such cases, it makes sense to develop the application code on a host computer and then download it to the target machine.

Parallelism arises at the beginning of cooperative processes in knowledge-based system development. There are two reasons: the system hardware can be parallel; and the problem can be parallel, as in real-time control situations. Software designed using Ada programming and the object-oriented technique is best suited for dealing with parallelism in implementation.

Unfortunately, the consensus in the current implementations of object-oriented prototypes is that they are too slow. Although many factors are involved, two are obvious. The first is class granularity. When the number of classes in a system is huge, overhead is considerable. The second is that the time spent on class searching is also considerable. The strategy for implementing the deliverable system is two-fold:

- Remove as much of the structure overhead as possible by optimizing the size of classes and the decision-aided domain

- Minimize the time required on class searching by using radix search trees to implement dynamic hashing technique.

3. Real-Time Control

An embedded real-time system is extremely expensive in CPUs (Central Computing Units) and has limited memory resources. The pilot aiding system is required to be implemented in this kind of architectural model. Thus the focus on resource allocation, response time, and communication problems is useful for real-time control of the system implementation.

a. Resource Allocation

The resource allocation is based upon software architecture to allocate processors dedicated to various inference engines and functional managers for the desired performance. The software architecture has to partition its application domain into several sub-domains those parts which are executed as stand-alone knowledge-based systems on each CPU. Each local memory contains its inference engine software and mapping table of global knowledge base. The global knowledge base is stored in the main global memory. The high speed data bus is used for communication.

b. Response Time

The time it takes to complete a task is unpredictable because it depends on the number, arriving rate, and difficulty level of tasks that can affect the response time. Decision times of the pilot aiding system are based on human performance. Interviews with pilots indicate that many problems could be considered at several different levels. Human response time was quite different at these different levels of reasoning. System response time and capabilities are suggested as follows:

- A response time of 10 ms (for rapid reflex-like responses) to about 20 minutes (for very complex decisions involving future possibilities)
- A capability of 5000 rules or equivalent knowledge structure in an embedded computer

•   The ability to handle many object-oriented classes (about 10,000)

•   The ability to continue functioning despite much uncertainty.

c. Communication

Real-time control needs a cooperative problem solver, therefore, an individual processor should be able to communicate and share data in order to achieve goals. A manager can utilize local memory as buffer to store temporarily a partially executed task. A new task could be executed until the stored problem is resumed. This queuing technique has been used for various real-time applications. The most important of them, communication, has been proved too expensive. Instead of processor interrupt, a logic interrupt is used to inform the status of the current state of the system.

A distinction is made, at the communication level, between processor interrupts and logic interrupts. A processor interrupt is essentially identical to the interrupts in conventional real-time operating systems. Logic interrupts refer to software interruptions of shifting from a low priority task to a higher one. Suspended tasks may have to be resumed. The decision to suspend, resume, or abort the current task is made by the task manager and is amenable to parallel processing.

H. THE SEVENTH PHASE: Maintenance

System maintenance is extremely important in terms of profitability. The system will continue to evolve for a long time after installation as improvements are added and the system is updated. It is useful to divide maintenance into two categories: systems update and system repair. For either update or repair, three main functions are involved.

1. Understanding the Existing System

This implies the need for good documentation and good traceability between the system requirements and the final system. The pilot aiding system can be easily understood by maintenance personnel. A maintenance and user's guide will be provided as part of the system for routine operation and normal maintenance.

2. Modifying the Existing System

This implies the need for a system design that can accommodate software, hardware, and system structure changes with minimum impact to the overall system. These updates are incorporated into the system beginning with a reversion of the requirements specification. The size of the change determines the development lifecycle that is followed. If the update dramatically changes the purpose and interface of the system, every stage of the system development lifecycle is repeated.

3. Revalidating the Modified System

This implies the need for a system structure that facilitates selective retest that makes retesting more thorough and efficient. It may be possible to repair the components that fail the retesting by reworking their internal implementation. If this does not work, the problem is resolved by a system redesign.

IV. SYSTEM ARCHITECTURE

The pilot aiding system is defined conceptually as being "object oriented." The proposed system shown in Figure 2 is the knowledge engineering design overview of the pilot aiding system. The three major subsystems are the executive system, intelligent component system, and knowledge base management system. The executive system is the top-level controlling module of the pilot aiding system. The problem solving processes of a knowledge-based system are performed by the different type of inference engines and a knowledge base management system (integrated with knowledge and data). The control flows of the pilot aiding system are concurrent. The operations of the intelligent component system are scheduled in a cooperative plan and the information is exchanged by updating the attributes of classes. The objected-oriented techniques are used to update dynamically the contents of knowledge structure.

To facilitate integration of the pilot aiding system into an aircraft, the avionics architecture must be an extended mission computer concept. However, the system architecture presented here is of a generic approach type. It can be augmented or modified to

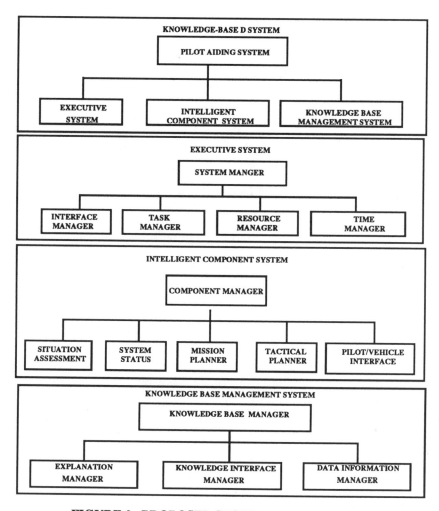

FIGURE 2. PROPOSED SYSTEM ARCHITECTURE

replace the functions performed in the current aircraft. The Pilot
Associate Program serves as a guide to describe the intelligent
components of the pilot aiding system that can be modified for
commercial applications. Thus, the pilot aiding system can be used
to solve a wide range of problems in military as well as in civilian
applications.

The system architecture description, outlined in this section, provides a baseline for further enhancement as the system development progresses. Some of the baseline requirements are as follows:

•   The cockpit layout must be able to enhance the pilot's situation awareness

•   The system design must be founded on an open architecture to maximize the modification capability with minimal side effects

•   The execution must be in real-time so that the system can respond to the pilot assuming control of the aircraft

•   The heterogeneous inference engines must perform as a cooperative problem solver

•   The global knowledge base must be integrated with a data base

•   The system must operate reasonably and safely on inaccurate or uncertain data inputs including temporal and spatial properties.

A.   Executive System

The Executive System (ES) consists of a system manager and four sub-managers. ES controls and manages the information flow and information processing. The information flow consists of external and internal communications. External communications are categorized by pilot interaction and avionic system data exchanges. The avionic system processes status data to/from the mission computer bus, tactical data to/from sensors, and communication data from data links. Internal communications are the two-way communication between the ES and the intelligent component system, and the knowledge base management system.

The ES interacts with the pilot inputs and the pilot aiding responses. These responses are information services, decision making, situation advisories, and emergency procedures for assisting the pilot's "situation awareness" to gain a tactical advantage and ensure mission success. The inputs and outputs are performed by a variety of ways including voice, tone, pictorial format, head-up display, and helmet mounted display. The outputs may also be sent to other avionic subsystems based upon the embedded system and the level of autonomy. The system manager and four sub-managers are described as follows.

1. System Manager

The System Manager (SM) initializes the system operation, receives the pilot requests, sends messages, and controls the SM sub-managers. The SM sub-managers are the interface manager, resource manager, task manager, and time manager. The SM is a controller that manages the timing and content of the information flow and channels it to the pilot. All information flow to the pilot is directed through the interface manager that is capable of providing a broad range of system status, explanatory information, and tactical messages.

An embedded system has very limited resources that must be utilized conservatively. This requires methods for allocating resources efficiently and determining what tasks need to be done. In particular, the resource manager must be concerned with how to expend optimally limited resources within mission constraints. Some capacities of system resources are reserved to respond to emergency mission needs or to a system failure. The task manager decodes the pilot's requests into tasks that are scheduled and monitored by the reasoning system. The time manager acts as a watch-dog. If the time of the maximum-length task is significant with respect to the time required to respond to urgent tasks, a preemptive policy is required. The preemptive scheme should be able to suspend or to abort a low-priority task that is currently running in favor of a high-priority task. The preemptive scheme should also be based on optimizing the system performance.

The SM allocates time and resources to a task and submits them to the intelligent component system for processing. The response is provided by the intelligent component and knowledge base management system.

2. Interface Manager

The user-interface development toolkits can provide a windowing system for menu layout that will be very helpful in the prototype design. The color display presents menus and selection options for interacting with the system. Whenever an option is selected, the appropriate software module is invoked.

In the final pilot aiding system, the Interface Manager (IM) is designed to be easily changeable without affecting the whole system. The primary function of the IM is to manage all the pilot requests, transfer input data to the knowledge base management system, and present information to the pilot. The IM does not interpret the pilot's intentions, but checks for valid inputs. It establishes an I/O queue for incoming requests or outgoing information. When the queue is full, any additional incoming requests are ignored. A message will be sent to the pilot indicating that the queue is full. For multiple requests, I/O queue policy is first-in and first-out (FIFO) in which the order the requests are presented is the same as the order of their arrival. Emergency requests will be assigned a higher priority to achieve faster service. Requests are transmitted from the queue to the task manager.

3. Task Manager

When a request is received from the queue, the Task Manager (TM) decomposes the request into a sequence of tasks according to the function domain. Each task has one of five states: *ready, pending, running, resuming,* and *terminating.* These states indicate the task execution states in the intelligent component system. The TM is aware of the task states but is not aware of the task context. The requirements of task management are as follows.

a. Task Concurrence

In a real-time environment, there is a need for concurrent execution of tasks with hardware support. This helps achieve desired performance in terms of speed and quality of the solution.

b. Task Priority

The priority assignment is performed by estimating the importance of a task based on two types of criteria. The first criterion is the suitability of the task in relation to the current mission. The second criterion is based on the number of other tasks that depend on the task being prioritized. The priority also depends on the tasks already under execution.

c. Task Interruption

The task interruption is a request to change the status state of a task. The decision to switch the ready, pending, running, resuming, or terminating status of the current task is dependent on resource availability and time constraints.

d. Task Scheduling

When the TM dispatches and controls task transitions from state to state, the transitions are based upon the task scheduling algorithm. Before the control is transferred from a ready state to a running state, the TM must know the time constraints, resource availability, and priority of the task. Termination can be either normal or unconditional. Normal termination allows a task to run until its completion. This can benefit system performance, because extra time will not be used for suspending and then resuming the task. The unconditional termination is to suspend the current task and start execution of the next scheduled task. The states between ready and termination are dependent on the scheduling mechanism.

4. Resource Manager

The Resource Manager (RM) controls the system resources
which are the knowledge base, memory, data bus, CPU, and I/O
processor of the pilot aiding system. The task manager provides a
simple and important task scheduling algorithm for overall system
operation. The system calls are embedded in the scheduling
algorithm. After a request from the task manager, the resource
manager will check the availability and health of the resources
required for the task. It also checks the status of the intelligent
components performing the task. The aircraft resource sharing,
memory management, fault isolation, re-configuration, and recovery
functions are dependent on the responses provided by the intelligent
component system as to the system status.

In practice, a decision about resource allocations involves a
tradeoff between granularity and overhead. Granularity refers to the
size of resources and the time duration that may be assigned to a
task. Overhead refers to the cost of the computational work required
to make an allocation. Therefore, overhead costs are proportional to
granularity costs. The smaller granularity is, the lower the overhead
is for assigning resources to a particular task.

When resource allocations are predictable, the situation is
static. Static allocation has a fixed number of intelligent components
dedicated to various concurrent tasks. However, because
unpredictable situations often occur dynamic resource allocation is
needed. Dynamic allocation works by reallocating resources to meet
the demands of the problem. It also enables designs to be

independent of a specific hardware configuration. Dynamic allocation makes it difficult to guarantee the availability of resources when needed or to dedicate a fixed amount of resources to a particular task.

Resource allocation can be viewed as: (1) distributing resources optimally; and (2) monitoring the health of resources. Strategies for resource allocation can be quite expensive because they require measuring constraints for the optimal assignment of resources, and then supervising processor utilization. Implementing a dynamic allocation strategy can be very expensive. In a real-time application, costs must be kept low. Resource management requirements are as follows.

a. Resource Constraint/Contention

The RM must be able to determine which constraints are rigid, which can be relaxed, and how the resources should be shifted among the tasks.

b. Task Interruption/Resumption

The RM must coordinate with the task manager to perform task interruptions when shifting resources occur to a higher priority task. Suspended tasks may have to be resumed later, after higher priority tasks have been executed. The abortion of current tasks may be required for reallocation of resources.

c. Flexible Allocation of Resources

The RM needs the capability to allocate finite resources (memory and computational power) flexibly to the tasks and the ability to fine tune the resource distribution for the desired performance.

d. Risk-Reduction Strategies

The RM must be able to handle the situations of: (1) reduction in capability due to partial system failures; and (2) demands exceeding the designed capability. Under these conditions, the system should still be able to provide the best possible performance.

5. Time Manager

The Time Manager (TM) must use a timing scheme which optimizes the performance of all tasks. A time-tag is assigned to each task. These time-tags are the basic data type in the TM. The TM also controls the logical clock time for synchronization and asynchronization operations in the pilot aiding system.

Within the system, functions are divided between two layers that are the system control layer and intelligent component control layer. The low-data-rate of the intelligent component control layer enables the intelligent components to perform their reasoning in the decision-making process. The high-data-rate used in the system control layer facilitates the direct interfaces between subsystems.

The timing logic is based on the system clock and is used to generate properly synchronized operations between the system control layer and the intelligent component control layer. The logical clock is required to implement the clock synchronization. Time granularity affects both the normal task execution time and the time required to suspend a task.

B.   Intelligent Component System

The intelligent component system consists of one component manager and several intelligent components. The component manager provides context sensitive scheduling, processing and control on all intelligent components in a cooperative processing environment. The context sensitivity is dependent on the context of the current situation both in time limitation and in task dependency. Each intelligent component consists of one or more inference engines in a hierarchical structure. In order to schedule the computations, assign time limits, monitor progress, control system resources, and solve cooperative problems, the following must be known:

- The current context and the determination of when significant context changes require re-scheduling
- The computational cost, including the cooperation of solution time for each of the tasks that may be invoked both in real-time and non-real-time to predict and monitor performance on a given problem set

• The actual relative time or time left to go, and not absolute time, on a scale relative to the events being controlled and monitored.

The following sections provide a description of the component manager and the intelligent components.

1.   Component Manager

The Component Manager (CM) provides common and consistent process scheduling for its intelligent components. The differences between task and component schedules are context sensitive and processing rates. The CM coordinates five components to carry out tasks in pursuit of a common result or goal. A cooperative control algorithm will be used to coordinate the actions of the individual components for achieving a goal. The cooperative control algorithm schedules and monitors the participating components. The CM reports the intelligent component system status to the task manager during operation.

The scheduling plan can be defined as a pilot request resulting in a set of tasks to attain a given goal, assuming those tasks are executed on the several simultaneously active intelligent components. The scheduling plan is based on the following observations:

• A pilot's request may involve multiple intelligent components to accomplish a tactical goal

• Each intelligent component has its own special reasoning capability to deal with a specified application domain. The cooperative control algorithm consists of three steps.

a. Step One. Defining Dependency

The pilot's request is associated with a set of tasks. These tasks describe the properties and behavior for achieving a tactical goal. Some tasks are dependent, so that the execution of these tasks cannot start until other tasks are completed. The degree of dependency associated with a goal determines the level of interdependency of the tasks required to complete that goal. The dependency factor will be the guideline for scheduling intelligent components in sequence or parallel. Otherwise tasks are concurrent processes.

The degree of dependence is measured by how much the success of one task depends on the success of the other tasks. Dependency involves the temporal, spatial, causal, and simultaneity relationships from task A to Task B in a one-way direction. Some of the preconditions for performance of an action may be attained through the actions of another component. The degree of dependence may range from none to total. Computation of the degree of dependency is based upon whether the preconditions supplied by other components are necessary or sufficient to complete a task. If a component relies upon other components to provide necessary preconditions, dependency is high. If the component relies only upon the supply of sufficient preconditions then dependency is low.

b. Step Two. Generating A Cooperative Plan

The cooperative plan is a general purpose dependency-satisfaction facility that enables a broad range of tasks to operate together. Given a set of tasks, and dependencies on task behaviors, a network of interrelated tasks is constructed to achieve some stated set of goals. Starting with some initial task, the iterative plan chooses some dependency by testing to see whether that dependency is satisfied. The plan continues onward in this fashion, repeatedly expanding, searching, and testing its plan until all task dependencies are met.

This plan can be extended by introducing the concept of a regional plan for utilizing regional task dependencies. Each node of the plan network may be associated with a regional plan that is constructed by the same method as the dependency-satisfaction process. Upon reaching a regional plan, a regionally defined task dependency must be chosen and checked to see whether that dependency is satisfied. If it is not, the plan requires either to backtrack or to modify so that the dependency will be satisfied.

When the time intervals are estimated for each of the tasks of a plan network, a time chart can easily be constructed for guiding the planning process. The time chart contains all the tasks of the plan.

c. Step Three. Assigning The Participating Components

The ultimate objective of the scheduling plan is to assign intelligent components to tasks, based on the constructed time chart, so that the pilot's request can be completed on time. The representation of the scheduling plan is a set of assignment triplets, (Task-id, Component-id, time-tag). The assignment is a simple process whose objective is the matching of a component's application domain with the contextual information of a task.

The step three, the final step, is monitored by the CM. This monitoring includes the policy of cooperative implementation. If a failure of the operations or timing constraints of any component occurs, an interrupt will be issued to suspend activities and to report the failure to the task manager. When intelligent components interact with the global knowledge base, read and write operations are continuously based on the time chart.

2. Intelligent Components

Each intelligent component consists of an inference engine, knowledge translation, and explanation facility. The inference engines implement the reasoning mechanisms for manipulating their local knowledge bases. The local knowledge bases are transformed from the global knowledge base by table mappings. The function of table mapping is to translate the global knowledge base to a format suitable for use by an inference engine. During the mapping process, intelligent components communicate with the knowledge base manager requesting the rights of reading and writing attribute values. The function of the explanation facility is to provide information to the intelligent components to generate advisory decisions back to the interface manager for display. The explanation facility also offers explanations of why adversary actions are being issued.

The software design of the inference engines is based on the object-oriented paradigm. The rule-based reasoning is designed to created an augmented AND/OR inference graph with uncertainty. The forward chaining and backward chaining are traversed over the nodes of the graph to draw conclusions. Each node is a class. The associations and operations are for node traversal and evaluation. The challenge in selecting an inference engine, is in deciding which type of reasoning or planning is best for the application. There are many different reasoning paradigms based on the application domain, such as evidential reasoning, qualitative reasoning, procedural reasoning, etc.

The inference engine software is independent on the knowledge provided by application domain. The domain's physical environment, its functional decomposition, the scope of domain constraints, and the cooperative control strategies may all play a role in the pilot aiding system. Therefore, the intelligent components are determined by choosing inference engines according to the types of reasoning for the application domain. The inference engines may be selected from the following commercial application domains to reduce pilot's workload.

a. Landing management can track signal beams produced by ground radio transmission when ceilings are nonexistent and virtually no runway markings are visible. Specifically, the precise flare maneuver and terrain-avoidance systems are required.

b. Mission management can improve flight performance in fuel economy situations. During a period of dense traffic, it also allows some leeway in planning arrival time and route.

c. On-board fault monitoring and diagnosis are being developed to detect aircraft health and correct pilot procedural errors.

d. Flight-path management can manipulate aircraft control surfaces and the propulsion system to lead the aircraft along a predetermined path. It also determines the safety of landing an aircraft under various weather conditions.

In the military application domain, the Pilot's Associate Program Statement of Work provides an overall functional description of the pilot aiding system. The application is limited to single-seat fighter scenarios. The function areas are: *situation assessment, systems status, mission planner, tactical planner,* and *pilot/vehicle interface.*

a. Situation Assessment

The Situation Assessment (SA) correlates available information to determine hostilities of the external environment. It will consider friendly and enemy aircraft, ground forces, terrain, weather, aircraft status, and available resources. This information will be obtained from sensor data, C^3I (Command, Control, Communication and Intelligence), track files, intelligence databases, and its own avionic systems. The available information may be incomplete or uncertain for determining the identity and intent of objects. Hostile and unknown objects will be prioritized based on the estimated threat level and intent.

The SA contributes to the pilot a prioritized list of the most important objects and keeps continuous track of the other objects. It will also provide a justification and explanation of its assessment when queried by the pilot. According to these assessments, the SA may alert other aircraft of the need for changes in tactics, maneuvers, and mission plans. In cooperation with other components, the assessment of threat vulnerabilities may help in formulating offensive strategies. It may also improve target recognition and cueing to assist weapon delivery.

Finding consistent, correct, and meaningful information is of paramount importance. Data often contains high level of noise and errors due to missing, erroneous, or extraneous data values. The requirements of the SA are:

•   The SA must be able to identify what information are overload, where assumptions have been made, and when the data are unreliable.

•   The SA must be able to cope with partial information

•   The SA must be able to compute confidence level of the information.

Identification means to classify the targets by analyzing formations, maneuvers, emissions, and tactics. The key problems are:

•   Classification requires reasoning about time for targets that change over time and for events that are ordered in time

•   Classification requires the integration of incomplete information

•   Classification must account for multiple possible future scenarios (hypothetical reasoning), and indicate sensitivity to variations in the input data

• Classification must be able to make use of diverse data, since information contributing to the identification of the target can be found in many places.

The SA has to deal with conflicting and uncertain information. These two information characteristics have usually been treated independently. Methods for dealing with conflicting information involve utilizing non-monotonic logic and truth maintenance systems. Methods for dealing with uncertain information include an approach describing how to combine confidence levels in probability reasoning, fuzzy reasoning, and evidence reasoning domains.

b. System Status

The System status (SS) is designed to monitor and diagnose the resources and to detect and isolate faults. Monitoring means to continuously interpret signals and to set off alarms when intervention is required. An alarm condition is often context-dependent. The SS must be able to recognize alarm conditions and detect false alarms when considering the context. The SS can analyze own ship information that includes positions, states, and capabilities. States are different combinations of conditions and configurations of aircraft equipment, and the status of its resources. The outputs of the SS include resource status, healthy conditions, weapons status, and fuel remaining to provide mission maneuverability and readiness.

The following contributions of the SS functions will reduce the probability of pilot and aircraft loss and increase mission effectiveness. They are:

- To assist the pilot under normal conditions and provide essential information on the attitude, state, and capability of the aircraft
- To identify, diagnose, anticipate, verify, and compensate the status of all on-board systems and their possible malfunctions
- To contribute information useful with checklists, recommend actions, and handle multiple malfunctions
- To give advisory information, provide emergency procedures, and report pertinent information to other components.

The SS determines the level of system faults and informs the Pilot on whether to continue the mission or not. Emergency advisory is an important operational aid to the pilot. Three levels of fault are categorized as follows:

- Fatal faults are those faults for which it is impossible to maintain the accuracy of information during processing
- Warning faults are those faults for which continued processing is possible with the substitution of a default value for the value found inaccurate
- Cautionary faults are those faults for which processing is continued with no changes. Cautionary faults are indicated when an assumed inconsistency has occurred.

Of significant importance to the reasoning process is the explicit enumeration of all possible faults of the aircraft. The model-based reasoning has more advantages than the rule-based approach.

A model-based fault diagnosis is to reason directly on the structure and function of a physical system. Performance of different mechanisms can drive analytical conclusions from empirical and model knowledge. Model-based reasoning is done by qualitative reasoning followed by quantitative reasoning. Functional operations are considered as methods in a set of hierarchical classes to check the mechanical behavior at various levels of a device.

c. Mission Planner

The Mission Planner (MP) monitors progress of the mission, evaluates the impact of route deviations, and replans when changes in the current situation occur. A critical factor in combat missions is the element of surprise. The pilot must mask his aircraft using the terrain and maneuver in a manner that optimizes resources if there is any hope of utilizing the element of surprise. If his aircraft is located by the enemy or his sensors detect a threat, then the route may need to be dynamically replanned so that interception or avoidance can be accomplished depending on mission goals. The importance of the mission planning problem derives from the fact that a plan cannot be altered while being executed. Besides, the unforeseen always occurs, therefore, "flexibility" is more desirable than "optimality."

Knowledge of how to utilize available resources within mission constraints is critical. The MP receives information about the aircraft fuel and weapons status. The MP will generate paths to and from target areas that will minimize fuel consumption and risk. The plan must be dependent on the constraints and state of the aircraft. The MP will also be able to reroute the aircraft to different paths given new constraints during any phase of the mission. The main functions of the MP offer the pilot alternative mission plans with explanations as to the advantages and disadvantages of each. This will enhance the pilot's situation awareness by presenting prioritized route options in response to the changing environment.

Mission planning is the creation of action steps that implement a plan to satisfy particular requirements. A plan is a program of actions that can be carried out to achieve goals. The key problems in formulating a plan are the following:

(1). In a full mission, a system designer cannot immediately assess the consequences of mission decisions. He must be able to explore mission possibilities tentatively.

(2). Constraints of a mission plan come from many sources. Usually there is no comprehensive method that satisfies all constraints with mission choices.

(3). In very large systems, a system designer must cope with the mission complexity by factoring the design into subgoals. He must also cope with interaction between the subgoals, since they are seldom independent.

(4). When a mission is complicated, it is easy to forget the reasons for some mission decisions and hard to assess the impact of a change to part of a goal. This suggests that a system design should record justifications for mission decisions and be able to use these justifications to explain decisions later.

(5). When global plans are being modified, it is important to be able to reconsider alternative possibilities. During redesign, the system designer needs to be able to see the big picture to escape from points in the mission goal that are only locally optimal.

(6). Many plans require reasoning about spatial relationships. Reasoning about distance, shapes and contours demands considerable resources. There are currently no well-defined ways to reason approximately or qualitatively about shape and spatial relationships.

The selection of knowledge-based technique to solve the planning problem is determined by the following considerations:

(1). General constraints are from sensor management, situation assessment, or system status that a plan should never violate. Replanning is required for a changed set of constraints. The original plans are refined, updated, adapted, modified, and used as templates.

(2). An initial situation given in terms of resource quantities and start states, on which the plan will be executed. Mission planning is able to plan with conflicting and uncertain information.

(3). A set of goal states and final states which are to be achieved and resources are used "wisely."

(4). Explanations of result are required for explaining whether one plan is better than another.

The heuristic method of "generate-and-test" is an approach that can be used to generate actions and test the constraint-satisfaction. "Wisdom" can be embedded in constraints or actions. Plans should also be allowed to include effective replanning. Judicious use of evidence combinations supports planning with incomplete information. Because all application-specific knowledge, from the "goals" and "constraints" to the "rules" and

"deductions" are expressed and justified, the planning process itself
is explainable.

d. Tactical Planner

The Tactical Planner (TP) sorts and prioritizes threats and
targets, selects appropriate countermeasures, aids in weapons
employment, and coordinates the wingman responsibilities. The
major function of the TP is to analyze the current and predicted
situation and provide the pilot with short term offensive and
defensive tactical options. The TP will derive these alternatives
based on: the information provided by other components about the
current situation, threat characteristics, pilot performance capability,
available resources, and the current mission goals and constraints.
The TP analyzes the current and predicted situation and provides the
pilot with short term offensive and defensive tactical options. The
TP will also advise and assist the pilot in critical situations by
recommending flight path changes, maneuvers, countermeasures,
and weapons employment. Recommendations on appropriate
wingman roles will also be provided. Some or all of these functions
may be automated, depending on the present level of autonomy that
the pilot is delegating to the system. The pilot could activate,
control, and override any or all of these functions based on the
changing course of events.

The TP provides justification or explanation of any plan it
proposes. The TP makes the pilot aware of tactical options available
and allows him to select the most appropriate one for accomplishing
his mission. This may provide the pilot with an option he may have

forgotten or had not considered. The TP will be able to compare simultaneously the set of possible tactics, maneuvers, countermeasures, and weapons selection and delivery options against the current situation, mission goals, and constraints.

The TP must evaluate the mission plans that achieve goals without consuming excessive resources or violating constraints. If there is a goal conflict, the TP establishes priorities. If current decision data are not fully known or change with time, then the TP must be flexible and opportunistic, since planning always involves a certain amount of prediction. The key problems are described as follows:

(1). Mission planning is sufficiently large and complicated that it cannot give all the consequences of actions taken. The TP must be able to act tentatively, to explore results of possible plans.

(2). If the details are overwhelming, the TP must be able to focus on the most important considerations.

(3). In a large complex mission, there are often interactions between plans for different subgoals. The TP must attend to these interactions and cope with constraint conflicts.

(4). Often the mission context is only approximately known, so that the TP must operate in an environment of uncertainty. This requires preparation for contingencies.

(5). If a mission has to carry out multiple plans, the TP must be able to coordinate these plans.

The solutions to the planning problems of MP and TP contain a set of actions, an assignment of resources to the actions, and a schedule of the sequence in the actions. Goal states, constraints, and utility functions are parts of the statement of a planning problem. Although they may be elaborated upon as part of solving the planning problem, the planner may uncover additional constraints while trying to construct the plan.

While this formulation of a plan is task-centered, the MP and TP should also support the dual resource centered view of the solution as a schedule of resource utilizations, where resources enable tasks. Resources are primitive, aggregate, and composite. Primitive resources are those things whose availability can be known a priori. Aggregate resources are created dynamically by those resources that are always composed of specific numbers of primitive resources. Composite resources are those resources that are composed of new resources, but whose composition is not known a priori.

Actions refer to things that happen during the execution of a plan, and tasks are the things that must be done to prepare a plan. The notion of an action can also be broadened to include "making a decision" and "applying a strategy." A specific action can be completely described by giving a list of its starting and ending goals (often given as times), a set of needs and the resources that are meeting them, and a set of other associated actions.

The MP and TP utilize various sources of knowledge: specific constraints, actions, rules, and deductions. Actions have constraints attached to them. Constraints that are stated in terms of predicates, consist of time restrictions, resource restrictions, and task restrictions. To see if some predicate is true, the rule is deduced. The same rules are also used to help figure out what could be changed in the plan element knowledge base to make some predicate TRUE or FALSE. When a constraint is violated, a system-generated difficulty report goes to conflict resolution. The conflict resolution can be built as an agenda procedure of resolving the difficulty. An agenda list includes standard control elements and data flows. There are a number of options open to the utilization of the agenda that must be brought to the attention of enumeration of the alternatives of a plan.

A schedule is a time interval assigned to each action. The planning problems are generally dynamic because the plan assigned time schedule needs to take all eventualities into account. Actions along the schedule complexity dimension include time intervals, partial precedence graphs, single-process conditional branches, single-process general recursive plans, multi-agent cooperative plans, and multi-agent adversarial plans.

e. Pilot/Vehicle Interface

The Pilot/Vehicle Interface (PVI) makes decisions, based on levels of autonomy, as to what information will be presented to the pilot and what will be processed by the avionic system. The levels of autonomy are determined by which functions the pilot performs and which are automated. The major difference between the PVI and the executive system interface manager concerns the context in which the information is processed. The PVI processes information in a context-sensitive manner, whereas the executive system interface manager processes information in a context free environment.

The primary functions of the PVI consists of: presenting information to the pilot, interpreting information from the pilot aiding system, and monitoring the pilot's performance. The determination of what information should be presented to the pilot is based on the importance and urgency of the information. The reasoning process is chosen from the inference mechanisms based on the design specification. All the intelligent components post and update information in a continuous manner. This information includes explanations, messages, and advisories that are prioritized according to their importance. The PVI must present this information to the pilot in a succinct manner. Also, the pilot's requests and commands must be concise. Thus, the PVI provides fast and efficient two-way communication between the pilot and the pilot aiding system.

The criteria used to interpret the pilot's intentions are the intent function and intent model. The intent function involves interpreting the pilot's intention and checking the consistency of the pilot's plans and goals with the pilot aiding system. The intent model is used to maintain the pilot's action process and consequence reasoning at several levels of abstraction. The pilot's action process is based on the structure of his beliefs, desires, and intentions. The error critics interpret the structure of pilot error. The consequence reasoning function models the consequences of pilot actions for verification and validation purposes.

Monitoring a pilot's performance provides a dynamic method for estimating his workload and capabilities. The measurements are based upon the current mission demands, expected activities, pilot behavior, and the general status of his environment. The intent function may decide to shift the pilot's role to an emergency procedure or to assist pilot in handling operations.

The pilot is aware of what is being done and reserves use of the override option.

C. Knowledge Base Management System

The knowledge base management system (KBMS) is based on a centrally shared knowledge base structure. The KBMS provides a comprehensive knowledge and data organization that facilitates introspection, location of relevant expertise for a problem, and the setting up of a cooperative plan to reach the goal. This design also provides inference engines a transparent translation for adapting the knowledge base to the different inference mechanisms. There can be one or more knowledge base partitions. The pros and cons of the knowledge translation design are as follows:

• Potential advantages of this approach include: minimal communication, consistency and validation checking, and association with the interface to many different types of inference engines

• Disadvantages may include: bottleneck conditions and reduced throughput if parallel hardware architecture is not employed.

Cooperative decisions require a knowledge/data base that has both a breadth and depth of knowledge. Breadth of knowledge refers to knowledge of many expertises, whereas the depth of knowledge refers to the levels of expertises' understanding. The knowledge base manager controls the cooperative policy by using a cooperative updating algorithm. The algorithm specifies the data flow among different intelligent components and the control scheme of the retrieving/updating operations in real-time.

One of the major features necessary in the pilot aiding system is a facility to generate useful explanations of its behavior. The issue of explanation is characterized by three different types of knowledge: strategic, structural and support messages. The explanation manager facilitates knowledge and data base structure to provide message formats which are independent of the application domain. The contents are based on the mission phase. The explanation manager allows the pilot to ask the pilot aiding system what it is currently doing, why it is doing so, and how it is reasoning.

The description of the KBMS is divided into a knowledge/data paradigm description and a knowledge/data management description. The knowledge/data paradigm description covers knowledge structure, application domain, and operations. The knowledge/data management description emphasizes functionality and covers the knowledge base manager, explanation manager, translation manager, and data information manager.

1. Knowledge/Data Paradigm

The utilization of an object-oriented technique to design a knowledge base management system is required to support reasoning and to store information. If numerous classes are involved, some of them may need to be put in secondary storage. The design phase consists of a two-stage process. The knowledge organization design comes first and the application source design next. These two stages are described as follows.

a   Knowledge Structure

The knowledge structure is represented as classes and relationships which have been described in the fourth phase of the knowledge engineering design of Section III. Relationships used to model connections among classes are formed as an associative class. Such relationships may be one to many, or many to many. Classes that have various properties, methods, and constraints participate in a number of relationships with other classes. There are various types of relationships that are categorized by generalization, aggregation, and association. These types are used to represent a knowledge structure, and are summarized below:

(1). Classes. Each class can be stored independently and included in a generalization, aggregation, and association hierarchy. Classes contain the following essential parts:

• Attributes, which reference to goals, fact (preconditions), an invocation condition, effects (postconditions), and properties of an individual class

• Constraints, which represent desired limitations of fact. An atomic formula of logic describes a statement that is true or false

• Relationships, which encompass the link list types that are representative of a semantic network or frame structure

• Methods, which respond to serve some desired procedures, algorithms, and forms of logic predicate.

(2). Association. Each association is a set of rules linked by "is-set-of" pointers. The association graph can be inferenced by forward chaining and backward chaining.

(3). Generalization. Each generalization is also a chain of classes linked by "is-class-of" pointers between the class and the subclasses. A subclass will inherit all the properties of the class. A subclass may come from more than one class.

(4). Aggregation. Each aggregation is a chain of classes linked by "is-part-of" pointers linking all parts of a class. It contains other classes and creates an aggregation hierarchy. The order of this hierarchy determines the presentation order. An aggregation class can share a subtree with another aggregation class, thus creating a lattice. The lattice of an aggregation class is the basis for data sharing and some database access techniques.

b. Application Domain Design

The sources of information, utilized to develop a knowledge data base, are based upon; the review of relevant documentation and operation manuals; discussions with the pilots; observation of training procedures; and utilization of the design team's technical and testing expertise. This information can be categorized as follows:

(1). Tracking information. The tracking information includes sensor data on the location and tracking of friendly and enemy forces.

(2). Threat classification data. The threat classification data is used for identification of threats according to emitter characteristics, and the location and patterns of their tracking.

(3). Intelligence data. The intelligence data includes information known before a mission as well as data reported throughout a mission regarding the location, activity, fuel, and weapons status of enemy and friendly forces.

(4). Own ship status data. The own ship status data reflects the dynamic status of the own ship information, such as tactical data, as well as diagnostic readings.

(5). Stored data. The stored data consists of records of the important actions taken on a mission detailing the kinds of actions taken and decisions made. Data include the rules of engagement, tactics, order of battle, and operations plan.

c. Methods On Structure

The followings are various methods on the knowledge structure for checking goals, matching facts, accessing procedures, and testing predicates:

(1). Checking constraints. The checking of constraints of attributes consists of: testing conditions, maintaining conditions, and the achieving of a goal. A confidence level is assigned based on the importance of each goal. A partial goal completion is represented as a probability value for uncertainty reasoning use.

(2). Updating facts. Updating of facts is done explicitly to place values into more than one class hierarchy. The best ways to implement the object-oriented requirements are:

    •   by deferred copy, which means a logical copy without replication

    •   by reference, which means the inclusion of a reference to the copy in the class.

(3). Accessing methods. The accessing methods are combined to determine the allowable classes and to select the desired procedures. The accessing methods are:

    •   through the name of an association class

    •   through the aggregation hierarchy

    •   through the generalization hierarchy.

(4). Testing Predicates. The constraint fields are used to set up a set of logical expressions describing a current situation that consists of some current goals or facts, or both. The predicates within the method fields are tested and deduced. The field also points to related classes for generating an information network.

2. Knowledge/Data Management

Once applications have been defined, inference engines are selected to accomplish the decision aiding purpose. In fact, choices about the knowledge representation are based upon the selection of an application. Knowledge-based design consists of global and local knowledge bases. Once the global knowledge base is organized the Knowledge/Data Management (KDM) is required to provide a linkage from the global to the local knowledge bases to support the selected inference engine. In real-time applications, inference engines are designed to provide sufficient results to answer the pilot's request. The KDM has two other tasks that are performed by the explanation manager and the data information manager. The first task is to generate the messages, explanations, and advisories. The second task is to check incoming data, load it into the knowledge base, and prepare status reports.

The KDM consists of a knowledge base manager, explanation manager, translation manager, and data information manager.

a. Knowledge Base Manager

The Knowledge Base Manager (KBM) controls the logical time clock, time frames, data and knowledge integrity, three sub-managers, and monitors the access rights of intelligent components. The three sub-managers are: the explanation manager, the translation manager, and the data information manager. Depending upon the requests of the intelligent components, the KBM implements a cooperative updating algorithm for issuing commands to transfer the designated knowledge base to the components. Provisions are made for a mapping method. The KBM also provides a status report by using the data information manager to manipulate data and generate a status summary.

The cooperative updating algorithm is operated automatically on the sequence of initialization, conflict checking, controlling, manipulation, and termination. They are:

(1). Initialization. After the component managers send a message to the KBM that indicates the location of a cooperative plan, the KBM reads the desired plan and the mapping table of the global knowledge partition. The component requests are received and put in a queue. When a request is ready to leave the queue, the logical clock will be reset to an initial value of zero.

(2). Conflict checking. A conflict check will be performed against a component's request. If there is a conflict, the request with the highest priority will be applied and the other one will be put in a queue. The rejected request is set to "pending" for at least one logical clock.

(3). Controlling. The KBM checks the queue and processes the pending requests with the present updating requests. If the request is ready for operation, the locations of the knowledge base will be associated with the component identification number of the cooperative plan. The association will be performed by a matching algorithm to give manipulation permission.

(4). Manipulation. When a component starts to work, the KBM resets the local time-out clocks and semaphore variables. As the logical time advances, the operations are continued until either the completion, or preset time-out clock ticks. After updating, the logical clock will be incremented by a time unit.

(5). Termination. During manipulation, if an intelligent component has to quit, the intelligent component manager will delete the ID of that intelligent component from the participating list. An interruption will occur to terminate processing. When processing is terminated normally, the final knowledge base is stored. If the termination is not normal, a message will be displayed. The cooperative updating process is completed when the knowledge/data has been replaced.

b. Explanation Manager

The primary function of the Explanation Manager (EM) is to provide information to the pilot. The EM determines the complexity of each explanation type and also identifies the knowledge base needed to support any desirable collection of explanation capabilities. The pilot's queries may include requests to:

- Identify objects, positions, flight path, and mission plan
- Provide justification for a specific decision
- Provide clarification or additional supporting information.

Under a low stress situation, the pilot will have time to obtain detailed explanations for plans and recommended actions. When response time becomes critical, only the essential information desired by the pilot will be presented. To prevent the EM from overloading the pilot with excessive explanations, the need to display explanation will be determined by pilot/vehicle interface based on the current situation and pilot workload.

The EM facilitates the use of intelligent component system results to support the pilot who may be facing difficult, uncertain decisions, for which solutions are not obvious. Explanations are procedure calls which are embedded in the executive statements of the intelligent components. The different types are: WHY, WHYNOT, HOW, WHAT, WHEN, WHERE, etc. The steps of an explanation procedure are:

- Retrace solution paths
- Examine specific portions of the knowledge base defined by suitable qualifiers and context

- Compare alternative solutions and explain the differences
- Identify knowledge that is missing, and facilitate consistent knowledge refinement
- Explain the effects that the modifying portions of the solution paths used will have upon the solutions generated.

c. Knowledge Translation Manager

The primary function of the Knowledge Translation Manager (KTM) is to bridge the gaps between the global and local knowledge base. The KTM can optimally control and coordinate intelligent components to access required knowledge. During the early stages of development, the global representations are required to maintain and store data in a uniform way. There is no way to support all the different inference mechanisms by using a global knowledge base. The KTM is designed to implement the function in two steps. The first step is to partition the global knowledge and the second step is to convert the local knowledge base into a global structure using the mapping method.

The global knowledge base is partitioned into three levels of sharing among intelligent components. They are fully shared, partially shared, and private partitions:

- A fully shared knowledge base allows any inference engine access to shared knowledge information. This may create conflict resolution problems due to private knowledge and shared knowledge incompatibilities

- A partially shared knowledge base permits access by selected components, and can be partitioned both vertically and horizontally

- The private knowledge base is dedicated to a particular component use.

The ability to share knowledge among multiple inference engines is based upon an understanding of the role that the knowledge has to play in solving particular problems, and in the resultant organization and representation structures.

To make a completely cooperative heterogeneous reasoning system, the KTM must know how the knowledge is represented in both parts and how to convert back and forth. The conversion consists of the following four steps:

- Knowledge identification which recognizes knowledge elements to be used in the special purpose inference engine

- Class mapping which associates each part of a knowledge element with a related class in the global knowledge base. A mapping table is generated for each local knowledge base

- Level sharing which establishes the communication path with the necessary cooperative components to exchange information

- Transformation which exports the mapping knowledge in current use into the inference engine local memory.

One advantage of mapping knowledge is that it enables knowledge engineers to flexibly express or extend the knowledge domain without having to worry about specific applications. The intelligent component only needs a limited space to store the mapping knowledge in local memory and can manipulate information freely within the global knowledge base. The read/write behaviors are based on the cooperative updating algorithm that is under the control of the knowledge base manager.

d. Data Information Manager

The primary task of the data information manager (DIM) is to analyze the data from external sources and summarize the data for report generation. The incoming sensor data and status data are loaded into classes of the object-oriented knowledge base. That allows dynamic, run-time checking and modifying of actions. The DIM is responsible for gathering and reporting information to the pilot in response to requests/queries.

For a dynamic knowledge base, the DIM will perform data security checking and trace the class-relationships of the knowledge base to update and access data values. The knowledge base is not substantially different from the object-oriented data model concept, with the following exceptions:

(1). The data model is the basic foundation of knowledge structure that is queried by a navigation search of class relationships. Data manipulations, which consist of deleting and creating a class, and changing attribute values in a class, are the object-message calling techniques.

(2). All queries are interpreted by the pilot/vehicle interface component and sent to the DIM for data processing. The processing includes an algorithm for statistical calculations. The responses from the DIM are sent back to the executive system interface

manager for display. The statistical report is implemented by filling
in values in a fixed format.

(3). Data updating operations can be viewed as consisting of the
same operations as intelligent components and must be performed
under a cooperative updating algorithm.

Addressing issues that relate to interfaces between DIM and
related modules, require a system view of the data flow. The input
data, from input devices, is applied through the interface manager to
the pilot/vehicle interface and then to the DIM. The DIM controls
attributes of classes, that are part of the knowledge base, to support
intelligent component reasoning operations and explanation manager
operations. Advisories, messages, and reports are verified by
pilot/vehicle interface and forwarded to the interface manager for
display.

V. KNOWLEDGE CONSTRUCTION

Attempting to enhance the capabilities of the pilot aiding
system requires a substantially large knowledge base. Therefore,
the Knowledge Construction (KC) becomes a top priority and time-
consuming task. The purpose of the KC is gathering expert
knowledge and transforming it into a library. The knowledge
gathering is influenced by the knowledge inherent abstraction and
complexity. The expert knowledge involves the pilots' quality,
skill, and training. The knowledge library is used for security and
reuse purposes and also to reduce the complexity of the system

development. The five primary functions of the knowledge library are:

- •	Answer knowledge requests which services requirements by performing deductive reasoning under meta-knowledge control, and provides explanations of reasoning and error recovery
- •	Perform knowledge transactions that perform knowledge request translations between the knowledge base system and the global knowledge base to support the inference engines of each intelligent component
- •	Perform knowledge maintenance that performs updates operations on the global knowledge base including modifications and extensions in a manner that assures consistency
- •	Perform context management that provides multiple knowledge-based systems a problem-specific view of shared knowledge base
- •	Perform access control that provides multiple knowledge base systems access to a dynamic shared knowledge base in a secure manner.

The application domain is divided into several major functional areas for building a pilot aiding system to meet the customer's requirements. The stepwise decomposition of functional areas and the refinement of mission goals is a gradual progression from the top level to the lower level, with greater detail at each level. The knowledge construction steps are:

- •	Specify the application domain clearly and precisely

•   Divide, connect, and check functions of the application domain by reexpressing it as an equivalent structure of properly connected subfunctions

•   Repeat the above step enough time to reach the subgoals

•   Record the solving procedure of each subgoal in detail

•   Store the knowledge source in an organized and manageable knowledge structure.

The final knowledge library structure must be consistent with the global knowledge base. This approach has been used by more practitioners, and over a longer period of time, than any other method. There is a gap between gathering expert knowledge and creating a global knowledge base. This gap can be bridged by the following processes of knowledge acquisition, knowledge base organization, and knowledge management:

A.   Knowledge Acquisition

The purpose of knowledge acquisition is to express a pilot's thoughts, experience, and rules-of-thumb in written form. The problem is that so many of a pilot's abilities are innate and difficult to document. It requires extensive interviewing and probing by a skilled interviewer to acquire the knowledge needed. The current techniques for achieving this are as follows:

1. Thinking Process

A key ingredient in the knowledge acquisition process is the detailed study of a pilot's thinking process. As part of his thinking process, a pilot will represent an event as either: auditory thoughts, visual thoughts, sensory memories, or feelings. One way to help categorize a thinking process is to have the pilot perform in a realistic simulation. The following techniques can help categorize the pilot's thinking process:

•   The choice of predicates used by the pilot to employ internal auditory dialogue for processing information

•   The use of visual memories to assist in problem solving

•   The employment of language methods to suggest the use of feelings in problem solving

•   The use of nonverbal cues to reveal conscious or unconscious thinking process.

2. Presenting Process

The degree to which the pilot's language accurately reveals his thinking process will vary depending upon the pilot's presentation style and the interviewer's skill in interpreting what the pilot is saying. Certain types of ambiguities occur in verbal reports. It is easy to get caught in such ambiguities and assign meaning to them that is not valid. The following language properties should be kept in mind by the interviewer to help clarify the most common ambiguities in language:

•   The language a pilot use may obscure key components of his thinking process

- Words that are vague, without a specific reference point require further elaboration, such as, "better," "easier," and "cheaper"
- Words sometimes condense complex processes into static entities that sound complete but are not
- Language sometimes implies a causal connection between events that is either insufficiently detailed or does not exist
- Several commonly used words define causal probabilities and necessities included: "can or can't," "should or shouldn't," "will or won't," "possible or impossible," "must or must not," and "necessary or unnecessary"
- Another group of commonly used words may erroneously imply universal principles, like "always," "never," "every," and "none."

The purpose of these queries is to insure that the information gathered from the pilot accurately reflects his thinking process. Other forms of verbal ambiguities may occur, but if the interviewer focuses on achieving a complete report of structures and their associated content, a reasonably adequate model of the pilot's problem solving process will result.

3. Recording Process

In the pilot aiding system, the knowledge base has to be the "best mind for the job" and cannot be of substandard intelligence. After a pilot has been interviewed, the results will be a collection of symbolic content presented in a long sequence of steps. It will also probably include comments and other interconnections reflecting the pilot's representational and thinking process. A pilot's thinking process is essentially equivalent to the if-then-else rule used to link information. By following the interviewing techniques presented here, the knowledge engineer can maximize the accuracy of information transferred to the knowledge base. The recording languages used are categorized into three groups. They are:

a. Pure natural language. The pure natural languages occur in every day technical writing. English has a richness in its grammar and many words have multiple definitions. Knowledge recorded in a natural language contains statements that are ambiguous. This blurs the clarity of logic expression quantification and mathematical operations that are required for accuracy.

b. Natural language with restricted syntax and semantics. It is a subset of the natural languages. This approach offers a reasonable combination of clarity and simplicity and is justifiably popular with users.

c. Natural language with an augmented symbol. The natural languages with an augmented symbol share the restrictions of group (b) and, in addition, make use of alphanumeric symbols such as mathematical notations and graphics.

4. Coding Process

The coding process involves an Object-Oriented Programming (OOP) that focuses on classes, time, space, events, conditions, predicates, functions, etc. Each class can be characterized as a configuration of attributes, restrictions, and methods satisfying a hierarchical structure and having specified relationships to each other. The recorded expert knowledge is encoded. The encoded representation has to be understandable, explainable, extendible, maintainable, and reusable.

An important consideration in selecting a programming language is that the language can support data abstraction, inheritance, and runtime method determination. Examples of OOP languages are Smalltalk and Eiffel. Unfortunately, pure OOP languages are not supported with an excellent programming environment. Recently OOP concepts have been incorporated into other languages including Flavors, Loops, CommonLoops, C++, Object Pascal, and Current Prolog. The Ada programming language is the best choice because it is a structured programming language and it supports object-oriented concepts. It also has a long-term development commitment from the DoD (Department of Defense) community.

B.   Knowledge Base Organization

The knowledge base organization is one of the most crucial features of a large knowledge base and depends on the physical hardware configuration. A knowledge base may be implemented in different ways, some of which are: centralized, loosely distributed, and tightly distributed organizations. They are described below with some detail.

1. Centralized Knowledge Base

The centralized knowledge base approach is based on a centrally located shared knowledge base that is controlled by a single management system. This approach provides to the knowledge-based system components, a transparent interface to one or more knowledge base partitions. The knowledge base manager controls intelligent components using an algorithm that facilitates scheduling and interacting by not requiring a detailed understanding of knowledge base component representation. Potential advantages of this approach include:

- Minimal communication to initiate a cooperative inference system
- Easier maintenance of consistency and validation.

2. Loosely Distributed Knowledge Base

The loosely distributed knowledge base partitions among multiple intelligent components require a knowledge base manager to control and coordinate them. The knowledge base partitions are under a local control structure, but are accessed through a single management function. Access privileges are granted to knowledge requests, thereby achieving a level of security and protection. The knowledge base manager decomposes requests and directs them to the intelligent components. A composite response consists of integrating full or partial results of intelligent components. The advantages of this approach are due to the exploitation of parallelism. The primary disadvantage is the variability of knowledge request activity before its occurrence.

3. Tightly Distributed Knowledge Base

The tightly distributed knowledge base approach is based on distributed, shared partitions that are external to the components of each knowledge base manager. These components have specific knowledge about their own single partition. Each component must communicate to a specific knowledge base manager for servicing knowledge requests, thus requiring each component to understand where specific shared knowledge resides. The major advantage may be the improved processing speed achieved by distributing knowledge request processing. Potential disadvantages include the need for different interface requirements in the knowledge base managers, as well as the complexity associated with maintaining consistent and valid knowledge base partitions.

C. Maintenance

Large scale knowledge requires storage, retrieval and archives. As a knowledge base grows in size and complexity, it becomes more difficult to maintain and expand. Even if the current knowledge base contents are consistent, an error or inconsistency may be introduced whenever an additional modification is made. One approach to this maintenance problem is to provide validation checking and a special editor. Additional attention should be given to maintaining consistency in a knowledge base. Keeping a historical record of events is necessary to find the occurrence of a mistake, the version of a class, or changes in a class.

1. Truth Maintenance

A knowledge base may not be free of conflicting or misleading information after it is specified by the experts. Truth maintenance detects inconsistencies within classes. The integrity of the maintained class semantics is crucial. When a pilot alters a procedure or specifies one that contradicts an earlier procedure, the original procedure must be changed and that may affect a previously valid relationships between classes. A contradiction can be created not only by explicitly revising true or false, but also by violating the allowable value of a variable. Inconsistency tests must be highlighted so that deficiencies can be discovered and remedied.

Inconsistency tests are built to validate the pilot's inputs and justify knowledge. Validation verifies the truth of the context. Justification ensures that whenever facts match the premise of the rule, the correct conclusions are arrived at. The test processes include a set of steps to check the boundary conditions, meaningless contexts and error conditions derived from the inference engines. These steps are:

• Determine the testing criteria, such as, accuracy, adequacy, reliability, robustness. Another criterion is that a knowledge-based system should act as a pilot and demonstrate deep knowledge rather than shallow, recipe-like, knowledge

• Specify the realms or sets of input data the knowledge-based system must correctly handle

•   Determine objective metrics for the selected criteria. This is a difficult step, and is often not done because of the difficulty in coming up with a meaningful objective measure or surrogate.

The objective of the testing procedures are:

•   To validate the pilot aiding system by an independent panel of experts not connected with the development effort

•   to develop the testing software to perform automatically on the specific knowledge changes which they must revalidate

•   To develop scenarios for testing the knowledge source, when the knowledge base is constructed

•   To perform consistency testing as the system is elaborated

•   To optimize the knowledge source to improve system performance, because interactions in working memory can cause substantial degradations in performance.

If performance is severely affected, a high-level redesign, such as a revised subproblem structure to limit the focus at any one time, may be needed.

2. Editor

The editor is used mainly by the person maintaining the knowledge base and the domain specialist. The editor display enables a user to see the overall structure of knowledge represented by a graphic relationship, and offers interaction options for rearranging the structure and updating the domain knowledge.

The editor must meet the following requirements:

- It must ensure that relationships between classes, such as "is-class-of," "is-set-of," and "is-part-of," are maintained
- It must distinguish two different references of generalization and aggregation between the classes
- It must keep track of the original relationships to distinguish between classes
- It must be able to trace the structures and contents of class being modified.

The editor is a tool of the system development environment. It is an essential part of knowledge construction. The Ada editor can be substituted without the graphics capability.

VI. SUMMARY AND CONCLUSIONS

The ideas presented in this article are an important design concept for developing an effective pilot aiding system. They represent an investigation into many important issues in the Pilot Associate, Crew Associate [42], and other intelligent assistant programs. The design philosophy discussed explores the level of autonomy for the baseline functional capabilities of a future pilot aiding system. A prototype design is recommended which provides a method of proving concepts, examining the requirements, and improving the system before the final production. Thus, many more iterations and design changes take place early in the lifecycle. Of major importance in the knowledge representation is the unification of frame-based, semantic-network, and rule-based representation into a uniform, flexible, efficient, and extendible scheme by the use

of object-oriented representation. A suggested technique of
knowledge acquisition provides a useful tool to transfer experts'
experience into a knowledge library which contains the necessary
knowledge source for a pilot aiding system use.

The system architecture consists of three basic parts: the
system control module, the knowledge base, and the inference
mechanism. The architecture can be sized to fit in a space vehicle,
fighter, bomber, and commercial aircraft. The avionics systems are
required to provide a connection to the pilot aiding executive system.
The knowledge base is organized into one global knowledge base
and several local knowledge bases. The knowledge translation
provides different views of local knowledge base to the specific
inference mechanism. In addition, the explanation manager and the
data information manager provide the pilot with on-line assistance
that displays summary information, command descriptions,
advisories, error messages, and emergency procedures.

The primary design goal of the inference mechanism gives
the advantage of the complete expressability of different reasoning
schemes. This allows inference engines to perform identification,
monitoring, planning, and prediction for assisting the pilot's
situation awareness. The intelligent components concurrently record
and update information continuously using their own inference
engines and local knowledge bases. They work cooperatively to
complete a single task. Once the task is done, a new task may occur
that gives rise to a new cooperative scheme among the intelligent
components. The ability to share a common interest among
intelligent components is made possible by a task scheduling

algorithm, a cooperative control algorithm, and a cooperative updating algorithm.

By exploiting available knowledge-based techniques, we can draw the following additional conclusions:

1. The prototype design and open architecture concepts make it easy for the system architecture to evolve and facilitate the integration of new modules. As in pilot aiding application domains, these concepts are crucial to the successful enhancement of the system.

2. Algorithms now exist and are quite extensively used in many control applications, particularly those for conventional logic control. Much of this information is in the form of numeric processes. The object-oriented design keeps these conventional algorithms as methods in a class and accesses them by messages through the class interface.

3. Commercial database management systems are inadequate because their data models lack the ability to present semantic information and their interface designs do not satisfy real-time requirements. The object-oriented techniques provide a new way to represent knowledge structure which is suitable for integrating knowledge and data.

4. There have been a number of attempts to solve cooperative decision problems by the use of Blackboard Architecture, Yellow Page Architecture, and Common Pages Architecture. This article

addresses a cooperative architecture that will decrease communication requirements and increase system performance.

5. The pilot aiding system will not be used for critical real-time applications until the system is proved without making catastrophic errors. Validation and verification are emphasized in the system design.

6. In future developments, knowledge-based techniques will permit increased autonomy and a widening scope of applications will be implemented in the pilot aiding system.

VII. REFERENCES

1.   A. B. Chambers and D.C. Nagel, "Pilots of the future: Human or Computer?" *Communications of the ACM*, Vol. 28, No. 11 (1985).
2.   A. A. Covrigaru, and R. K. Lindsay, "Deterministic Autonomous Systems," *AI Magazine*, Vol. 12, No.3, pp. 110-117 (1991).
3.   Air Force Systems and Command, "The Pilot's Associate Program," *Executive Summary*, Depart. of the Air Force System Command, Wright-Patterson Air Force Base, OH (1985).
4.   M, L.Wright, M. W. Green, G. Fiegl, and P. F. Cross, "An Expert System for Real-Time Control", *IEEE Software*, pp. 16-24 (1986).

5. D. L. Nichols, and R. S. Evans "System Specification for ADA Avionics Real-Time Software (AARTS) Project," AARTS-SSS-002 (1987).

6. L. D. Pohlmann, et al., "Avionics Expert Systems Definition Study," *Final Report*, AFWAL-TR-86-1190 (1986).

7. J. R. Weitzel and L. Kerschberg, "Developing Knowledge-Based System: Reorganizing the System Development Life Cycle" *Communications of the ACM*, Vol. 32, No. 4 (1989).

8. R. Budde, et al. (Eds), *Approach to Prototyping*, Springer-Verlag, NY (1984).

9. B. W. Boar, *Application Prototypying: A Requirements Definition Strategy for the 80's*, John Wiley & Son, NY (1987).

10. H. H. Chin, and G. H. Gable, "An Application of Artificial Intelligence to Aircraft Weapon Delivery Systems," *Proceedings of the AIAA Computers in Aerospace IV Conference*, pp. 440-449 (1983).

11. H. H. Chin, "Understanding Natural Language Commands", *Proceedings of Spatial Information Technologies for Remote Sensing*, IEEE, pp. 106-119 (1984).

12. H. H.Chin, "Knowledge-Based System of Supermaneuver Selection for Pilot Aiding," *Journal of Aircraft*, Vol. 26, No. 12, pp. 1111-1117 (1989).

13. J. A. Schirs, et al., "Expert Systems Combat Aid to Pilots," *Final Report*, AFWAL-TR-84-1109 (1984).

14. A. Terriac, et al., "Adaptive Tactical Navigation Study," *Final Report*, AFWAL-TR-85-1036 (1985).

15. H. H. Chin, "Knowledge Data Management System for an AEW Environment," *The Third Air Force/NASA Symposium on Multi-disciplinary Analysis and Optimization*, San Francisco (1990).

16. H. H. Chin, "Intelligent Information System: for Automation of AEW Crew Decision Processes," *Applications of Artificial Intelligence IX*, SPIE, Vol. 1468, pp. 235-244 (1991).

17. R. Szkody, and Ostgaard J. C., "Architecture Specification for PAVE PILLAR Avionics," AFWAL-TR-87-1114 (1987).

18. P. Friedland (Ed.), "Special Section on Architectures for Knowledge-Based Systems" *Communications of the ACM*, Vol. 28, No. 9 (1985).

19. A. Barr, P. R. Cohen and E. A. Feigenbaum (Eds.), *The Handbook of Artificial Intelligence*, William Kaufman Inc., Los Altos, CA (1982).

20. Hoffmann, R. R.,"The Problem of Extracting Knowledge of Experts from the Perspective of Experimental Psychology," *AI Magazine*, pp. 53-67 (1987).

21. Prerau, D. S., "Knowledge Acquisition in the Development of a Large Expert System." *AI Magazine*, pp. 43-57 (1987).

22. M. Minsky, "A Frame Work for Representing Knowledge," *The Psychology of Computer Vision*, P. H. Winston (Ed.), McGraw-Hill, NY (1975).

23. R. J. Brachman, "The Future of Knowledge Representation," *The Eighth National Conference on Artificial Intelligence*, pp. 1082-1092, Menlo Park, CA (1990).

24. H. H. Chin, "Rule-Based Evidential Reasoning System," *The 1992 Long Island Conference on Artificial Intelligence and Computer Graphics*, pp. 1-13 (1992).

25. J. W. Schmidt, and C. Thanos (Eds), *Fundamentals of Knowledge Base Management Systems*, Springer-Verlag, NY (1988).

26. G. Castore, "Validation and Verification for Knowledge-based Control Systems," *SOAR Conference*, NASA/JSC, Houston, TX (1987).

27 C. Culbert, G. Riley, and R. T. Savely, "An Expert System Development Methodology which supports Verification and Validation," *The Fourth IEEE Conference on Artificial Intelligence Application* (1988).

28. P. H. Winston and B. K. P. Horn, *Lisp*, The 3rd Edition, Addison-Wesley (1989).

29. R. J. Abbott, "Rule-Based Systems," *Communications of the ACM*, Vol. 28, No. 9, pp. 921-932 (1985).

30. M. P. Georgeff, et al., "A Procedural Logic," *Proceedings of the Ninth International joint Conference on Artificial Intelligence* , pp. 516-523, LA (1985).

31. N. J. Nilsson, "Probabilistic Logic," *Artificial Intelligence*, No. 28 (1986).

32. L. A. Zadeh, "Fuzzy Probabilities and Their Role in Decision Analysis," *Proceedings of The 4th MIT/ONR Workshop on Command, Control, and Communication*, MIT, pp. 159-179 (1981).

33. G. Shafer, *A Mathematical Theory of Evidence*, Princeton University Press, Princeton, NJ (1976).

34. C. V. Ramamoorthy, and S. Shekhar, " A Cooperative Approach to Large Knowledge-Based Systems," *Sixth International conference on Data Engineering,* Los Angeles, pp. 346-352 (1990).

35. H. P., "Blackboard Systems: Blackboard Application System, Blackboard System from a Knowledge Engineering Perspective," *AI Magazine,* pp. 82-106 (1986).

36. D. P. Miranker and D. A. Brant, "An Algorithmic for Integrating Production Systems and Large Database," *Sixth International Conference on Data Engineering,* pp. 353-360, LA (1990).

37. M. Stefik and D. G. Bobrow, "Object-Oriented Programming," *AI Magazine,* Vol. 6, No.4, pp. 40-62 (1986).

38. D. Maier, et al., "Development of an Object-Oriented DBMS," *Proceedings of ACM OOPSLA Conference,* NY (1986).

39. C. Booch, "Object-Oriented Development," *IEEE Trans. Software Engineering,* Vol. SE-12, No.2, pp. 211-222 (1986).

40. A. Kaneko, et al., "Logical Clock Synchronization Method for Duplicated Database Control," *First International Conference On Distribute Computing System,* pp. 601-611 (1979).

41. W. Suwa, A. C. Scott, and E. H. Shortliffe, "An Approach to Verifying Completeness and Consistency Checking of Expert Systems with Rule-Based Expert System," *AI Magazine,* pp. 16-21 (1982).

42. L. A. Lansky, "Distributed Reasoning Dynamic Environments", *Final Report,* SRI/Grumman Crew Associate Program (1988).

Techniques for Optimal Sensor Placement for On-Orbit Modal Identification and Correlation of Large Aerospace System Structures

Daniel C. Kammer

Department of Engineering Mechanics
University of Wisconsin
Madison, Wisconsin 53706

I. INTRODUCTION

Proposed Large Space Structures (LSS) will require accurate analytical models for performing on-orbit loads analysis and control system design and simulation. A modal identification must be performed to obtain modal parameters which can be compared with pretest analytical results using test-analysis correlation techniques [1-3]. Based upon the results of the correlation analyses, the analytical models are updated such that they more accurately predict the test results [4-7]. Due to the size and flexibility associated with an LSS, modal testing cannot be economically or accurately performed in a ground vibration test. The structure must be tested on orbit.

Testing a structure the size and complexity of an LSS would be difficult enough on the ground. Testing on orbit will present a myriad of new problems not usually encountered during a ground vibration test [8]. One of the Key problems which will be faced by the designers of on-orbit modal identification experiments is the placement of the sensors which will take the data [9]. Actuator placement is also of great importance, however, actuator locations in many cases will be fixed such that the existing hardware is used for excitation during testing. For example, RCS jets used for attitude control may also be used to excite an LSS during an on-orbit modal test. This chapter will only consider the sensor placement problem, assuming that the actuator locations are already designated.

In general, sensor resources will be scarce due to cost and weight considerations. In a ground vibration test, it is quite common to place several hundred sensors, usually accelerometers, on the structure to obtain all of the data required to identify the dynamically important modal parameters. During the course of the test, sensors can be easily moved if the appropriate data is not being obtained. In stark contrast, for the case of on-orbit modal identification, a relatively small number of sensors must be placed such that the experiment is able to identify a relatively small number of selected target modes. In addition, sensors cannot be easily moved, if at all. The designer of the on-orbit modal identification experiment cannot afford to use engineering judgement alone to place sensors. Systematic sensor placement methods must be developed and employed.

A vast literature exists concerning the placement of both sensors and actuators. Most of the work has been performed by the control dynamics community to address both parameter identification and control. Many of the references found in the literature deal with the placement of sensors in distributed parameter systems [10-16]. Some of the techniques used to place sensors include maximizing error sensitivity [10], maximizing the determinant of an information matrix [11], minimizing the trace of an estimate error covariance matrix [12-14], and maximizing a measure of observability [15]. An even larger number of references deal with sensor placement for spatially discrete systems. For example, references [17-22] place sensors for structural control by minimizing a cost or an objective function. Vander Velde and Carignan [23] place sensors considering possible failures. Sensor placement for the purpose of detecting dynamic changes in multivariable systems was investigated by Basseville et al. in references [24, 25]. A relatively smaller amount of literature deals with placement of sensors for structural parametric identification, for example references [26-28].

In contrast, the literature is generally lacking in the area of sensor placement techniques which are directed specifically at the problem of modal identification, and even more so in regards to on-orbit modal identification [29-32]. This chapter presents in detail the Effective Independence (EfI) method of sensor placement which was introduced by the author in reference [29]. The technique approaches the sensor placement problem from the standpoint of a structural dynamicist who must use the modal parameters identified during an on-orbit test to perform test-analysis correlation and analytical model updating. It is vital that the targeted test mode shapes are linearly or spatially independent. In fact, the sensors should be placed such that the partitions are as independent as possible. The Effective Independence method attempts to maximize this independence by selecting sensor locations which contribute significantly to the determinant of a corresponding Fisher Information matrix.

The following sections describe, in detail, the EfI method and its application to examples. Section II describes test-analysis correlation techniques and illustrates the need for spatial independence of the target modes. Section III presents the theory behind the EfI method. Sections IV and V discuss the effects of sensor noise and analytical model error, respectively.

II. TEST-ANALYSIS CORRELATION

It has become common place in the aerospace industry to require a test-validated analytical model for all spacecraft. Test-analysis correlation is the process by which an analyst determines how well test modal parameters agree with modal parameters predicted by an analytical model. If the agreement is good, the analytical model can be used with confidence to predict dynamic loads and design control systems.

Test and analysis frequencies can be compared directly, however, mode shape comparison is a little more difficult. Perhaps the most simple technique for comparing test and analysis mode shapes is given by the Modal Assurance Criterion (MAC) [33]. The ijth term in the MAC matrix is given by

$$MAC_{ij} = \frac{\left(\Phi_{ti}^T \Phi_{fsj} \right)^2}{\left\| \Phi_{ti} \right\|^2 \left\| \Phi_{fsj} \right\|^2} = cos^2 \theta_{ij}$$

(2.1)

where Φ_{ti} is the ith test mode shape, Φ_{fsj} is the jth analytical mode shape partitioned to the sensor locations, $\|\cdot\|$ represents the Euclidean norm of the enclosed vector, and θ_{ij} is the angle between the mode shapes. It is important to note that the test modal coefficients are obtained only at the sensor locations while the analytical modes must be partitioned to the sensor locations in order to use the MAC comparison.

The MAC is a measure of the parallelism of the ith test mode and jth analytical mode. If the mode shapes are the same and thus parallel, $MAC_{ij}=1.0$. If the mode shapes differ significantly, MAC_{ij} will be a small number which can in fact be zero if the two modes are orthogonal in an ordinary sense. The advantage of the MAC measure of mode shape correlation is that it does not require a reduced analytical model which possesses only the degrees of freedom corresponding to the sensor locations. Test engineers quite often use MAC during the course of a ground vibration test to check the quality of their data. In order for MAC to produce valid and meaningful results, the sensors must be placed such that they render the test target mode shapes linearly independent. The analytical target modes partitioned to the sensor locations must also be linearly independent. If this is not the case, there will be repeated shapes within the MAC comparison which will not be spatially distinguishable. The MAC will indicate that the repeated shapes are the same mode.

The remaining test-analysis correlation techniques mentioned in this chapter require a reduced representation of the analytical model which usually comes in the form of a finite element model (FEM). The reduced representation is called a Test-Analysis-Model (TAM). It contains only the degrees of freedom which correspond to sensor locations, the remaining degrees of freedom are eliminated during the reduction process. The undamped equations of motion for the FEM representation of an LSS can be partitioned in the form

$$\begin{bmatrix} M_{oo} & M_{oa} \\ M_{ao} & M_{aa} \end{bmatrix} \begin{Bmatrix} \ddot{u}_o \\ \ddot{u}_a \end{Bmatrix} + \begin{bmatrix} K_{oo} & K_{oa} \\ K_{ao} & K_{aa} \end{bmatrix} \begin{Bmatrix} u_o \\ u_a \end{Bmatrix} = \begin{Bmatrix} 0 \\ F_a \end{Bmatrix} \tag{2.2}$$

where the subscript a denotes the degrees of freedom which are to be retained in the TAM representation and subscript o denotes the degrees of freedom to be omitted in the reduction. Note that it is assumed that external loads are only applied at the a-set degrees of freedom. The TAM representation is generated by relating the FEM displacement vector to the TAM or a-set degrees of freedom using a transformation matrix T

$$u_{FEM} = T u_a = \begin{bmatrix} D^T & I \end{bmatrix}^T u_a \tag{2.3}$$

in which D is a smaller transformation matrix relating the o-set degrees of freedom to the a-set. The FEM mass and stiffness matrices from Eq. (2.2) are then reduced to the TAM level using the relations

$$M_{TAM} = T^T M_{FEM} T \qquad K_{TAM} = T^T K_{FEM} T \tag{2.4}$$

The most commonly used reduction method for TAM generation is the Guyan or static reduction [34]. The Guyan transformation is generated by solving the static portion of the equation corresponding to the o-set partition of Eq. (2.2) producing the relation

$$u_o = -K_{oo}^{-1} K_{oa} u_a = D_s u_a \tag{2.5}$$

The static transformation T_s to be used in the relations in Eq. (2.4) is given by

$$T_s = \begin{bmatrix} -K_{ao}K_{oo}^{-1} & I \end{bmatrix}^T$$

(2.6)

This reduction method ignores the dynamic terms associated with the *o-set* degrees of freedom. In order for test-analysis correlation using a TAM to provide accurate and useful information concerning the correlation between the test and FEM modal parameters, the TAM must accurately represent the FEM target modes. Therefore, use of the Guyan reduction requires that the TAM degrees of freedom or *a-set* contains all degrees of freedom which possess significant kinetic energy in the target modes. The kinetic energy for each FEM degree of freedom in each target mode is given by [1]

$$k_{li} = \Phi_{li} \sum_{j=1}^{n} M_{FEMij} \Phi_{lj}$$

(2.7)

where Φ_{li} is the *i*th entry in the *l*th target mode, n is the total number of degrees of freedom in the FEM representation, and M_{FEMij} is the *ij*th term from the FEM mass matrix. The kinetic energy expression in Eq. (2.7) is essentially an itemized generalized mass. It is also important to note that the *a-set* partition of the FEM target modes must be full column rank or the resulting static TAM will not predict the complete target mode set.

In many cases, the static TAM produced by the Guyan reduction technique produces an accurate representation of the FEM target modes which can be used in test-analysis correlation. This has especially been the case in ground vibration tests of conventional spacecraft where large numbers of sensors have been available. More recently, O'Callahan [35] improved upon the static TAM by generating a frequency independent approximation for the dynamic terms omitted in the Guyan reduction process.

However, in the case of on-orbit identification of LSS, sensor numbers will be very restricted, therefore the experiment designer cannot afford to place sensors at all of the degrees of freedom possessing large amounts of kinetic energy as required by the static TAM. More advanced TAM generation techniques must be used to produce accurate test-analysis-models for on-orbit identification and correlation of LSS. The Modal TAM method was introduced by Kammer [36] to address the problem of restricted sensor number. The method uses the Modal expansion equation for the target modes partitioned according to the *o-* and *a-sets* already defined

$$\begin{Bmatrix} u_o \\ u_a \end{Bmatrix} = \begin{bmatrix} \Phi_{fo} \\ \Phi_{fa} \end{bmatrix} q \tag{2.8}$$

where q is a vector of target mode responses. The second partition of the vector equation (2.8) can be solved for the modal response vector using the relation

$$q = \left[\Phi_{fa}^T \Phi_{fa} \right]^{-1} \Phi_{fa}^T u_a \tag{2.9}$$

Note that this computation also requires that the sensor configuration renders the target mode partitions linearly independent. If Eq. (2.9) is now substituted into the first partition of relation (2.8), an expression relating the o- and $a\text{-set}$ displacement vectors is obtained as

$$u_o = \Phi_{fo} \left[\Phi_{fa}^T \Phi_{fa} \right]^{-1} \Phi_{fa}^T u_a = D_m u_a \tag{2.10}$$

The modal transformation matrix T_m can be formed using Eq. (2.3) and the modal TAM mass and stiffness matrices can then be generated using the expressions in Eq. (2.4). It is important to note that the modal reduction method is exact for the modes used in the reduction process. This means that if the eigenvalue problem associated with the Modal TAM mass and stiffness matrices is solved, it will exactly predict the FEM target mode frequencies and mode shapes partitioned to the sensor locations.

It is easy to see the advantage of using the Modal TAM in a case where there is only a relatively small number of sensors available. However, in most cases, the number of target modes used in the reduction process will be smaller than the number of sensor locations available. While the Modal TAM will exactly compute the FEM target modes used in the reduction, if non-target modes are computed using the TAM representation, either by going beyond the target mode frequency range or in the case where non-target and target modes are interspersed in frequency, the resulting mode shapes and frequencies will be very inaccurate when compared with the corresponding FEM parameters. It has been proposed that the Modal TAM's poor representation of the non-target or residual modes causes it to be too sensitive to small participations of the test residual modes in the extracted test target mode shapes. Errors of this type will always be present because test modal identification techniques are never exact.

In order to improve the residual mode representation of the Modal TAM, the Hybrid TAM representation was devised [37]. Briefly, the method takes the exact representation of the target modes produced by the modal reduction and augments it with residual mode approximation of the static reduction. Using this approach, an approximation of the complete FEM displacement vector can be written as

$$\hat{u}_{FEM} = T_m P_T u_a + T_s P_R u_a \tag{2.11}$$

where P_T and P_R are oblique projectors onto the target mode and residual mode column spaces, respectively, given by

$$P_T = \Phi_{fa} \Phi_{fa}^T M_{mTAM} \qquad P_R = I_s - P_T \tag{2.12}$$

in which I_s is an identity matrix with order equal to the number of sensors. The terms $P_T u_a$ and $P_R u_a$ therefore represent the *a-set* displacement vector restricted to the target modes and residual modes, respectively. Using the second expression in Eq, (2.12), the Hybrid TAM transformation equation can be written as

$$\hat{u}_{FEM} = \left[T_s + (T_m - T_s) P_T \right] u_a = T_h u_a \tag{2.13}$$

This transformation has been shown to produce TAM representations which produce exact predictions of the target modes and improved predictions of residual modes. Further details can be found in reference [37].

Once a reduced mass matrix has been generated which is consistent with the sensor locations, several more test-analysis mode shape correlation computations can be performed to assess the accuracy of the analytical model. The first that will be mentioned is the test mode orthogonality check

$$O = \Phi_t^T M_{TAM} \Phi_t \tag{2.14}$$

If the test modes are normalized with respect to the TAM mass matrix and if the mass matrix is an accurate representation of the mass distribution of the test article, the test modes will be orthogonal with respect to the TAM mass matrix and the orthogonality matrix will be an identity matrix. In practice, the terms on the diagonal give a measure of how well the TAM mass distribution agrees with the test mode shapes. It is the state of the practice to accept a value of 0.90 or larger as good test-analysis correlation. Off-diagonal terms larger than 0.10 indicate significant coupling between the test modes produced by inaccuracies in the TAM mass matrix.

Test and analysis mode shapes can be compared directly using a cross-orthogonality check

$$C = \Phi_t^T M_{TAM} \Phi_{TAM} \tag{2.15}$$

where Φ_{TAM} are the TAM generated mode shapes. In general, the cross-orthogonality matrix will not be square unless there are equal numbers of test and TAM mode shapes. Assuming that the mode shapes are again mass normalized and that there are more TAM modes than test modes, the largest term in the ith row of the cross-orthogonality matrix identifies the TAM mode which most closely resembles the ith test mode shape. If the ith test mode is perfectly consistent with the TAM mass matrix and mode shapes, there will be a single nonzero value with a magnitude of 1.0. If the largest value in each row is greater than or equal to 0.90 and the remaining terms are less than or equal to 0.10, the test and TAM mode shapes are said to possess good correlation. Once again, the test modal partitions must be linearly independent for this computation to be valid.

One final test-analysis correlation technique will be discussed here. The method uses the effective mass matrix to compare test and analysis mode shapes [1]. The effective mass matrix M_E is computed using the relation

$$M_E = \left(\Phi^T M_{TAM} \Phi_r \right)^{\wedge 2} \left[diag\left(\Phi_r^T M_{TAM} \Phi_r \right) \right]^{-1} \tag{2.16}$$

where Φ are either the test or TAM mode shapes, M_{TAM} is the TAM mass matrix, Φ_r are rigid body mode shapes generated with respect to a point of the structure which is fixed to ground, and $[diag( )]$ is the diagonalized TAM rigid body mass matrix. The ijth term within M_E represents the fractional contribution of the ith mode shape to the jth rigid body mass of the total structure. If all the modes are present in modal matrix Φ, each of the six columns in M_E will sum to 1.0. Modes with large effective mass values are usually of a global nature and thus dynamically important. Test and analysis mode shapes can be compared by computing the effective mass matrix for each and then comparing to find discrepancies. Unfortunately, effective mass is identically zero for structures which are not constrained. In its basic form it thus has limited application in the case of LSS.

This section is certainly not complete in its listing of test-analysis correlation techniques or even methods for generating test-analysis-models. The important point is that all of the methods discussed, and otherwise, require that the sensors are located such that the resulting test target mode partitions are linearly independent. If not, no technique will be able to distinguish and thus correlate the offending modes.

III. SENSOR PLACEMENT THEORY

As demonstrated in the previous section, sensors must be placed on the LSS such that the target modes are spatially differentiable. The target mode set should include all mode shapes which have significant response in the sensor output. Assuming displacement sensors, in the noise-free case, the sensor output is given by

$$u_s = \Phi_s q \tag{3.1}$$

in which u_s is the output displacement vector and Φ_s is the matrix of target modes partitioned to the sensor locations. Note that velocity or acceleration output could also be used without changing the derivation presented in this section. The linear independence requirement for the target modes implies that, if the target mode partitions are known, at any time t, Eq. (3.1) can be solved in a least-squares sense to obtain an estimate of the modal response q given by

$$\hat{q} = \left[\Phi_s^T \Phi_s \right]^{-1} \Phi_s^T u_s \tag{3.2}$$

Therefore, placing sensors to maintain target mode spatial independence can be cast in the form of an estimation problem. Estimation theory can be used to obtain a measure of sensor configuration goodness directly related to the accuracy of the target mode response estimate. A measure of the contribution of each candidate sensor location to the independence of the target modes can also be derived. Before proceeding, it is important to note that the linear independence requirement imposed upon the target mode partitions in the case of modal identification is more restrictive than the usual observability requirement found in the control dynamics literature [38]. The target modes must be absolutely identifiable [10].

The sensor placement problem addressed in this section can be restated in the following manner; given s initial candidate sensor locations and m available sensors, how should m optimal locations be selected such that there is as much independent information concerning the response of the target modes in the sensor output as possible. Unfortunately, the true target mode partitions in Eqs. (3.1) and (3.2) are not known a priori, therefore, the prelaunch FEM must be used to design the sensor configuration. Due to the unavoidable presence of noise, the output equation (3.1) must be generalized to a static Fisher model [39]

$$u_s = \Phi_{fs} q + v \tag{3.3}$$

where Φ_{fs} represents the FEM target modes partitioned to the initial candidate sensor locations and v is a vector of sensor noise. The initial candidate set of sensor locations must be large enough to contain all of the important dynamics corresponding to the target modes. At the same time, the candidate set must render the target modes as initially independent as possible. The final selected sensor configuration will only be as good as the initial candidate set. The kinetic energy distribution, computed for each of the target modes using Eq. (2.5), can be used to help determine a good candidate sensor set.

It is assumed that the sensor noise is a stationary random observation disturbance with zero mean and positive definite covariance intensity matrix R such that

$$E\left[v(t)v(\tau)^T \right] = R\delta(t-\tau)$$

(3.4)

in which E represents the expectation operator. The unbiased estimate of the target mode response produced by the Fisher model estimator, W_F, is given by

$$\hat{q}_F = W_F u_s = \left[\Phi_{fs}^T R^{-1} \Phi_{fs} \right]^{-1} \Phi_{fs}^T R^{-1} u_s$$

(3.5)

The associated covariance matrix of the estimate error is of the form

$$\Sigma = E\left[(q - \hat{q}_F)(q - \hat{q}_F)^T \right] = \left[\Phi_{fs}^T R^{-1} \Phi_{fs} \right]^{-1} = Q^{-1}$$

(3.6)

where Q represents the Fisher Information Matrix (FIM) [40]. For any given set of sensor locations, the estimate produced by the estimator in Eq. (3.5) results in the smallest error covariance matrix for all possible estimators. This implies that the Fisher model estimator is efficient. The smallest error covariance matrix results in the best estimate. Thus, for each subset of m sensors selected from the initial candidate set, the estimate produced by Eq. (3.5) is the best estimate. The optimum set of m sensor locations, out of all possible m-sensor configurations, will correspond to the smallest of all possible error covariance matrices. Conversely, the optimum sensor configuration will produce the largest FIM. Thus, a suitable matrix norm of Q must be maximized as the sensor configuration is selected. For example, references [12] and [27] suggest the trace norm as the most useful and physically meaningful measure of the size of the FIM. On the other hand, reference [41] states that the determinant of the FIM for all linear unbiased estimators is largest for the best estimate.

The only assumption that has been made regarding the noise covariance matrix R is that it is positive definite. This only implies that none of the sensors is perfect, which is physically realistic. Each of the sensor locations can possess its own noise statistics due perhaps to differences in environment based upon geometric location within the structure. Matrix R can also be fully populated corresponding to noise correlation between sensors. The inverse of the noise covariance matrix can thus be decomposed in the form

$$R^{-1} = \gamma \beta^{-1} \gamma^T \tag{3.7}$$

in which γ is a matrix of orthonormal eigenvectors and β is the corresponding matrix of positive eigenvalues. The square root of the inverse of R can then be expressed as

$$\left[R^{-1}\right]^{1/2} = R^{-1/2} = \gamma \beta^{-1/2} \gamma^T \tag{3.8}$$

where $R^{-1/2}$ is defined by $[R^{-1/2}R^{-1/2}]=R^{-1}$ and the diagonal matrix $\beta^{-1/2}$ is easily computed. The FIM can thus be conveniently written in the form

$$Q = \Phi_{fs}^T R^{-1} \Phi_{fs} = \Phi_{fs}^T \left[R^{-1/2}R^{-1/2}\right]\Phi_{fs}$$

$$= \left[\Phi_{fs}^T R^{-1/2}\right]\left[R^{-1/2}\Phi_{fs}\right] = \overline{\Phi}_{fs}^T \overline{\Phi}_{fs} \tag{3.9}$$

in which

$$\overline{\Phi}_{fs} = R^{-1/2}\Phi_{fs} = \gamma \beta^{-1/2} \gamma^T \Phi_{fs} \tag{3.10}$$

represents a set of noise-modified target modes at the candidate sensor locations.

The FIM can be decomposed into a contribution from each candidate sensor location, Q^i, such that

$$Q = \sum_{i=1}^{s} \overline{\Phi}_{fs}^{iT} \overline{\Phi}_{fs}^i = \sum_{i=1}^{s} Q^i \tag{3.11}$$

where $\overline{\Phi}_{fs}^i$ is the ith row of the noise-modified target mode matrix partition associated with the ith candidate sensor location. It is apparent from Eq. (3.11) that as sensors are added to, or deleted from the initial candidate set, information is added to, or deleted from the FIM. Sensor locations which do not contribute significantly to the information matrix, and thus the independence of the target modes, can be deleted from the candidate set.

The initial candidate sensor location set is selected such that Q is positive definite. For k target modes, Q can be decomposed using its k orthonormal eigenvectors Ψ and real positive eigenvalues λ such that

$$\Psi^T Q \Psi = \lambda \qquad \text{and} \qquad \Psi^T \Psi = I \tag{3.12}$$

The k orthonormal eigenvectors Ψ represent orthogonal directions in a k-dimensional space which will be referred to as Absolute Identification Space. Forming the matrix product

$$G = \left[\overline{\Phi}_{fs} \Psi \right] {}^{\wedge 2} \tag{3.13}$$

in which symbol $^{\wedge 2}$ represents a term-by-term square of the enclosed matrix, produces a matrix G in which each row contains the square of the components of the rows of $\overline{\Phi}_{fs}$ in terms of the coordinate system defined by the columns of Ψ. Each column of G sums to the corresponding eigenvalue of Q and, thus, the ith term within a column represents the contribution of the ith sensor location to the eigenvalue. Post-multiplying G by the inverse of the matrix of eigenvalues λ yields

$$F_E = \left[\overline{\Phi}_{fs} \Psi \right] {}^{\wedge 2} \lambda^{-1} \tag{3.14}$$

where now each direction within absolute identification space is of equal importance such that the ith term in the jth column of the $s \times k$ matrix F_E represents the fractional contribution of the ith sensor location to the jth eigenvalue. Adding the terms within each row of F_E produces

$$E_D = \left[\sum_{j=1}^{k} F_{E1j} \quad \sum_{j=1}^{k} F_{E2j} \quad \cdots \quad \sum_{j=1}^{k} F_{Esj} \right]^T = \left[\overline{\Phi}_{fs} \Psi \right] {}^{\wedge 2} \lambda^{-1} \{1\}_k \tag{3.15}$$

where F_{Eij} represents the jth term in the ith row of matrix F_E and $\{1\}_k$ is a column vector of 1's with dimension k. Column vector E_D will be referred to in this chapter as the Effective Independence Distribution of the candidate sensor set.

At this point in the discussion, it is proposed that the ith term in the vector E_D is the fractional contribution of the ith candidate sensor location to the linear independence of the target mode set. This statement can be proven to be true by considering the matrix

$$E = \overline{\Phi}_{fs}\Psi\lambda^{-1}\Psi^T\overline{\Phi}_{fs}^T \tag{3.16}$$

With the help of Eqs. (3.14) and (3.15) and some matrix algebra, the EfI distribution vector E_D can be shown to be the diagonal of matrix E. Using Eqs. (3.12), E can be written as

$$E = \overline{\Phi}_{fs}Q^{-1}\overline{\Phi}_{fs}^T \tag{3.17}$$

which becomes

$$E = \overline{\Phi}_{fs}\left[\overline{\Phi}_{fs}^T\overline{\Phi}_{fs}\right]^{-1}\overline{\Phi}_{fs}^T \tag{3.18}$$

using the definition of the information matrix. The form for matrix E can be immediately recognized as an orthogonal projector [42] onto the space spanned by the noise-modified target mode partitions. The rank of the projector is equal to the number of linearly independent target modes, k. It is clear from the expression in Eq. (3.18) that $E^2=E$ which implies that the projector is an idempotent matrix. A well known characteristic of idempotent matrices is that their trace is equal to their rank, i.e. $tr(E)=k$. Therefore, each term on the diagonal of E, and thus also in the EfI distribution vector E_D, represents the contribution of the corresponding sensor location to the rank of the projector or the linear independence of the noise-modified target mode partitions $\overline{\Phi}_{fs}$, as proposed.

The physical meaning of the EfI distribution can be made more clear by considering a k-dimensional ellipsoidal surface associated with absolute identification space. This Absolute Identification Ellipsoid is represented by the relation

$$x^T\lambda^{-1}x = 1 \tag{3.19}$$

where x is a k-dimensional vector and λ is the previously defined matrix of FIM eigenvalues. The principal axes of the ellipsoid have directions corresponding to the orthonormal eigenvectors Ψ_i and lengths corresponding to the square root of the associated eigenvalues λ_i. The larger the eigenvalue λ_i, the more identifiable or independent the target modes are when viewed along the direction Ψ_i in identification space. Thus the distance from the center of the ellipsoid to any point on its surface is a measure of how independent the target modes are along the corresponding direction.

If, as candidate sensor locations are deleted from the initial set, one of the eigenvalues becomes zero, the absolute identification ellipsoid collapses and the the target mode partitions are no longer independent. It is important to note here that the volume of the ellipsoid is proportional to the square root of the determinant of the information matrix Q [43]. Following the derivation presented in reference [44], the determinant of the information matrix can be considered as formally analogous to the definition of information [45]. As sensor locations are deleted from the initial candidate set, the determinant of the information matrix must, therefore, be maintained as large as possible. From this brief discussion, it becomes obvious that the initial candidate sensor set should be selected in some fashion to produce an initial identification ellipsoid which is as large and round as possible.

Further study of the form of the projector given in Eq. (3.16) shows that the diagonal terms in E, and thus also the terms in the EfI distribution E_D, can be written as

$$E_{ii} = E_{Di} = \overline{\Phi}_{fs}^{i} \Psi \lambda^{-1} \Psi^{T} \overline{\Phi}_{fs}^{iT} = \rho_i^T \lambda^{-1} \rho_i \qquad (3.20)$$

in which ρ_i is a vector containing the ith row of $\overline{\Phi}_{fs}$ expressed in Ψ coordinates. Therefore, the ith term in the EfI distribution represents the ellipsoidal norm of the corresponding row of the noise-modified target mode partition expressed in Ψ coordinates using metric λ^{-1}. If each of the rows of $\overline{\Phi}_{fs}$ are plotted as vectors in identification space, it is proposed that the tips of the vectors will lie within or on the identification ellipsoid implying that

$$0.0 \le E_{Di} \le 1.0 \qquad (3.21)$$

If $E_{Di}=0.0$, the ith row of $\overline{\Phi}_{fs}$ is null and the target modes are not even observable from the corresponding sensor location. This sensor location could be immediately discarded without any loss of information. It is apparent from the quadratic form of E_{Di} in Eq. (3.20) that the EfI value of a sensor location can never be less than zero. In contrast, if $E_{Di}=1.0$, the tip of the ith row vector plotted in identification space will touch the surface of the identification ellipsoid. It is further proposed that the corresponding sensor location is absolutely vital to the linear independence, and thus the absolute identification of the target modes. Elimination of this sensor would cause the identification ellipsoid to collapse and a maximal loss of information.

Before proving the validity of these proposals, a simple example, excluding noise, is presented to illustrate the point. Consider two target modes with coefficients at three candidate sensor locations and the corresponding Fisher Information matrix given by

$$\Phi_{fs} = \begin{bmatrix} 1 & 0 \\ 1 & 1 \\ 1 & 1 \end{bmatrix} \qquad Q = \begin{bmatrix} 3 & 2 \\ 2 & 2 \end{bmatrix}$$

The FIM eigenvectors and eigenvalues are

$$\Psi_1 = \{-0.788 \quad -0.615\}^T \qquad \lambda_1 = 4.562$$

$$\Psi_2 = \{0.615 \quad -0.788\}^T \qquad \lambda_2 = 0.438$$

Using Eq. (3.15), the EfI distribution is given by the vector

$$E_D = \{1.0 \quad 0.5 \quad 0.5\}^T$$

As proposed, the EfI values corresponding to the three sensor locations lie between 0.0 and 1.0. It is important to note that the sum of the terms within E_D is always equal to the number of target modes, two in this case. With a value of 1.0, the first sensor is vital to the independence of the target mode partitions and thus cannot be deleted from the candidate sensor set. The corresponding row vector of Φ_{fs} plotted in identification space would touch the identification ellipsoid. This result is obvious from an inspection of the modal partitions presented above. In contrast, the other two sensors are of equal and lesser importance. The corresponding row vectors lie entirely within the identification ellipsoid and either sensor could be deleted from the candidate sensor set while still maintaining the independence of the modal partitions. This is also obvious from inspection. This simple example supports the proposed ideas. It also shows that the EfI distribution can be effectively used to rank the importance of sensor locations in the initial candidate set.

A more mathematically rigorous proof of these propositions will now be presented. First, it will be shown that if a row vector of the noise-modified target mode partitions touches the identification ellipsoid when plotted in identification space, i.e. $E_{D\,i}=1.0$, the corresponding sensor location must be retained in the sensor configuration. For simplicity in notation, let r_i be a column vector containing the ith row of $\overline{\Phi}_{fs}$ and assume that this row touches the identification ellipsoid such that

$$E_{Di} = r_i^T \Psi \lambda^{-1} \Psi^T r_i = r_i^T Q^{-1} r_i = \langle r_i, Q^{-1}r \rangle = 1.0 \tag{3.22}$$

where $<,>$ represents the usual inner product for column vectors. If all sensors are retained, the information matrix Q is assumed to be nonsingular. If the ith sensor location is vital to the linear independence of the target modes, the elimination of the ith term Q^i from the sum in Eq. (3.11) will result in a singular matrix. Define the matrix B such that

$$B = Q - Q^i = Q - r_i r_i^T \tag{3.23}$$

Premultiplication by the inverse of the information matrix produces

$$H = Q^{-1}\left[Q - Q^i\right] = I_k - Q^{-1} r_i r_i^T \tag{3.24}$$

Note that H has the same rank as matrix B. Taking the trace (tr) of each term in Eq. (3.24) results in the expression

$$tr(H) = tr(I_k) - tr\left(Q^{-1} r_i r_i^T\right) \tag{3.25}$$

It is known from matrix algebra that

$$\langle r_i, Q^{-1}r \rangle = tr \rangle Q^{-1} r_i, r_i \langle = tr\left(Q^{-1} r_i r_i^T\right) \tag{3.26}$$

where $>,<$ denotes the outer product. Using Eqs. (3.22) and (3.26), Eq. (3.25) indicates that $tr(H)=k-1$. Transposing the expression for H and introducing the matrix P_i yields

$$H^T = I_k - r_i r_i^T Q^{-1} = I_k - P_i \tag{3.27}$$

The following relation can then be derived using P_i and the condition in Eq. (3.22)

$$P_i^2 = P_i P_i = r_i r_i^T Q^{-1} r_i r_i^T Q^{-1} = r_i \langle r_i, Q^{-1} r_i \rangle r_i^T Q^{-1} = r_i r_i^T Q^{-1} = P_i \tag{3.28}$$

This expression implies that P_i is an idempotent matrix and thus the trace of

P_i is equal to its rank (rk). There exits a one-to-one relation between idempotent matrices and projectors [42], therefore, P_i is recognized as an oblique projector onto the space spanned by r_i (range space) along the corresponding null space of P_i. This is illustrated by the expression

$$P_i r_i = r_i r_i^T Q^{-1} r_i = r_i \langle r_i, Q^{-1} r_i \rangle = r_i \tag{3.29}$$

Equation (3.27), therefore, indicates that H^T is the complimentary oblique projector onto the null space of P_i along its range space. Matrix H^T is thus also idempotent, yielding the final result

$$tr\left(H^T\right) = rk\left(H^T\right) = rk(H) = k - 1 = rk(B) \tag{3.30}$$

which implies that the $k \times k$ matrix B is singular and thus the sensor location corresponding to r_i and Q^i cannot be discarded from the candidate sensor set. This proves that any sensor location with EfI value $E_{Di}=1.0$ is vital to the independence of the target modes and must be retained in the sensor configuration.

The question remains as to whether a sensor location can possess an EfI value $E_{Di}>1.0$. This would imply that the corresponding vector r_i extends beyond the identification ellipsoid when plotted in identification space. Assuming this to be the case, then

$$\langle r_i, Q^{-1} r_i \rangle = r_i^T Q^{-1} r_i = \sigma_i > 1.0 \tag{3.31}$$

Starting again with the matrix B introduced in Eq. (3.23), note that it may also be expressed as

$$B = \Phi_{fsi}^T \Phi_{fsi} \tag{3.32}$$

where Φ_{fsi} represents the original target mode partitions with the ith row deleted. The expression presented in Eq. (3.32) implies that B is at least positive semidefinite and thus the eigenvalues of B are all greater than or equal to zero. Forming the matrices H, H^T, and P_i using Eqs. (3.24) and (3.27), and noting that B, H, and H^T all have the same sign-definiteness because of the positive definite nature of Q^{-1}, the matrix product $P_i r_i$ is now given by

$$P_i r_i = r_i r_i^T Q^{-1} r_i = r_i \langle r_i, Q^{-1} r_i \rangle = r_i \sigma_i \tag{3.33}$$

This indicates that σ_i is an eigenvalue of matrix P_i. The corresponding eigenvalue of H^T, and thus also H, according to Eq. (3.27) is given by

$$\lambda_{H^T_i} = 1.0 - \sigma_i < 0.0 \qquad\qquad (3.34)$$

Matrix H thus has a negative eigenvalue and is not positive semidefinite. However, this implies that B is also not positive semidefinite which contradicts the result presented in Eq. (3.32) that B had to at least be positive semidefinite. This proves that no vector r_i can extend beyond the surface of the identification ellipsoid.

The effective independence distribution can, therefore, be used to rank candidate sensor locations based upon their contribution to the rank of the target mode partition matrix and thus the independence of its columns. The EfI value for each sensor location has been shown to lie in the range between 0.0 and 1.0. In an iterative fashion, low ranking sensor locations can be eliminated from the initial candidate set and the target mode partitions. The iterative process results in a rapid reduction of the initial set of candidate locations to a sensor configuration containing the allotted number of sensors which will give the best results for identification and correlation of the target modes. It is very important to understand that when a sensor is deleted from the candidate set, the EfI values of the remaining locations change because the E_D vector must always add up to the value k. The relative importance of candidate sensor locations may also change. The implication of this is that the initial candidate set cannot be truncated to the allotted number of m sensors in a single step. Vital sensor locations may be discarded resulting in unacceptable information loss. In order to get the best approximation to the true optimal solution, the initial candidate set should be ranked, the lowest ranked sensor location should be deleted from the target mode partitions, and then EfI values must be computed for the new candidate set and ranked again. A single sensor should be discarded at each iteration. Compared to most methods found in the literature which border on exhaustive search, the iterative EfI approach is very fast and efficient.

As suggested, the iterative sensor placement method proposed here is sub-optimal in that the derived sensor configuration is not guaranteed to have the largest Fisher Information matrix determinant. However, it is believed that the EfI solution will be close to the optimum configuration, derived using a much more time consuming optimization algorithm such as the genetic algorithm which searches for the m-sensor configuration with the largest FIM determinant. This belief is fortified by the following theorem [46] which shows that deleting the lowest EfI ranked sensor location produces the smallest change in the FIM determinant. Before stating the theorem, the following lemma will be useful.

Lemma: Let $C \in R^{n \times m}$, $D \in R^{m \times n}$, and I_p be a $p \times p$ identity matrix, then $det(I_n - CD) = det(I_m - DC)$. The proof can be found in the appendix of reference [47].

Theorem: For all $Q = \overline{\Phi}_{fs}^T \overline{\Phi}_{fs} \in R^{k \times k}$ and Q positive definite, let $r_i \in R^{k \times 1}$ contain the ith row of $\overline{\Phi}_{fs}$ and $B = Q - r_i r_i^T$, then $det(B) = det(Q)(1.0 - E_{Di})$ where $0.0 \le E_{Di} \le 1.0$.

$(Proof):$ $det(B) = det(Q - r_i r_i^T) = det(Q[I_k - Q^{-1} r_i r_i^T])$. Since Q and $[I_k - Q^{-1} r_i r_i^T]$ are both square matrices, $det(B) = det(Q) det(I_k - Q^{-1} r_i r_i^T)$

$= det(Q) det(1.0 - r_i^T Q^{-1} r_i) = det(Q)(1.0 - E_{Di})$ where the lemma and Eq. (3.22) have been used. The range of values for E_{Di} has been previously discussed.

Thus, the Effective Independence sensor placement method iteratively deletes candidate sensor locations which have the smallest impact upon the value of the Fisher Information matrix determinant. The accuracy of the estimation or the goodness of the sensor configuration can be monitored during the iteration process by tracking the determinant of the FIM, which is a measure of the amount of information contained in the sensor output concerning the response of the target modes. Tracking the condition number of the FIM in the spectral norm produces a measure of the robustness of the sensor configuration to errors in the measurement matrix $\overline{\Phi}_{fs}$ due to uncertainty in the FEM representation. This will be discussed thoroughly in a later section. It is also important to note that all of the results presented in this section are totally valid for the case where either there is no sensor noise, or where there is no knowledge of the noise covariance matrix. In either of these cases, merely substitute the target mode partitions Φ_{fs} for the noise-modified mode partitions in the equations of interest.

A. SIMPLE BEAM EXAMPLE

As a first example, a simple unconstrained beam representation of an LSS with a concentrated mass at the midpoint three quarters the mass of the beam itself will be considered for sensor placement. The finite element representation, illustrated in Figure 1, was constructed using 22 grid points and 21 elements. Each grid possesses a transverse and a rotational degree of freedom. The first ten mode shapes and frequencies were computed including two rigid body modes and eight elastic modes.

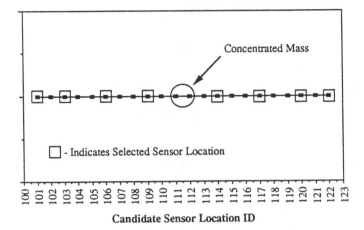

Candidate Sensor Location ID

Fig. 1. Simple beam model with central concentrated mass.

A typical beam bending mode is illustrated in Figure 2. The first seven elastic mode shapes were selected as target modes to be identified during a modal test. All 22 transverse displacement degrees of freedom were considered in an initial candidate sensor set. Figure 3 presents the fractional contribution to the target mode independence of each of the initial 22 sensor locations. This is essentially a ranking of the importance of prospective sensor locations to the success of the modal survey. As expected for this case, the beam endpoints are the most important locations. Note, however, that none of the locations is vital to independence for the initial set.

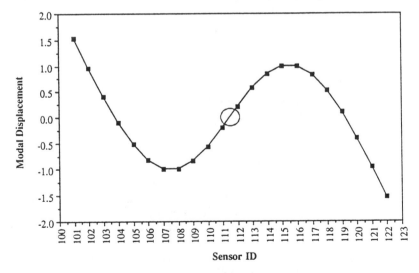

Fig. 2. Typical elastic free-free beam bending mode.

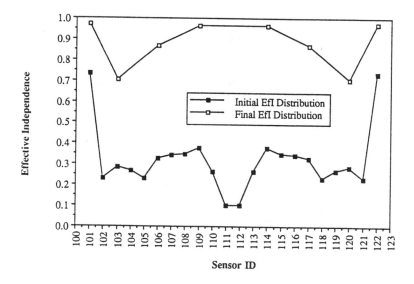

Fig. 3. Fractional contribution of sensor locations to linear independence.

Fourteen iterations were used to reduce the initial candidate sensor set to a final configuration of eight sensors which maintains the determinant of the FIM and the linear independence of the target modes. The FIM determinant is pictured in Figure 4 at each iteration. The eight selected sensor locations are illustrated in Figure 1, while the corresponding EfI distribution for the selected configuration is pictured in Figure 3. Note that

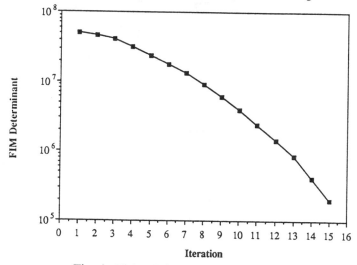

Fig. 4. Fisher Information matrix by iteration.

the EfI values for the sensor locations in the final configuration are larger than their initial values in the original candidate set. As sensors are deleted, the remaining locations become more important and, in some cases, relative importance of sensors has changed. For instance, in the initial EfI distribution, sensor 116 is more important that sensor 120, however, as sensors were deleted, location 120 surpassed 116 in importance and was selected for the final configuration while 116 was deleted.

The value of the FIM determinant for the final configuration was 1.96×10^5 while the corresponding condition number was 8.02 indicating that the target mode partitions are nicely independent. As a comparison, engineering judgement was used to place eight sensors on the beam. Starting at the ends, a sensor can be placed once every three locations resulting in a perfectly even distribution over the beam's length. However, the corresponding FIM determinant and condition number are given by 4.46×10^4 and 23.87, respectively. The EfI sensor configuration clearly outperforms the configuration based on engineering judgement, even in this case of a simplistic structure. The EfI method for sensor placement has another advantage in that the distribution for the final configuration indicates the cost of losing a sensor. This information can be used to place backup sensors.

B. SPACE STATION EXAMPLE

A second, more complicated example will now be considered for sensor placement. It consists of an early version of the Space Station. The finite element model representation, containing approximately 3,800 degrees of freedom, is illustrated in Figure 5. Fifteen target modes were selected for modal identification and test-analysis correlation. The target mode set is comprised primarily of main truss bending modes, although, there is also a large amount of photovoltaic array participation in each of the mode shapes. Table I. lists the target modal frequencies and descriptions. The initial candidate sensor location set was selected based upon the kinetic energy distribution in the target modes computed using Eq. (2.7). The highest ranked 187 degrees of freedom were designated as candidate sensor locations resulting in at least 50% of the total kinetic energy for each of the target modes being contained in the candidate set. One hundred and sixty-seven iterations were used to reduce the initial candidate set to the final configuration of 20 sensor locations which are illustrated in Figure 5.

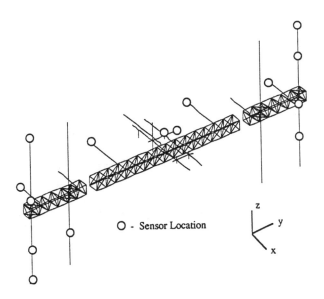

Fig. 5. Early version Space Station FEM representation.

Table I. Selected Space Station target modes.

Mode	Freq. (Hz.)	Description of Main Truss
21	0.124	First-order bending about x
29	0.144	First-order bending about z
33	0.158	First-order bending about z
34	0.160	First-order bending about x
38	0.331	Second-order bending about x
43	0.488	Second-order bending about z
44	0.513	Second-order bending about x
53	0.653	Third-order bending about z
55	0.722	Third-order bending about x
56	0.792	Port side torsion
67	1.169	Third-order bending (z)/torsion
68	1.242	Fourth-order bending (x)/torsion
69	1.257	Port side bending (x)/torsion
70	1.334	Torsion
71	1.352	Starboard side bending (z)/torsion

Table II. lists the selected sensors by FEM grid number and location description. Figure 6 shows that the FIM determinant varies smoothly with iteration number for this example. The EfI distribution for the final sensor configuration is presented in Figure 7.

Table II. Selected Space Station Sensor Configuration.

Grid No.	Dof	Location
290	23	Port outboard EPS radiator
294	2	Stbd outboard EPS radiator
319	23	Stbd station radiator
324	23	Port station radiator
8006	1	Stbd outboard PV array
8009	3	Stbd outboard PV array tip
8016	1	Stbd outboard PV array
8019	12	Stbd outboard PV array
8055	3	Port outboard PV array
8061	1	Port outboard PV array tip
8071	1	Port outboard PV array
8074	23	Port outboard PV array tip
8097	1	Port inboard PV array
30104	2	HAB mounted SSRMS
39835	3	Aft nodes-module cluster

The corresponding FIM determinant and condition number are given by 4.82×10^{-19} and 69.06, respectively. It is important to note that even though the determinant value is small, this does not necessarily mean that the FIM is close to being singular and the target modes are close to being dependent. The relatively small condition number indicates that the partitions are nicely independent. The value of the determinant, by itself, is not an absolute measure of the goodness of the sensor configuration. However, the determinant can serve as a basis for comparing the goodness of one sensor configuration with respect to another. For example, if the top 20 kinetic energy degrees of freedom are selected as the final sensor configuration, the FIM determinant is 2.18×10^{-42}, while the corresponding condition number is 3.06×10^{6}. Clearly, the EfI based sensor configuration is far superior to the one selected using modal kinetic energy.

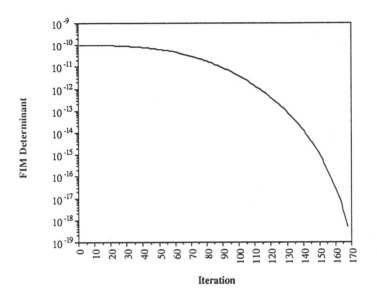

Fig. 6. Fisher Information matrix by iteration.

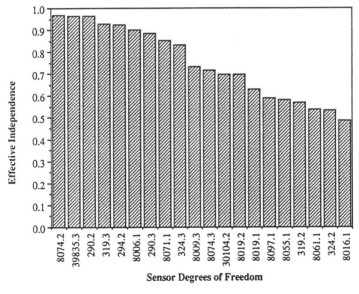

Fig. 7. Effective Independence distribution for selected sensor configuration.

IV. EFFECTS OF SENSOR NOISE[1]

In general, there will always be a certain amount of noise present in output data. If the experiment designer has no knowledge of the sensor noise statistics, the best approach is to use the noise-free target mode partitions in the formulations presented in the previous section, as was done in the two examples. However, in many cases, the designer may have a priori knowledge of the sensor noise covariance matrix, perhaps due to past experiences with similar structures and environments. Knowledge of the sensor noise statistics and distribution can be invaluable in designing the modal identification experiment such that the chance of on-orbit success is greatly enhanced.

One of the drawbacks of the straightforward application of the EfI sensor placement method is that it gives no indication as to the number of sensors required for the best modal identification, other than the mathematical minimum of k sensors required to maintain linear independence of k target modes. An important consideration, which will have a great impact upon the number of sensors required for accurate on-orbit identification, is signal-to-noise ratio (SNR). It is well known that the precision of a measurement is limited by the SNR in the output data. Therefore, while it is important to maintain an adequate level of SNR in the physical responses measured by the sensors, it is also important to maintain a desired minimum SNR in the target modal response space. The number of sensors used in the identification has no impact upon SNR levels at individual sensor locations in physical space, but it has a profound effect upon SNR levels in modal space. The analysis presented in this section derives a criterion for the number of sensors required to maintain a desired minimum level of SNR in target modal space. In a similar application, Juang and Pappa [48] applied the idea of a SNR to determine the order of the realized state matrix A in the presence of noise when using the Eigensystem Realization Algorithm (ERA) for modal identification.

Using the approach of Hoerl and Kennard [49], the square of the Euclidean distance between the estimate and the actual target modal response vector at time t can be expressed in the form

$$d^2 = \left(\hat{q} - q\right)^T \left(\hat{q} - q\right) = \hat{q}^T \hat{q} - \hat{q}^T q - q^T \hat{q} + q^T q \tag{4.1}$$

Considering the efficient unbiased estimator W_F presented in Eq. (3.5) and the output relation in Eq. (3.3), the first term in expression (4.1) can be written as

---

[1] Figures and portions of the text contained in this section were taken from reference [30]

$$\hat{q}^T \hat{q} = q^T q + q^T W_F v + v^T W_F^T q + v^T W_F^T W_F v \tag{4.2}$$

Similarly, the other expressions in Eq. (4.1) can be written in terms of q, W_F, and v. When these relations are substituted back into Eq. (4.1), the expression for d^2 becomes

$$d^2 = v^T W_F^T W_F v \tag{4.3}$$

Taking expected values, and due to the zero mean value of the measurement noise, Eq. (4.2) produces

$$E(\hat{q}^T \hat{q}) = q^T q + E(v^T W_F^T W_F v) \tag{4.4}$$

The mean square value of the magnitude of the estimate is, therefore, equal to the sum of the square of the Euclidean norm of the true target mode response and the expected value of the square of the Euclidean distance between the estimate and the true response vector.

The expression for d^2 in Eq. (4.3) is a quadratic form, thus the expected value can be expressed as [50]

$$E(d^2) = E(v^T W_F^T W_F v) = tr\left[\left(W_F^T W_F\right) E(vv^T)\right] = tr\left[\left(W_F^T W_F\right) R\right] \tag{4.5}$$

Substituting the expression for the estimator produces

$$E(d^2) = tr\left[R^{-1} \Phi_{fs}\left(\Phi_{fs}^T R^{-1} \Phi_{fs}\right)^{-1}\left(\Phi_{fs}^T R^{-1} \Phi_{fs}\right)^{-1} \Phi_{fs}^T \right] \tag{4.6}$$

The well known identity $tr\{AB\}=tr\{BA\}$ results in the expected value of d^2 being given by

$$E(d^2) = tr\left[\left(\Phi_{fs}^T R^{-1} \Phi_{fs}\right)^{-1}\right] \tag{4.7}$$

Using this result, Eq. (4.4) becomes

$$E(\hat{q}^T \hat{q}) = q^T q + tr\left[\left(\Phi_{fs}^T R^{-1} \Phi_{fs}\right)^{-1}\right] = q^T q + tr\left[Q^{-1}\right] \tag{4.8}$$

The last term in Eq. (4.8) is the contribution of the measurement noise to the expected value of the response estimate in the target modal space.

Defining α as the difference between the square of the Euclidean norm of the modal response vector estimate and the corresponding true value, its expected value is then given by

$$E(\alpha) = E(\hat{q}^T \hat{q}) - q^T q = tr[Q^{-1}]$$

(4.9)

Prior to launch, analysis is performed to simulate the response of the LSS to proposed inputs to be applied during the modal identification experiment. The response of each target mode can be predicted over the identification period T. Thus, a predicted true mean square response of all the target modes can be computed based upon the simulation using the expression

$$\left(q^T q\right)_{ave} = \frac{1}{T} \int_0^T \left(q^T q\right) dt$$

(4.10)

Given the noise covariance intensity matrix R and the anticipated response of the target modes, the problem of how many sensors are required to maintain a desired minimum SNR in the target modes will now be addressed.

A noise-to-signal ratio can be defined as the root-mean-square (RMS) noise divided by the RMS signal. The ratio ε^2 can then be defined

$$\varepsilon^2 = \frac{E(\alpha)}{\left(q^T q\right)_{ave}} = \frac{tr[Q^{-1}]}{\left(q^T q\right)_{ave}}$$

(4.11)

and identified as the square of the noise-to-signal ratio associated with $q^T q$. The mean square value of the Euclidean norm of q predicted by simulation can be expressed as

$$\left(q^T q\right)_{ave} = \sum_{i=1}^{k} \overline{q_i^2} \geq k \overline{q^2}_{min}$$

(4.12)

where $\overline{q^2}_{min}$ represents the minimum mean square response among all the target modes during period T. Writing the trace of the Fisher Information Matrix in terms of its eigenvalues

$$tr\left[Q^{-1}\right] = \sum_{i=1}^{k} \lambda_i^{-1} \leq k\lambda_{min}^{-1}$$

(4.13)

where λ_{min} represents the smallest eigenvalue of Q, the relation for ε^2 can be expressed in the form

$$\varepsilon^2 = \frac{\displaystyle\sum_{i=1}^{k} \lambda_i^{-1}}{\displaystyle\sum_{i=1}^{k} \overline{q_i^2}} \leq \frac{\displaystyle\sum_{i=1}^{k} \lambda_i^{-1}}{k\overline{q^2}_{min}} \leq \frac{\lambda_{min}^{-1}}{\overline{q^2}_{min}}$$

(4.14)

Equation (4.14) produces an upper bound for the square of the noise-to-signal ratio in $q^T q$ during on-orbit identification given by

$$\left(\varepsilon^2\right)_{max} = \frac{\lambda_{min}^{-1}}{\overline{q^2}_{min}}$$

(4.15)

The derivation of this bound assumes that all the target modes possess the minimum predicted mean square response, therefore, Eq. (4.15) represents the upper bound of the squared noise-to-signal ratio for the least excited mode. The lower bound on the corresponding SNR is thus given by

$$\sigma_{min} = \left[\lambda_{min} \overline{q^2}_{min}\right]$$

(4.16)

Given the predicted minimum mean square response in the target modes and the minimum eigenvalue of the Fisher Information Matrix, Eq. (4.16) yields a lower bound for the minimum SNR that can be expected for the target modes during the identification. From this result, a criterion can be derived which can be applied during the course of the generalized EfI sensor placement analysis. This criterion can be used to determine the number of sensors which are required to maintain a desired minimum signal-to-noise ratio in the target modes given a predicted value for the minimum mean square target mode response. At each iteration of the EfI method, the minimum eigenvalue of the Fisher Information Matrix must satisfy the condition

$$\lambda_{min} > \frac{\sigma_{min}^2}{\overline{q^2}_{min}}$$

(4.17)

This condition can be easily checked during the EfI analysis after the truncation of each sensor. If the condition in Eq. (4.17) is violated, the sensor deleted during the iteration must be retained to maintain the desired minimum SNR in all the target modes, thus fixing the number of required sensor at the level prior to the current iteration.

The application of the theory set forth in this section will be demonstrated using the simple LSS beam representation introduced in Section III. In order to demonstrate the effect of measurement noise upon the placement of sensors using the EfI method, two different sensor noise distribution cases will be considered. Each of the noise distributions is illustrated in Figure 8. Noise case 1 assumes the variance in the sensor noise increases linearly towards the center of the structure while case 2 assumes that the variance is large at the tips of the structure and decreases linearly towards the center. For simplicity, the covariance intensity matrix is assumed to be diagonal in each case. As in previous examples, all 22 grid point locations in the FEM are included in the initial candidate sensor location set and the first seven elastic mode shapes will be selected as target modes.

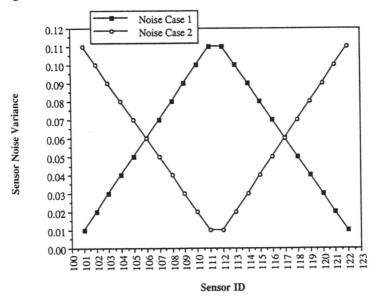

Fig. 8. Sensor noise variance by station within structure.

The generalized EfI procedure begins by generating the noise-modified target mode shapes. The noise covariance intensity matrix R acts as a weighting matrix for the sensor locations. Sensors possessing a high noise variance are suppressed while sensor locations possessing little noise are emphasized. Figure 9 illustrates this trend for the second order bending target mode in which the noise-free mode and the noise-modified mode for cases 1 and 2 are plotted. In noise case 1, the mode shape has been modified such that, with respect to the noise-free shape, the modal coefficients have increased at outboard sensor locations where the noise variance is low and decreased at locations towards the center of the structure where the variance is much larger. In case 2, the opposite trend appears, consistent with the second noise distribution.

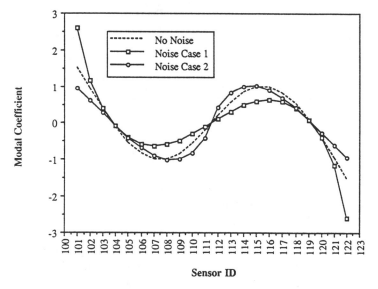

Fig. 9. Effect of sensor noise on second order bending mode shape.

The EfI distribution vector E_D for all 22 initial candidate sensor locations is plotted versus station in the beam in Figure 10 for the noise-free case, noise case 1, and noise case 2. The trends that were apparent in the mode shapes illustrated in Figure 9 are also seen in Figure 10. In case 1, the EfI method tends to rank the outboard sensors higher with respect to the noise-free ranking because they possess less noise. Locations towards the center are ranked lower which is consistent with the increased noise variance levels. The ranking for case 2 again illustrates the opposite trend consistent with higher noise levels outboard and less noise at the center of the structure.

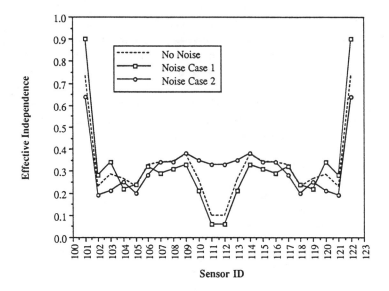

Fig. 10. Effect of sensor noise on EfI distribution.

Fourteen iterations were performed to reduce the initial candidate sensor location set to eight optimum locations for target mode identification. Figure 11 presents the EfI distribution vector plotted versus station within the beam for the eight optimum sensor locations selected by the EfI method for each noise case. The eight optimum sensors for the noise-free case and noise case 1 are identical, however, as in previous results, the outboard locations are ranked higher and the central locations lower in case 1 with respect to the noise-free ranking. In case 2, a one-to-one comparison of EfI ranking values is not possible because different sensor locations were retained. The tip sensor locations were still selected but the remaining six locations are concentrated in the central portion of the structure reflecting the lower noise levels in this area.

This simple example can also be used to investigate the effect of measurement noise upon the number of sensors required to perform an on-orbit identification of the target modes. An on-orbit test is simulated by applying a unit impulse to the left end of the beam representation. For demonstration purposes, no damping is included in the simulation. Accelerometers are used to sense the response of the structure which is simulated for 30 seconds. Simulation predicts that the first mode has the minimum RMS acceleration response given by $\left(\overline{q^2}_{min}\right)^{1/2} = 0.839$.

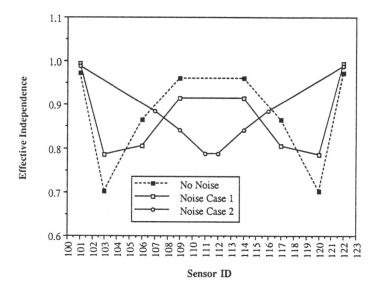

Fig. 11. Effective Independence distribution for 8 selected sensor locations.

An uncorrelated uniform noise distribution was assumed with a standard deviation of 0.222 in/sec^2. This noise level corresponds to an average 100:1 physical acceleration SNR over all 22 candidate sensor locations due to the seven target modes. Applying the criterion set forth in Eq. (4.17), the number of sensors required to maintain a desired level of SNR in the modal space over all seven target modes was computed for the proposed excitation. The results are presented in Figure 12. Up to a desired modal SNR level of 4.0, only seven sensors are required. As the desired SNR level increases, additional sensors are needed until at an SNR level of 10.0, all 22 sensors must be retained. If higher SNR ratios are desired, larger inputs must be provided.

The example presented here shows that the noise covariance intensity matrix can be thought of as a sensor weighting matrix. Sensors with little noise are weighted more heavily in the analysis than sensors with larger noise levels. Independent of the noise problem, other weighting matrices may be of interest. Using an identity weighting matrix, the general EfI sensor placement method reduces to an approach which ignores noise. This produces a sensor configuration which renders the noise-free target modal partitions spatially independent. This property enhances the computation of test-analysis-models using the advanced techniques presented in Section II.

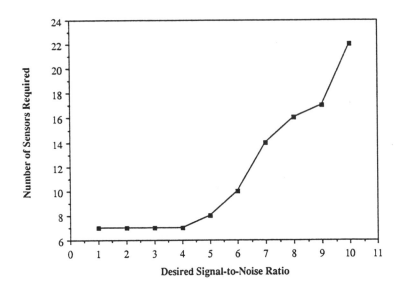

Fig. 12. Minimum number of sensors required to maintain SNR.

The FEM mass matrix is a symmetric positive definite matrix that can also be used as a weighting matrix in the EfI approach. Locations with large mass values will be weighted more heavily than locations possessing little mass. This weighting matrix could be used to select a sensor configuration to be used in conjunction with a statically reduced TAM. The EfI method for sensor placement using mass weighting has a distinct advantage over the usual approach in static reduction, where all large kinetic energy degrees of freedom are selected, in that the method maintains the linear independence of the target modal partitions which is vital to any reduced representation. It is important to note, however, that mass weighting will in general produce a sensor configuration which possesses less information concerning the target modal response than the configuration derived using the identity weighting.

V. EFFECTS OF ANALYTICAL MODEL ERROR[2]

A common thread running throughout the optimal sensor placement literature, including the EfI method presented in this chapter, is the use of

---

[2] Figures and portions of the text contained in this section were taken from reference [31]

some type of analytical model to perform the placement analysis. This analytical representation usually comes in the form of a finite element model and it will always possess errors due to incorrect input parameters, incorrect modeling techniques, unmodeled dynamics, or small nonlinearities. These errors must be considered in the sensor placement procedure. The problem addressed in this section arises because the analytical model which is to be updated based upon the on-orbit test results is being used to determine the sensor configuration which in turn dictates the modal test results. The error present in the prelaunch FEM can result in degradation and even failure of the on-orbit modal identification test. The designer of an on-orbit test must, therefore, have some form of guarantee that the experiment will work based upon sensor placement results derived from a prelaunch FEM. This section presents an error analysis from within the framework of the EfI method. A lower bound will be derived for the number of sensors required to guarantee the spatially independent identification of the true on-orbit target modes using prelaunch FEM results. This analysis assumes that there is no noise, however, all of the results are applicable to the general case by substituting the noise-modified target mode partitions where appropriate.

The effect of prelaunch analytical model error upon the sensor placement analysis is studied by assuming that the sensor partition of the FEM target mode shapes can be expressed in the form

$$\Phi_{fs} = \Phi_{rs} + \delta_s$$

$$(5.1)$$

where Φ_{rs} represents the real target mode shapes at the sensor locations and δ_s is the corresponding matrix of errors in the prelaunch FEM target modes. This general form for mode shape error is very useful because it can accommodate errors due to incorrect input parameters, incorrect modeling, unmodeled dynamics, or small nonlinearities. Subtracting δ_s from both sides of Eq. (5.1) and premultiplying each side by its corresponding transpose produces

$$\Phi_{rs}^T\Phi_{rs} = \Phi_{fs}^T\Phi_{fs} - \Phi_{fs}^T\delta_s - \delta_s^T\Phi_{fs} + \delta_s^T\delta_s$$

$$(5.2)$$

Making the definitions

$$Q_r = \Phi_{rs}^T\Phi_{rs} \quad ; \quad Q_f = \Phi_{fs}^T\Phi_{fs} \quad ; \quad \Delta = -\Phi_{fs}^T\delta_s - \delta_s^T\Phi_{fs} \quad ; \quad D = \delta_s^T\delta_s$$

Eq. (5.2) becomes

$$Q_r = Q_f + \Delta + D \qquad\qquad (5.3)$$

in which Q_r represents the information matrix corresponding to the real target mode shapes, Δ is a symmetric matrix representing the uncertainty in the analytical modes, and D is a symmetric matrix, not to be confused with the definition in Section II, representing the information in the mode shape errors δ_s.

Information matrix Q_f, where the subscript has now been added to emphasize the fact that it is derived from the FEM representation, is positive definite by design, matrix Δ is in general indefinite, and without loss of generality, it will be assumed that the unknown error information matrix D is also positive definite. The EfI sensor placement strategy maintains the positive definiteness and determinant of the analytical model information matrix Q_f. In reality, the experiment designer wishes to maintain the positive definiteness and determinant of the true information matrix Q_r. In Section III, it was shown that a k-dimensional identification ellipsoid is associated with positive definite Q_f. Assuming initially that Q_r and D are also positive definite, the real identification ellipsoid can be associated with Q_r and an error ellipsoid can be associated with D.

The objective here is to determine conditions under which the prelaunch FEM can be used to make intelligent decisions concerning sensor placement for identification of the real target modes. Subtracting D from both sides of Eq. (5.3) results in matrix N

$$Q_r - D = Q_f + \Delta = N \qquad\qquad (5.4)$$

which will be called the net information matrix. Examining the left side of Eq. (5.4), if Q_r-D is positive definite (>0), then Q_r>D and the real identification ellipsoid contains the error ellipsoid. This implies that in each direction in identification space, there is more information in the real modes than there is in the analytical mode shape errors. If Q_r-D is positive semi-definite (≥ 0), the error ellipsoid touches the real identification ellipsoid, and finally, if Q_r-D is indefinite, the error ellipsoid pokes through the real identification ellipsoid.

It can be seen from Eqs. (5.3) and (5.4), and the identified forms of the matrices, that N>0 is a sufficient condition for the positive definiteness of the real information matrix Q_r. Thus, during the course of an EfI sensor placement analysis, if after each iteration, the net information matrix could be generated and determined to be positive definite, it could be guaranteed that the sensor just eliminated was not vital to the independence of the real target modes. A positive definite net information matrix for any sensor configuration guarantees that the sensors will be able to independently identify the real target modes on-orbit.

While the condition $N>0$ is sufficient for the identification of the real target modes, it is not necessary. However, it is proposed that this condition is necessary for the analytical representation to provide positive or useful information concerning sensor placement for the identification of the real modes. The FEM is then said to contain positive net information concerning the real target modes Φ_{rs}. Considering the smallest eigenvalue μ_{min} of N with corresponding eigenvector ψ_{min} possessing unit length, rearranging Eq. (5.4) and then pre- and postmultiplying by ψ^T_{min} and ψ_{min}, respectively, yields

or

$$\psi^T_{min}Q_r\psi_{min} = \mu_{min} + \psi^T_{min}D\psi_{min} \tag{5.5}$$

$$g_r = \mu_{min} + h \tag{5.6}$$

Scalar g_r is the square of the Euclidean norm of the vector $\zeta_r = \Phi_{rs}\psi_{min}$. The ith term within ζ_r is the projection of the ith row of Φ_{rs} onto the unit direction in identification space represented by ψ_{min}. Therefore, g_r is a measure of the amount of information contained in the real target modes along ψ_{min}. Likewise h is a measure of the amount of information contained in δ_s.

If N is positive definite, $\mu_{min}>0$, the information contained in the analytical mode shape partitions contributes in a positive manner to the information contained in the real target mode partitions along ψ_{min}. For this case, there is also an ellipsoid associated with the positive definite net information matrix. If N becomes positive semi-definite after the deletion of a sensor, $g_r = h$. The net information ellipsoid collapses and the information contained in the real target modes along ψ_{min} is totally due to the error in the corresponding FEM mode shapes. The analytical representation possesses no knowledge of δ_s, therefore it contributes no information toward the independent identification of the real target modes. If N becomes indefinite, $\mu_{min}<0$, the information contained in the model that is used to place sensors for identification of the analytical target modes actually detracts from the independent identification of the real target modes. At this point, the FEM should no longer be used to place sensors.

With the error theory in place, it remains to determine the bounds on a suitable norm of the error δ_s such that the net information matrix is positive definite. Over the past decade, dynamicists working in the area of robust control system design have been concerned with determining bounds on the norm of a perturbation matrix Δ such that if matrix A is stable (all eigenvalues have negative real parts), $A+\Delta$ will remain stable. The results obtained for this stability problem are directly applicable to the determination of the positive definiteness of the perturbed matrix $N=Q_f+\Delta$. Several References [51-54] have considered the case in which A is of a general form and Δ is allowed to be complex and unstructured. However, there is an absence of results specifically addressing the more restricted case of real symmetric A and structured real Δ which is of interest here.

For the complex unstructured case, Qiu and Davison [54] present a necessary and sufficient bound for the norm of Δ. Assuming A to be a normal matrix $(A^T A = AA^T)$, if Δ is bounded by its L^2, or spectral norm, $A+\Delta$ is stable if and only if

$$\|\Delta\|_2 = \kappa_{max}(\Delta) < \frac{1}{\left\|(sI-A)^{-1}\right\|_\infty}$$

(5.7)

in which $\| \cdot \|_2$ denotes the 2-norm, κ_{max} denotes the maximum singular value, and $\| \cdot \|_\infty$ is the H^∞ norm [54] given by

$$\left\|(sI-A)^{-1}\right\|_\infty = \sup_{\omega \geq 0}\left\{\left\|(j\omega I-A)^{-1}\right\|_2 : \omega \in R\right\}$$

(5.8)

where "sup" denotes the supremum and $j = (-1)^{1/2}$. Application of this tight bound to the determination of the positive definiteness of $N=Q_f+\Delta$ in which both Q_f and Δ are real and symmetric results in the condition

$$\kappa_{max}(\Delta) < \lambda_{min}(Q_f)$$

(5.9)

where $\lambda_{min}(\cdot)$ denotes the smallest eigenvalue of Q_f. Unfortunately, unlike the general case, while Eq. (5.9) is sufficient, it is not a necessary condition for the positive definiteness of N. This is due to the fact that in the case of interest, the perturbation Δ is structured, i.e.

$$\Delta = -\Phi_{fs}^T \delta_s - \delta_s^T \Phi_{fs}$$

Only the modal error matrix δ_s itself is unstructured. Therefore Eq. (5.9) gives a conservative condition on Δ for the positive definiteness of N.

Rather than Δ, a condition for the positive definiteness of N must be derived in terms of a suitable norm of δ_s. The selected measure of modal error size must have physical meaning to the structural dynamicist. For the present case, a condition will be derived in terms of $\|\delta_{max}\|$ which is the

Euclidean norm of the largest modal error vector, i.e. $\left\|\delta_{max}\right\| = \max_i\|\delta_{si}\|$ where δ_{si} is the ith column in δ_s. Therefore, in order to use the condition in Eq. (5.9), $\kappa_{max}(\Delta)$ must be related to $\|\delta_{max}\|$. Using the usual properties of matrix norms and the definition of Δ yields

$$\|\Delta\|_2 \le \left\|\Phi_{fs}^T\delta_s\right\|_2 + \left\|\delta_s^T\Phi_{fs}\right\|_2 \le 2\left\|\Phi_{fs}\right\|_2\|\delta_s\|_2$$

(5.10)

However,

$$\left\|\Phi_{fs}\right\|_2 = \left[\lambda_{max}\left(\Phi_{fs}^T\Phi_{fs}\right)\right]^{1/2} = \left[\lambda_{max}\left(Q_f\right)\right]^{1/2}$$

therefore

$$\|\Delta\|_2 \le 2\left[\lambda_{max}\left(Q_f\right)\right]^{1/2}\|\delta_s\|_2$$

(5.11)

Considering the modal error information matrix $D=\delta_s^T\delta_s$, where as above $\|\delta_s\|_2=[\lambda_{max}(D)]^{1/2}$, and using the trace identities

$$tr(D) = \sum_{i=1}^k \lambda_i(D) = \sum_{i=1}^k \|\delta_{si}\|^2$$

produces the inequality

$$\lambda_{max}(D) \le k\|\delta_{max}\|^2$$

(5.12)

where k is again the number of target modes. Finally, Eqs. (5.11) and (5.12) can be combined to give the desired relation

$$\|\Delta\|_2 \le 2k^{1/2}\left[\lambda_{max}\left(Q_f\right)\right]^{1/2}\|\delta_{max}\|$$

(5.13)

A sufficient condition for the positive definiteness of the information matrix N is thus given by

$$\|\delta_{max}\| < \frac{1}{2k^{1/2}} \frac{\left[\lambda_{max}\left(Q_f\right)\right]^{1/2}}{cond\left(Q_f\right)}$$

(5.14)

where $cond(\bullet)$ is the condition number in the spectral norm. Assuming a maximum value for target mode error, for example 10% of the Euclidean norm of the largest analytical target mode, a user can easily compute the right side of Eq. (5.14) and check for positive definiteness of N after each sensor is deleted during the sensor placement analysis. The form of the right side of Eq. (5.14) is physically reasonable because it is inversely proportional to the number of target modes, i.e. as the number of target modes increases, more information is needed in the analytical target modal partitions to identify the real target modes. It is also inversely proportional to the condition number of Q_f, implying that orthogonality of the analytical target mode partitions Φ_{fs} promotes the identification of the real target modes.

The drawback of the positive definiteness condition given by Eq. (5.14) is that for many cases it can be very conservative. More research by both the controls and structural dynamics communities on stability conditions for real structured perturbations of real symmetric matrices is needed. However, the derived bound can provide some meaningful insight into the number of sensors required for the guaranteed identification of a set of real target modes given a sensor configuration based upon a prelaunch finite element model with an assumed level of error.

The simple LSS beam representation will again be used to demonstrate the ideas presented in this section. In this case, the FEM representation will consist of the uniform free-free beam without the concentrated mass. The actual on-orbit structure is simulated by the same beam, but with the large concentrated mass at the center. This represents a significant error in the analytical representation for the purpose of demonstration. Eigenvalue solutions for each full order model produced two rigid body modes and twenty elastic mode shapes. In this example, five of the elastic FEM mode shapes were selected as target modes. Table III lists the analytical target mode frequencies, the corresponding real structure modal frequencies, and the related cross-generalized mass (CGM) values. The CGM value for the ith analytical/real target mode pair is computed using the relation

$$C_i = \Phi_{fsi}^T M_{FEM} \Phi_{rsi}$$

in which M_{FEM} is the analytical mass matrix. The mode shapes are normalized such that a value of $C_i = 1.0$ corresponds to perfect shape correlation between the ith analytical target mode and the corresponding real mode, as discussed in Section II. Values below 0.8 indicate significant differences between finite element model and real mode shapes.

Table III. FEM and real target modal frequencies and CGM values.

Target Mode	FEM Freq. (Hz.)	Real Freq. (Hz.)	CGM
3	4.89	3.83	0.88
5	26.17	21.23	0.85
7	64.04	56.27	0.88
8	88.99	76.24	0.91
10	150.73	124.91	0.76

As in previous examples, the initial candidate sensor location set was assumed to contain all 22 transverse displacement degrees of freedom. Based upon the analytical mode shapes, the EfI method was used to eliminate 17 sensors, one during each iteration, resulting in five sensor locations which maintain the determinant of the associated Fisher Information Matrix. In this example, the real target modes are known, therefore the net information matrix N can be computed at each iteration and checked for positive definiteness. If the net information matrix is positive definite, the FEM is providing positive or beneficial information to the identification of the real target modes and thus another sensor can be deleted based upon the results of the sensor ranking. Figure 13 illustrates magnitudes of the eigenvalues of the net information matrix for all 17 iterations. At the seventh iteration, after six sensors have been deleted from the initial candidate set, one of the eigenvalues becomes negative and thus the net information matrix becomes indefinite. A second eigenvalue becomes negative at the twelfth iteration.

Beyond the seventh iteration, there exists a direction in identification space along which the FEM detracts from the identification of the real target modes. This direction is along the eigenvector ψ_{min} corresponding to the minimum eigenvalue as presented in Eqs. (5.5) and (5.6). The parameters in Eq. (5.6) are illustrated in Figure 14 at each of the 17 iterations for the example sensor placement analysis. Parameter g_r represents the amount of information in the real target modes along ψ_{min}, h represents the amount of information contained in the analytical mode shape errors, and λ_{min} represents the contribution of the analytical representation.

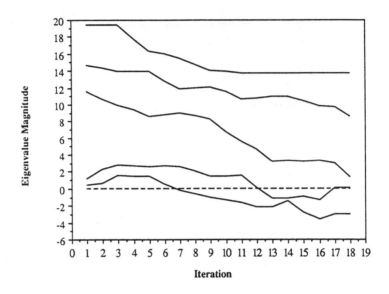

Fig. 13. Eigenvalues of net information matrix N.

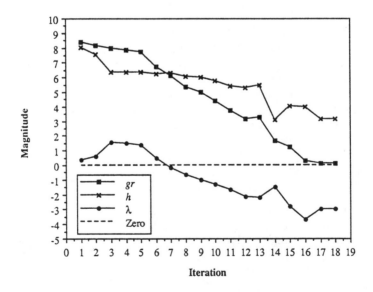

Fig. 14. Information contributions along net information eigenvector ψ_{min}.

At the seventh iteration, the FEM contribution becomes negative and the information within the real target modes becomes less than the information in the modal error vectors.

Figure 15 presents the value of the determinant vs. iteration for the FEM Fisher Information matrix which is a measure of the volume of the corresponding identification ellipsoid. This curve is based upon EfI sensor placement using the analytical target modes. Note that the method tends to maintain the value of the determinant as in previous examples. The figure also illustrates the value vs. iteration of the determinant of the FIM corresponding to the real target modes assuming the sensor placement analysis was based upon the real target modes. The third curve in the figure presents the determinant of the real FIM when the sensor truncation analysis is based upon the analytical target modes. This curve represents the actual on-orbit results using a sensor configuration which is derived based upon maintaining the independent information in the prelaunch FEM target modes. After the seventh iteration, corresponding to an indefinite net information matrix, the determinant of the real FIM falls of rapidly and irregularly. In this regime, the "goodness" of the sensor configuration for the identification of the analytical target modes actually detracts from the independent identification of the real target modes resulting in large amounts of information loss after further sensor deletions.

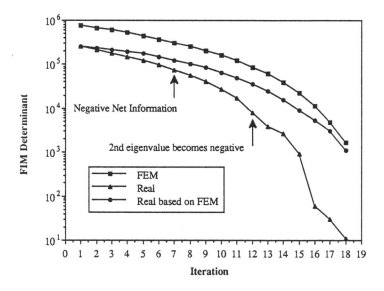

Fig. 15. Fisher Information matrix determinants.

This is especially evident after the twelfth iteration where a second eigenvalue of N becomes negative.

Unfortunately, during an actual prelaunch sensor configuration design analysis, the analyst will not have any knowledge of the real target modes. The sensor placement will be based solely on the FEM. However, the designer will want to place sensors in sufficient numbers to guarantee the identification of the real target modes on orbit. Assuming a level of modal error based upon engineering judgement, Eq. (5.14) can be used in conjunction with the EfI sensor placement method to determine the iteration at which the prelaunch finite element model can no longer be used for sensor placement. This approach yields the number of sensors required to guarantee on-orbit independent identification based upon prelaunch FEM analysis. Figure 16 illustrates the number of sensors required for on-orbit identification of the real target modes based upon the five analytical target modes selected for the uniform beam example assuming varying levels of maximum modal error. The plot indicates that if the length of the maximum error vector is 15.0% of the length of the largest analytical target mode shape, the FEM will provide no useful information for sensor placement.

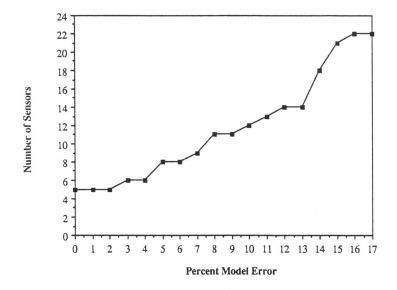

Fig. 16. Number of sensors required for $N > 0.0$.

Fifteen percent error is certainly not unexpected in prelaunch analytical representations. However, the analysis used to generate Figure 16 assumes that all of the modal error vectors are shorter than 15.0% of the longest analytical target mode. Therefore, percentage error corresponding to analytical modes with shorter Euclidean lengths could be greater than 15.0%. Also, as indicated previously, the condition for positive net information given by Eq. (5.14) can be very conservative in many cases.

VI. CONCLUSION

A method of sensor placement for modal identification has been presented which is based upon ranking the contribution of each candidate sensor location to the linear independence of the target modal partitions. In an iterative fashion, locations which do not contribute significantly to the independent information contained within the target modes are removed. Within a relatively small number of iterations, the initial candidate set of sensor locations can be reduced to the allotted number. The method, called Effective Independence, results in a sensor truncation process that maintains the determinant of the Fisher Information matrix which minimizes the covariance matrix of the estimate errors, thus giving the best state estimates of the target modes.

An advantage of the EfI method is that it is computationally nonintensive compared with borderline exhaustive search techniques found in the literature. The method also offers the benefit of physical insight into the ranking and ultimate selection of sensor locations. It is important to note that the EfI method does not directly determine how many sensors are required to identify the target modes in the presence of analytical model uncertainty and sensor noise. However, criteria have been presented, in the context of the EfI approach, such that the required number of sensors can be determined to maintain the desired minimum signal-to-noise ratio in modal space, and guarantee the independent identification of the true target modes on orbit.

REFERENCES

1. J. C. Chen and J. A. Garba, "Structural Model Validation Using Modal Test Data," *Joint ASCE/ASME Mechanics Conference*, Albuquerque, NM, pp. 109-137 (1985).

2. C. C. Flanigan, "Test/Analysis Correlation Using Design Sensitivity and Optimization," *Society of Automotive Engineers*, Warrendale, PATP871743 (1987).

3. D. C. Kammer, B. M. Jensen and D. R. Mason, "Test-Analysis Correlation of the Space Shuttle Solid Rocket Motor Center Segment," *Journal of Spacecraft and Rockets* **26**(4), pp. 266-273 (1989).

4. A. Berman, F. Wei and K. V. Rao, "Improvement of Analytical Dynamic Models Using Modal Test Data," *21st AIAA Structures, Structural Dynamics, and Materials Conference*, New York, NY, pp. 809-814 (1980).

5. A. Kabe, "Stiffness Matrix Adjustment Using Mode Data," *AIAA Journal* **23**(9), pp. 1431-1436 (1985).

6. D. C. Kammer, "An Optimum Approximation for Residual Stiffness in Linear System Identification," *AIAA Journal* **26**(1), pp. 104-112 (1988).

7. S. W. Smith and C. A. Beattie, "Optimal Identification Using Inconsistent Modal Data," *32nd AIAA/ASME/ASCE/AHS/ASC Structures, Structural Dynamics, and Materials Conference*, Baltimore, MD, pp. 2319-2324 (1991).

8. E. Denman, E. Hassleman, T. Sun, J. N. Juang, F. Udwadia, V. Venkaya and M. Kamat, "Identification of Large Space Structures on Orbit," Air Force Rocket Propulsion Laboratory, Technical Report, TR-86-054 (1986).

9. T. A. Kashangaki, "On-Orbit Damage Detection and Health Monitoring of Large Space Trusses - Status and Critical Issues," *32nd AIAA/ASME/ASCE/AHS/ASC Structures, Structural Dynamics and Materials Conference*, Baltimore, MD, pp. 2946-2957 (1991).

10. A. Le Pourhiet and L. Le Letty, "Optimization of Sensor Locations in Distributed Parameter System Identification," *Identification and System Parameter Estimation*, in *Optimization of Sensor Locations in Distributed Parameter System Identification*, (Rajbman), North-Holland Co., Amsterdam, pp. 1581-1592 (1978).

11. Z. H. Qureshi, T. S. Ng and G. C. Goodwin, "Optimum Experimental Design for Identification of Distributed Parameter Systems," *International Journal of Control* **31**(1), pp. 21-29 (1980).

12. T. K. Yu and J. H. Seinfeld, "Observability and Optimal Measurement Locations in Linear Distributed Parameter Systems," *International Journal of Control* **18**(4), pp. 785-799 (1973).

13. S. Omatu, S. Koide and T. Soeda, "Optimal Sensor Location for Linear Distributed Parameter Systems," *IEEE Trans. on Automatic Control* **AC-2**(4), pp. 665-673 (1978).

14. Y. Sawaragi, T. Soeda and S. Omatu, *"Modeling, Estimation and Their Application for Distributed Parameter Systems," Lecture Notes in Control & Information Science,* in *Modeling, Estimation and Their Application for Distributed Parameter Systems,* **11,** Springer-Verlag, Berlin, pp. (1978).

15. R. E. Goodson and M. P. Polis, "Identification of Parameters in Distributed Systems," *Distributed Parameter Systems,* in *Identification of Parameters in Distributed Systems,* (W. H. Ray and D. G. Lainiotis), Marcel Dekker, New York, pp. (1978).

16. C. S. Kubrusly and H. Malebranche, "Sensors and Controllers Location in Distributed Systems - A Survey," *Automatica* **21**(2), pp. 117-128 (1985).

17. J. N. Juang and G. Rodriguez, "Formulation and Application of Large Structure Sensor and Actuator Placement," *2nd VPI&SU/AIAA Symposium on Dynamics and Control of Large Flexible Spacecraft,* Blacksburg, VA, Virginia Polytechnic Inst. and State Univ., pp. 247-262 (1979).

18. G. A. Norris and R. E. Skelton, "Selection of Dynamic Sensors and Actuators in the Control of Linear Systems," *Journal of Dynamic Systems, Measurement, and Control* **111**, pp. 389-397 (1989).

19. H. Baruh and K. Choe, "Sensor Placement in Structural Control," *Journal of Guidance, Control, and Dynamics* **13**(3), pp. 524-533 (1990).

20. M. L. DeLorenzo, "Sensor and Actuator Selection for Large Space Structure Control," *Journal of Guidance, Control, and Dynamics* **13**(2), pp. 249-257 (1990).

21. K. B. Lim, "A Method for Optimal Actuator and Sensor Placement for Large Flexible Structures," *AIAA Guidance, Navigation, and Control Conference* (1991).

22. A. E. Sepulveda and L. A. Schmit Jr., "Optimal Placement of Actuators and Sensors in Control-Augmented Structural Optimization," *International Journal for Numerical Methods in Engineering* 32, pp. 1165-1187 (1991).

23. W. E. Vander Velde and C. R. Carignan, "Number and Placement of Control System Components Considering Possible Failures," *Journal of Guidance, Control, and Dynamics* 7(6), pp. 703-709 (1984).

24. M. Basseville, A. Benveniste, G. Moustakides and A. Rougee, "Detection and Diagnosis of Changes in the Eigenstructure of Nonstationary Multivariable Systems," *Automatica* 23, pp. 479-489 (1987).

25. M. Basseville, A. Benveniste, G. Moustakides and A. Rougee, "Optimal Sensor Location for Detecting Changes in Dynamical Behavior," *IEEE Transactions on Automatic Control* AC-32(12), pp. 1067-1075 (1987).

26. P. C. Shah and F. E. Udwadia, "A Methodology for Optimal Sensor Locations for Identification of Dynamic Systems," *Journal of Applied Mechanics* 45, pp. 188-196 (1978).

27. F. E. Udwadia and J. A. Garba, "Optimal Sensor Locations for Structural Identification," *JPL Workshop on Identification and Control of Flexible Space Structures*, pp. 247-261 (1985).

28. M. Salama, T. Rose and J. Garba, "Optimal Placement of Excitation and Sensors for Verification of Large Dynamical Systems," *28th AIAA/ASME/ASCE/AHS Structures, Structural Dynamics, and Materials Conference*, Monterey, CA, pp. 1024-1031 (1987).

29. D. C. Kammer, "Sensor Placement for On-Orbit Modal Identification and Correlation of Large Space Structures," *Journal of Guidance, Control, and Dynamics* 14(2), pp. 251-259 (1991).

30. D. C. Kammer, "Effects of Noise on Sensor placement for On-Orbit Modal Identification of Large Space Structures," *to appear in the Journal of Dynamic Systems, Measurement, and Control.*

31. D. C. Kammer, "Effect of Model Error on Sensor Placement for On-Orbit Modal Identification of Large Space Structures," *to appear in the Journal of Guidance, Control, and Dynamics.*

32. T. W. Lim, "Sensor Placement for On-Orbit Modal Testing," *32nd AIAA/ASME/ASCE/AHS/ASC Structures, Structural Dynamics and Materials Conference*, Baltimore, MD, pp. 2977-2985 (1991).

33. D. J. Ewins, *Modal Testing: Theory and Practice*, Research Studies Press LTD., Letchworth, England (1984).

34. R. J. Guyan, "Reduction of Mass and Stiffness Matrices," *AIAA Journal* 3(2), pp. 380 (1965).

35. J. O'Callahan, "A Procedure for an Improved Reduced System (IRS) Model," *7th International Modal Analysis Conference*, Las Vegas, NV, Union College, pp. 17-21 (1989).

36. D. C. Kammer, "Test-Analysis Model Development Using an Exact Modal Reduction," *International Journal of Analytical and Experimental Modal Analysis* 2(4), pp. 174-179 (1987).

37. D. C. Kammer, "A Hybrid Approach to Test-Analysis Model Development for Large Space Structures," *Journal of Vibration and Acoustics* 113(3), pp. 325-332 (1991).

38. P. C. Hughes and R. E. Skelton, "Controllability and Observability of Linear Matrix-Second-Order Systems," *Journal of Applied Mechanics* 47, pp. 415-420 (1980).

39. F. C. Schweppe, *Uncertain Dynamic Systems*, Prentice-Hall, Englewood Cliffs, NJ (1973).

40. D. Middleton, *An Introduction to Statistical Communication Theory*, McGraw-Hill, New York, NY (1960).

41. V. V. Fedorov, *Theory of Optimal Experiments*, Academic, New Yory, NY (1972).

42. A. Ben-Israel and T. N. E. Greville, *Generalized Inverses: Theory and Application*, John Wiley and Sons, New York, NY (1974).

43. R. H. Battin, *An Introduction to the Mathematics and Methods of Astrodynamics*, AIAA, New York, NY (1987).

44. C. D. Johnson, "Optimization of a Certain Quality of Complete Controllability and Observability for Linear Dynamical Systems," *Transcripts of the American Society of Mechanical Engineers, Series D* **91**, pp. 228-237 (1969).

45. C. E. Shannon, "Mathematical Theory of Communication," *Bell System Technical Journal* **27**, pp. 379, 623 (1948).

46. L. Yao, Personal Communication, Dept. of Electrical and Computer Engineering, Univ. of Wisconsin (1991).

47. T. Kailath, *Linear Systems*, Prentice Hall, Englewood Cliffs, NJ (1980).

48. J. Juang and R. S. Pappa, "Effects of Noise on Modal Parameters Identified by the Eigensystem Realization Algorithm," *Journal of Guidance, Control, and Dynamics* **9**(3), pp. 294-303 (1986).

49. A. E. Hoerl and R. W. Kennard, "Ridge Regression: Applications to Nonorthogonal Problems," *Technometrics* **12**(1), pp. 69-82 (1970).

50. J. V. Beck and K. J. Arnold, *Parameter Estimation in Engineering and Science*, John Wiley & Sons, New York, NY (1977).

51. W. H. Lee, "Robustness Analysis for State Space Models," Alphatech, Inc., Report TP-151 (1982).

53. R. K. Yedavalli, "Perturbation Bounds for Robust Stability in Linear State Space Models," *International Journal of Control* **42**, pp. 1507-1517 (1985).

54. D. Hinrichsen and A. J. Pritchard, "Stability Radii of Linear Systems," *Systems and Control Letters* **7**(7), pp. 1-10 (1986).

54. L. Qiu and E. J. Davison, "New Perturbation Bounds for the Robustness Stability of Linear Space Models," *25th IEEE Conference on Decision and Control*, Athens, Greece, pp. 751-755 (1986).

Investigation on the Use of Optimization Techniques for Helicopter Airframe Vibrations Design Studies

T. Sreekanta Murthy

Lockheed Engineering and Sciences Company
Hampton, Virginia, 23666

I. INTRODUCTION

All helicopters are prone to vibrations which can seriously degrade ride quality, reduce service life, and limit maximum speed in forward flight. Considerable progress has been made over the past forty years in reducing vibrations in helicopters. However, the level of vibration reduction achieved in newer helicopter designs has been, for the most part, either insufficient or only marginally acceptable in meeting the increasingly stringent vibration requirements which are being imposed. Even though excessive vibrations have plagued virtually all new helicopter developments, until recently, there has been little reliance on the use of vibration analyses during design to limit vibration. With only a few exceptions, helicopters have been designed to performance requirements while relying on past experience to "tinker out" excessive vibrations during the ground and flight testing phases of development. Most often, excessive vibrations have been reduced through the use of add-on vibration control devices at the expense of significant weight penalties associated with using them. Recently, however, there has emerged a consensus within the helicopter industry on the need to account for vibrations more rigorously during the design process. This need has resulted in the subject of helicopter vibrations receiving considerably increased attention in recent years [1,2]. The goal (unofficially) set down by the industry is to achieve the vibration levels associated with jet aircraft, the so-called "jet smooth" ride [3]. To achieve this goal will require the development of advanced design analysis methodologies and attendant computational procedures which properly and adequately take into account vibration requirements during all phases of the design process.

Recent research activities in the United States concerning the development of advanced design analysis methodologies to limit vibrations are focused in three major areas: (1) rotor system design, (2) control system design, and (3) airframe structural design. Various types of vibration analyses methodologies are evolving to support rotor and airframe design work. Rotor aeroelastic analysis codes are being implemented to evaluate design analysis of rotors. Methods for employing active control systems to suppress vibrations in both the rotor and the airframe are being studied. Design optimization analysis methods are being formulated to select rotor and airframe design parameters which yield low inherent vibrations.

At Langley Research Center, a study was undertaken to investigate the use of formal, nonlinear programming-based, numerical optimization techniques for airframe vibrations design work. Considerable progress has been made in that study since its inception in 1985. The purpose of this paper is to present a unified summary of the experiences and results of that study.

The paper begins with a background section which discusses the vibratory loads of interest, current vibration reduction approaches, and relevant research activities. Key tasks involved in the optimization of airframe structures are described. The formulation and solution of airframe optimization vibration problems are discussed. The implementation of a new computational procedure based on MSC/NASTRAN and CONMIN in a computer program system called DYNOPT for the optimization of airframes subject to strength, frequency, dynamic response, and fatigue constraints is described. Major steps of the airframe design process are outlined. An optimization methodology which appears suited to airframe design work is described. Considerations needed in formulating optimization problems for both new and existing airframes are discussed. Finally, numerial results from the application of the DYNOPT program to the Bell AH-1G helicopter are presented and discussed.

II. BACKGROUND

A. Predominant Vibratory Loads

The predominant source of vibration in a helicopter arises from the oscillatory airloads acting on the blades of the main rotor. These loads are transmitted from the rotor through the hub and into the airframe where they produce objectionable vibrations. In steady-state level flight, the loads acting on the individual blades sum in such a way that the resultant forces and moments transmitted to the airframe occur at integer multiples of the product of the rotor rotational speed Ω and number of blades N. Thus, the airframe is subjected to steady-state rotor-induced forces which occur at the discrete frequencies $N\Omega$, $2N\Omega$, $3N\Omega$, ... , $nN\Omega$. The dynamic characteristics of the rotor and the airframe and the way in which these two systems are coupled at the rotor hub determine the manner in which the helicopter airframe responds to the dynamic loads. Because the magnitude of the harmonic airloads generally decreases with increasing harmonic number, the lower harmonics of the loads occurring at $N\Omega$ (and sometimes $2N\Omega$) are usually more important with respect to vibrations than the higher harmonics.

B. Approaches to Vibration Reduction

Many approaches to the solution of the vibration problem have been proposed and studied. The most common approaches employed to reduce helicopter airframe vibrations include: (1) the modification of the main rotor system by altering blade stiffness and mass properties to bring about reduction in the magnitude of the resultant hub loads [4-6]; (2) the use of active and passive vibration control devices to absorb and/or isolate the forces transmitted from the rotor hub to the airframe [7-8]; and (3) the modification of the airframe structure to ensure that the natural frequencies of the airframe are well separated from the predominant rotor excitation frequencies to avoid resonance and to reduce the dynamic responses of the airframe under rotor-induced loads [9-10]. Among these approaches, the approach involving main rotor modification requires extremely complex multi-disciplinary design trade-off studies for which optimization approaches are being developed, and the approach involving the use of active and passive vibration control devices has weight penalties. The approach involving airframe structural modification is gaining renewed attention in the design community, and efficient and practical methods to perform airframe structural modification are being sought.

C. Airframe Design for Vibration Reduction

The requirement for low vibratory response of the airframe necessitates: (1) Insuring that none of the major airframe natural frequencies is close to the predominant rotor exciting frequencies to avoid resonance; and (2) Reducing the magnitude of the dynamic response of the airframe under the combined action of the exciting forces and moments at the frequencies of interest. Low vibration design of an airframe requires knowledge of the airframe dynamic characteristics in terms of both its frequency response characteristics and its frequencies and mode shapes. In practice, airframe design involves repetitive structural design modifications to obtain the desired dynamic characteristics. The identification of the necessary structural modifications typically requires extensive analyses using large finite element models of the airframe structure and multi-dimensional searches in design variable space to determine the optimum sizes of the structural members. Airframe design is primarily based on engineering judgement and involves a tedious trial-and-error modification process. Selection of the best airframe that meets all design requirements, in particular the vibration requirements, is a difficult task. Therefore, there is a need for a systematic procedure which considers vibration requirements during the airframe design process by properly accounting for the various multi-disciplinary interactions that influence the design modifications. It would appear that structural optimization tools, if properly brought to bear by the design engineer, would go a long way toward achieving the goal of an analysis capability for designing a low vibration helicopter.

D.   Recent Research In Airframe Optimization

The airframe structural optimization approach for helicopter vibration reduction has not been addressed much in the past. The reported work is contained primarily in references 11-19. References 11-15 address the vibration reduction problem by modifying the airframe structure to tune the natural frequencies of the airframe and/or to reduce the responses under dynamic loads. Although the word "optimization" is used in these references, the work described there addresses the use of ad hoc methods as the basis for making structural modifications without the use of any formal optimization techniques. The methods described include those based on considerations such as the Vincent circle trace and the strain energy in a member. The use of nonlinear mathematical programming methods to tune airframe frequencies has only recently gained attention, and references 16-17 describe what are apparently the first applications of that method to finite-element models of airframes. Computer codes for using the nonlinear programming optimization methods for airframe vibration reduction are beginning to be developed by the rotorcraft industry (see, for example, [18-19]). Clearly, there is a need for further research to explore more fully the potential of optimization approaches for vibration reduction in helicopter airframes.

As mentioned earlier, the objective of this research is to investigate the use of formal, nonlinear programming-based, numerical optimization techniques for airframe vibrations design work. The primary focus of this research study is directed toward: (1) identification and examination of key tasks involved in the application of optimization techniques to helicopter airframes; (2) development of practical computational procedures for optimization; (3) development of suitable optimization methodology which would be compatible with the airframe design process; and (4) application of the formulated computational procedures for optimization to real airframe structures. Some of the results from this airframe optimization research activity are reported in References 20-24.

III. KEY TASKS IN AIRFRAME OPTIMIZATION

The basic idea in airframe structural optimization for vibration reduction is to design the airframe in a way that the vibratory responses in the areas of interest are minimized. The application of the nonlinear mathematical programming approach typically involves formulation of the vibration problem to find a minimum value of an objective function, under a specified set of structural response constraints, and with prescribed bounds on the structural design variables. A solution to the optimization problem is sought via an iterative approach consisting of a sequence of computational tasks involving finite element analysis, sensitivity analysis, approximate analysis, and design change computations. The specific considerations required to accomplish these computational tasks, particularly with reference to the airframe optimization computations, are identified and discussed below.

A. Formulation of the Problem

A key task in the successful application of optimization techniques to helicopter airframes is the formulation of the optimization problem, which includes the establishment of a relevant set of design variables, an objective function, and constraints. An optimization problem is generally expressed in the form:

Minimize the objective function	$F(b)$	(1)
subject to the constraints	$g(b) \leq 0$	(2)
and bounds on design variables	$b_l \leq b \leq b_u$	(3)

where b_l and b_u are the lower and upper bounds on the design variables b.

In the formulation of the optimization problem, the objective function is typically the total weight of the airframe structural members expressed as a function of the design variables. The expression for the constraint functions are formulated based on the allowable limits on the vibration levels or equivalent structural forced response displacements at selected locations in the structure. The expression for the constraint functions can also be formulated to specify the allowable ranges in which the natural frequencies of the structure are to be placed to avoid resonances with the frequency of the excitation forces. The design variables are normally chosen as the cross-sectional sizes of the structural members which are to be varied within certain prescribed bounds to seek optimum values for the variables. A major issue in the formulation of the optimization problem arises from the fact that there is no unique way of formulating a given vibration reduction problem as an optimization problem because the design variables can be defined in several different ways, and because the objective functions and constraints can be written in many different forms. The formulation is also dependent on which phase of the design process is being addressed and on whether the airframe is a new design or an existing one which is to be modified. Because different formulations will yield different solutions to a vibration reduction problem, the various considerations needed in the formulation of alternative optimization problems need to be examined.

B.   Finite Element Analysis

After formulating an optimization problem, an important task in the optimization solution is the finite element analysis of the airframe structure to compute structural responses which are subsequently used in the numerical evaluation of the objective and constraint functions. In the present study, the MSC/NASTRAN program [25] was chosen for the finite element analysis task because it is the code of choice for structural analysis in the rotorcraft industry. The finite element analysis computations require considerable effort in managing the repetitive execution of large-sized finite element models and careful organization of the large amount of data which is associated with such models. Companion design models [26] need to be developed for the optimization computations. A design model may be defined as an organized collection of design variables, constraints, objective function, bounds on design variables, and linking of design variables. A design model is to optimization as a finite element model is to structural analysis. It should be noted that the development of a well-organized design model may require as much effort as the development of the underlying finite element model.

C.   Sensitivity Analysis

Design sensitivity analysis consists of the computation of the sensitivity coefficients of the objective and constraint functions with respect to changes in the design variables. In the airframe optimization computation, the sensitivities of weight, natural frequencies, forced response displacements, and dynamic stresses are needed. The MSC/NASTRAN (Version 65) program [25] has the capability to provide sensitivities of the natural frequencies and static stress constraint functions. However, the sensitivity analysis for the constraints on the forced response displacements and weight are not available in the program and therefore had to be implemented via an user-developed DMAP (Direct Matrix Abstraction Program).

D.   Approximate Analysis

Use of large finite element models of airframes in the repetitive structural analyses required during optimization iterations is computationally inefficient. Therefore, a crucial task in the successful application of optimization techniques to airframe structures is the development and the use of an efficient and accurate approximate analysis technique for evaluating the objective and constraint functions. Three of the most common types of approximate analysis techniques [27] - two based on a Taylor's series expansion and the other based on a hybrid constraint approximations - were chosen for use in the airframe optimization solution. It should be noted that while these approximate analysis techniques have been used in the solution of static optimization problems with some success, the accuracy and reliability of these techniques in static applications are still under investigation. These techniques have yet to be evaluated in dynamics applications, in particular to airframe dynamics problems.

IV.   FORMULATIONS OF THE VIBRATION OPTIMIZATION PROBLEM

Expressions for natural frequency, steady-state forced response displacement, and dynamic stress constraint functions, and the sensitivity derivatives of these functions, are presented and discussed here.   In general, the objective and constraint functions can be used interchangably in formulating an optimization problem. Because the expressions for the objective and constraint functions have the same form for a given type of response, in the discussion below only the constraint expressions are given.

A.   Natural Frequency Constraints

As discussed earlier, constraints on airframe natural frequencies are required to ensure that the frequencies are well-separated from the main rotor excitation frequencies to avoid resonance. The rotor-induced forcing frequencies are discrete frequencies $nN\Omega$ (where n is an integer; N is the number of blades; Ω is the rotor speed). The natural frequency constraints can be written as:

$$\omega_{il} \leq \omega_i \leq \omega_{iu} \tag{4}$$

where ω_{il} and ω_{iu} are the lower and upper bounds on the ith natural frequency ω_i of the airframe. The airframe natural frequencies are determined by solving the eigenvalue equation

$$[K - \lambda M] \Phi = 0 \tag{5}$$

where K and M are the stiffness and mass matrices of the structure, λ is the eigenvalue and is equal to the square of the frequency ω, and Φ is the mode shape vector. The sensitivity derivatives of the frequency constraints are obtained by differentiating Eq. (5) with respect to the design variables vector b. After simplifying the differentiated terms by making use of the symmetry and orthogonality of matrices K and M, and assuming unit generalized mass, the resultant expression for the sensitivity derivative is given by:

$$\partial\lambda/\partial b = \Phi^T [\partial K/\partial b - \lambda \ \partial M/\partial b] \ \Phi \tag{6}$$

B.   Steady-State Forced Response Constraints

In addition to natural frequency constraints, constraints are required on the forced response amplitudes of the airframe. The forced response amplitudes are obtained by solving the equation:

$$M(b) \ \ddot{X}(b) + C(b) \ \dot{X}(b) + K(b) \ X(b) = F(t) \tag{7}$$

where M, C and K are the mass, damping and stiffness matrices, F is a vector of steady-state harmonic forces acting on the top of the main rotor shaft, and X is a vector of harmonic response displacements. The forced response constraints can be written as:

$$X - X_a \le 0 \tag{8}$$

where X is the amplitude of displacement at a specified location in the airframe and X_a is the allowable value. The sensitivities of the forced responses are obtained by differentiating Eq. (7) with respect to the design variables b and leads to the expression:

$$[-\Omega^2 M(b)+i \Omega C(b) + K(b)] \partial X/\partial b =$$

$$[-\Omega^2 \partial/\partial b M(b) + i\Omega \partial/\partial b C(b) + \partial/\partial b K(b)] X \qquad (9)$$

where X is the response vector at the design b. The partial derivatives of the matrices M, C and K with respect to b are determined using either explicit analytical differentiation or finite difference techniques.

C.   Dynamic Stress Constraints

Because vibration and fatigue are closely related problems, the airframe structural dynamic design considerations are formulated to also insure that the dynamic stresses are within acceptable limits by proper sizing of the structural members. The dynamic stress constraint is formulated as:

$$\sigma_{il} \le \sigma_i \le \sigma_{iu} \qquad (10)$$

where σ_i is the computed mean dynamic or fatigue stress and σ_{il} and σ_{iu} are the lower and upper bound stress for the ith structural member. The dynamic stresses are computed from the relation:

$$\sigma = S X \qquad (11)$$

where S is the stress-displacement transformation matrix computed in a finite element analysis. The sensitivity of the stress constraint can be written as:

$$\partial\sigma/\partial b = S \partial X/\partial b + \partial S/\partial b X \qquad (12)$$

where $\partial X/\partial b$ is the derivative of the forced response displacement, and $\partial S/\partial b$ is the derivative of the stress-displacement transformation matrix with respect to the design variables.

V. DYNOPT COMPUTER PROGRAM

A computer program called DYNOPT (DYNamics OPTimization) was developed to carry out the optimization computations based on the nonlinear mathematical programming approach. The DYNOPT code features a unique operational combination of the MSC/NASTRAN (Version 65) finite element structural analysis code [25] extended to include the calculation of steady-state forced response and dynamic stress sensitivities, and the CONMIN optimizer [28]. The computational steps used in the DYNOPT program are illustrated in Figure 1. The operations in the finite element analysis are: (1) stiffness and mass matrix assembly; (2) static analysis; (3) frequency analysis; and (4) steady-state forced response and dynamic stress analysis. The operations in the sensitivity analysis are: (1) static sensitivity; (2) frequency sensitivity; (3) forced response displacement and dynamic stress sensitivity; and (4) weight sensitivity. The static and frequency sensitivity analysis are performed using solution sequences 51 and 53 in the MSC/NASTRAN program. The forced response displacement, dynamic stress, and weight sensitivity modules are newly developed using the Direct Matrix Abstration Program (DMAP) language of NASTRAN. For repetitive objective and constraint function evaluations, three different approximation techniques have been incorporated into the DYNOPT program. These approximation techniques are based on the use of direct, reciprocal, and hybrid forms of the Taylor's series expansion in the design variables. Depending on the optimization problem and degree of nonlinearity of the objective and constraint functions, any of these approximation techniques are chosen. The method of feasible directions available in the CONMIN optimizer program is used for design change computations. The various computational steps are organized into several independent modules in the DYNOPT program. Each module of the DYNOPT program is organized to perform the necessary computations upon reading the appropriate input data, and to operate on a database to store and retrieve data generated and stored at intermediate steps of the computation.

The DYNOPT program also includes several FORTRAN programs for: (1) identification and numbering of the objective and constraint functions, (2) organizing and transfering data between programs and restarting the CONMIN optimizer after the NASTRAN reanalysis, and (3) updating the NASTRAN input data for changes in element cross-sectional sizes and material properties corresponding to the design variable changes computed in the optimizer. Using interactive commands of the computer operating system, the appropriate modules in the DYNOPT program can be selected for execution depending on the types of objective and constraint functions specified for the optimization problem.

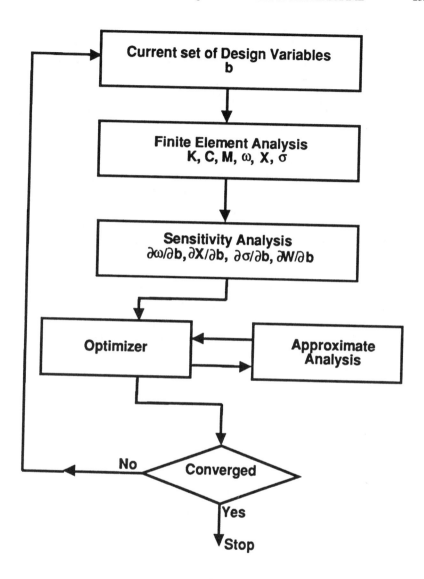

Figure 1. Computational Steps Used in the DYNOPT Program

Figures (1 through 8) and (11 through 16) adapted / reprinted from the authors' paper in the Journal of Aircraft, Volume 28, Number 1, January 1991, Copyrighted © by the American Institute of Aeronautics & Astronautics, Used with Permission.

VI.   AIRFRAME   DESIGN   PROCESS

Before proceeding to the application of the DYNOPT program to a helicopter airframe, it is helpful to recognize the different phases of a typical airframe design process. The primary phases of design are: 'Conceptual design', 'Preliminary design', 'Detailed design', and 'Ground and Flight test'. In the conceptual design phase, candidate configurations of an airframe are evaluated through trade-off studies on weight, aerodynamics, mission, performance, and stability. In the preliminary design phase, the configurations emerging from the conceptual design phase are worked out in greater detail, including layout of major structural members and selection of materials. In the detailed design phase, airframe members are sized based on strength, vibration, weight, and crash-worthiness requirements and the structural integrity is checked for various load cases within the flight envelope. In the ground and flight test phase, modifications are made to the airframe, if necessary, to enable the helicopter to satisfy the design requirements.

The practical application of optimization techniques to airframe vibration design work will require attention to the specific nature of the work performed in each of the phases of the airframe design process. This is necessary to ensure that any optimization-based design procedures which are developed are compatible with current engineering design practice. An optimization methodology which appears suited to the aforementioned design process is described in the appendix. A discussion of the major considerations needed in the application of optimization techniques in the various phases of an airframe design process is also given in the appendix.

VII.   APPLICATION   OF   DYNOPT   TO   A   HELICOPTER   AIRFRAME

This section summarizes the application of the the DYNOPT program to the Bell AH-1G helicopter airframe structure. The numerical results obtained from the application of the program are discussed to illustrate some of the essential computational tasks involved in applying the optimization methodology (as described in the appendix) in both the preliminary and detailed design phases of an airframe design.

A.   AH-1G   Helicopter   Airframe   Structure

The structure of the Bell AH-1G helicopter airframe [29] with its skin panels removed is shown in Figure 2a. The airframe structure is composed of several major components - fuselage, tail boom, vertical fin, landing gear, main rotor pylon, main rotor shaft, wings and wing-carry through structure. The gross weight of the AH-1G helicopter is 8399 lbs. This weight is composed of structural weight, non-structural weight, and useful weight items. The weight of the primary structure is about 1000 lbs.

B. Analysis Models of AH-1G Airframe

Three different finite element models of the airframe are available (Fig. 2). However, in the optimization studies discussed here only two of these models were used: the elastic-line (or 'stick') model [14] shown in Figure 2b and the detailed (or 'built-up') model [29] shown in Figure 2c. In the airframe stick model (Fig. 2b), the fuselage, tail boom, wings and rotor shaft structure were modeled with beam elements. The MSC/NASTRAN finite element model consists of 42 beam elements, 13 scalar spring elements and 12 rigid bar elements. There are 56 grid points in the model for a total of 336 degrees of freedom. A consistent mass representation is employed to model the weight of the primary structure in the airframe. A finite element analysis was carried out to determine the natural frequencies and mode shapes of the stick model of the airframe.

The fuselage and wing structures in the airframe built-up model (Fig. 2c) were modelled primarily with rods, shear panels and membrane elements. The tail boom, vertical fin and tail rotor shaft were modelled with beam elements in the same manner as they were in the stick model. The MSC/NASTRAN finite element model of the airframe consists of a total of 2954 finite elements which includes: 2001 rods, 197 beams, 340 shear panels, 243 triangular membranes, 160 quadrilateral membranes and 13 scalar spring elements. There are 504 grid points for a total of 3024 degrees of freedom. The natural frequencies computed using the stick model are within 10% of the frequencies of the built-up model for the modes of interest here.

C. Design Models

Two different design models of the airframe (see Figs. 3 and 4) were developed for the optimization studies using the DYNOPT program. The model shown in Figure 3 might be appropriate for use in preliminary design while the model shown in Figure 4 might be appropriate for use in detail design. In the discussions that follow, these models are referred to as the preliminary design model and the detailed design model, respectively.

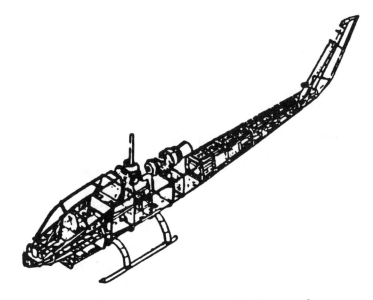

(a) Airframe Structure with Skin Panels Removed

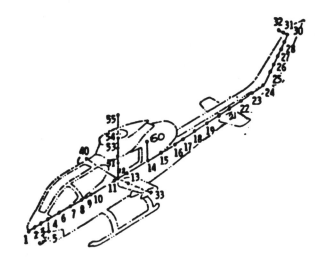

(b) Elastic-line Model

Figure 2. Bell AH-1G Airframe Structure and Finite Element Models

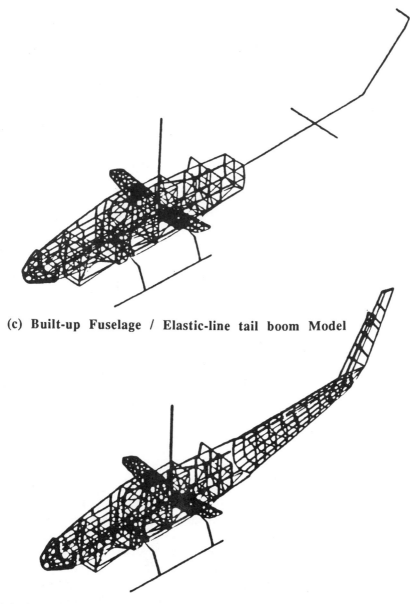

(c) Built-up Fuselage / Elastic-line tail boom Model

(d) Complete Built-up Model

Figure 2 (Continued). Bell AH-1G Airframe Structure and Finite Element Models

The preliminary design model (Fig. 3) takes as design variables the overall depth (d) of the cross-section of the primary structure at several stations along the airframe. The design model chosen reflects the several different types of cross-sections which comprise the primary structure of the airframe. The locations of structural members acting as stiffeners are indicated by solid dots while the location of flanges, webs, and skins are indicated by solid lines. In the design model, the depth of these sections in the fuselage and the tailboom was allowed to vary while holding fixed the sizes of the stiffeners, flanges, webs, and skins. The design model has a total of 46 design variables. An empirical relationship between the design variables of the design model (Fig. 3) and the element section properties of the finite element model (Fig. 2b) was established to update the NASTRAN bulk data deck during optimization iterations. For the numerical studies, the design variables were bounded to within ±50% of their initial values.

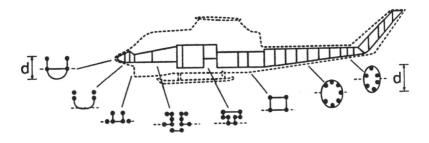

Figure 3. Preliminary Design Model Used for Optimization Studies

The detailed design model (Fig. 4) was developed to allow optimization of the sizes of the many individual structural members comprising the airframe structure. The development of this model required a more detailed consideration of the structure of the fuselage. Figure 4 shows details of some of the fuselage design variables, such as the panels and stiffeners located on either side of the fuselage. The design variables consisted of the thicknesses of the outer skin of each of the panels (t) and the cross-sectional areas of the stiffeners (A). A total of 191 design variables were used in this model, out of which 108 were independent design variables after using design variable linking. These design variables were related to the element properties of the built-up finite element model (Fig. 2c) of the airframe.

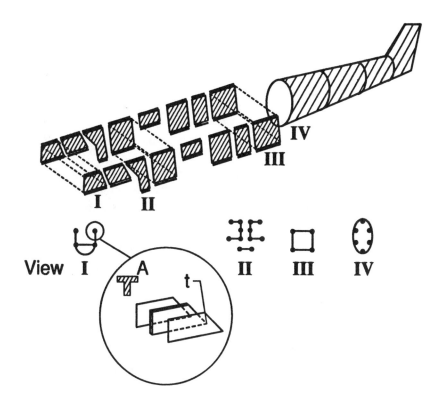

Figure 4. Detailed Design Model Used for Optimization Studies

D. Optimization Using the Preliminary Design Model

Three different optimization problems were formulated using the preliminary design model of the airframe. These optimization problems are used to demonstrate the application of the DYNOPT code for tuning the airframe natural frequencies, and for reducing dynamic stresses in the structural members.

1. First Problem

In the first optimization problem, the objective function was the weight of the primary structure of the airframe. Constraints were imposed on the natural frequencies corresponding to the pylon pitching (ω_1), pylon rolling (ω_2), first vertical bending (ω_3), second vertical bending (ω_4), and torsional (ω_5) modes. The first few natural frequencies and the corresponding mode shapes are shown in Figure 5. The objective function and the constraints are given by Eqs. (13-15):

$$F = \Sigma \rho_i A_i L_i \qquad i=1,46 \qquad (13)$$

$$g_{il} = \omega_{i1} - \omega_i \leq 0, \qquad i=1,5 \qquad (14)$$

$$g_{iu} = \omega_i - \omega_{iu} \leq 0, \qquad i=1,5 \qquad (15)$$

where ρ_i is the material density, A_i is the area of the cross-section and L_i is the length of the beam element. Subscripts l and u indicate lower and upper bounds on the natural frequencies. In Eqs. (14-15), the lower bounds on the frequencies are 2.5, 3.5, 5.0, 15.0, and 20 Hz and the upper bounds are 3.5, 4.5, 11.0, 20.0 and 25.0 Hz.

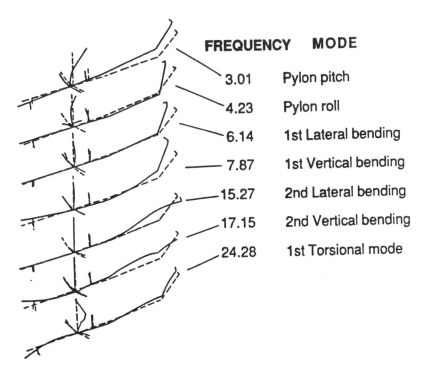

FREQUENCY	MODE
3.01	Pylon pitch
4.23	Pylon roll
6.14	1st Lateral bending
7.87	1st Vertical bending
15.27	2nd Lateral bending
17.15	2nd Vertical bending
24.28	1st Torsional mode

Figure 5. Natural Frequencies and Mode Shapes of the Elastic-line Model

The sensitivity coefficients for the frequency constraints were computed using the DYNOPT program. Figure 6 shows the distribution of sensitivity coefficients for the constraints on the first and second vertical bending mode frequencies. The sensitivity coefficients indicate that the design variables in the rear fuselage and most of the tail boom would be effective in changing the frequency of the first vertical bending mode. The figure also indicates that the design variables in both the fuselage and tail boom would be effective in changing the frequency of the second vertical bending mode. The weight gradients of the airframe were also computed and are shown in the figure. It is

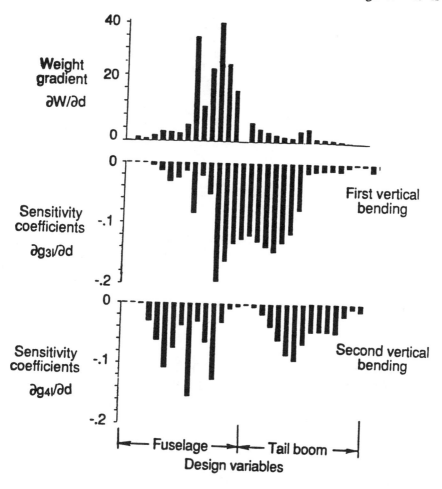

Figure 6. Sensitivity of Weight and Natural Frequency

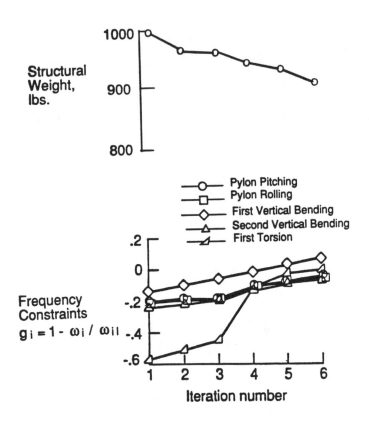

Figure 7. Optimization History for the Preliminary Design Model (First Problem)

seen that the design variables in the central and rear fuselage structures would have a significant effect in changing the weight of the airframe.

The history of the objective function and constraints is plotted in Figure 7. During the optimization iterations, the frequency constraints were satisfied and the optimizer computed the design changes by reducing the value of the objective function. In the first iteration, the airframe structural weight reduced from 1000 lbs. to 987 lbs. In the second iteration, the frequency constraints were still within bounds and the airframe weight reduced to 986 lbs. In this iteration, the weight reduction was small compared with the previous iteration because of the decreased step size used in the optimizer. In the third iteration, the airframe weight was reduced to 937 lbs. In this iteration, the constraint on the frequency of the first vertical bending mode reached its lower bound and the constraints on the other frequencies were still within their bounds. In the fourth, fifth, and sixth iterations the constraints on pylon pitching, rolling, first and second vertical bending and the torsion modes became active. Many of the design variables in the rear fuselage region reached their lower bounds. The weight of the airframe reduced to 900 lbs. in the sixth iteration. Subsequent iterations indicated insignificant changes in the design variables and the computations were terminated. A comparison of the initial and final values of the design variables is shown in Figure 8. It can be seen that the design variables were reduced in the rear fuselage and forward tail-boom structure whereas they increased in the forward fuselage and rear tail-boom structure.

2. Second Problem

In the second optimization problem, the objective function was taken to be the weight of the primary structure and the constraints were imposed on the dynamic stresses in the structural members (Eq. 10). The dynamic stresses in the structural members were computed for the case of a vertical excitation force at the rotor hub having a magnitude of 1,000 lbs. acting at a frequency of 10.8 Hz. (2/rev). The upper bound stress constraints in the tail-boom elements (see fig. 2b) connected by grid points (16-17), (18-19), (20-21), (22-23), (24-25), and

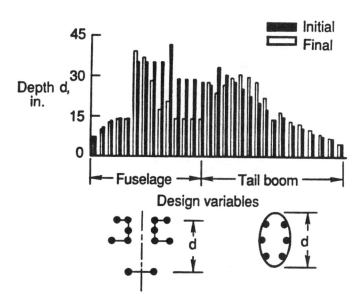

Figure 8. Initial and Final Design of the Primary Structure

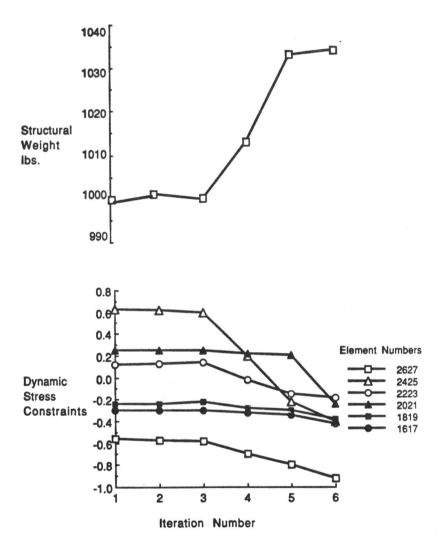

Figure 9. Optimization History for the Preliminary Design Model (Second Problem)

(26-27) are 10.0 psi., 10.0psi., 5.0psi., 5.0 psi., 4.0psi., and 5.0psi., respectively. The bending stresses corresponding to those elements at the initial design are 6.91psi., 7.44psi., 6.28psi., 5.64psi., 6.50psi., and 2.22psi., respectively.

The iteration history of the objective function and constraints for this problem are shown in Figure 9. In Figure 9, the element numbers are designated by the grid point number of their end points. For example, the element connecting grid points 26 and 27 is designated element 2627. In the first two iterations, the structural weight and the dynamic stresses reduced slightly because of the small step size in these iterations. In the 3rd and 4th iterations, however, there are considerable changes in the structural weight and stresses due to the larger step size in the optimizer. In the final iteration, the structural weight is 1034 lbs, which is 34 lbs. higher than the initial weight. The design variables increased by 8 to 11% of the initial design in the rear fuselage and 3 to 4% in the center fuselage and tail-boom region. This increase in the design variables (and hence the structural weight) is a result of the optimizer seeking a feasible design space to satisfy the dynamic stress constraints. It should be noted that optimization control parameters such as the step size, push-off factor (that pushes the violated constraints into feasible region) and the participation coefficient (indicating the degree to which the design is to be pushed to the feasible region) had a significant influence in the optimization solution.

3. Third Problem

In the third optimization problem, the objective function was again the structural weight, but the constraints on the natural frequencies and some of the constraints on the dynamic stresses from the last two problems were simultaneously imposed on the airframe. Therefore, this problem is similar to the first problem with 4 additional constraints on the dynamic stresses. It should be remarked that the natural frequency and dynamic stress constraints were imposed independently of one another in the previous two optimization problems. The upper bound stress constraints in tail-boom elements (see Fig. 2b) connected by grid points (16-17), (18-19), (19-20), and (21-22), are 5.0psi., 5.0psi., 5.0psi., and 5.0psi., respectively. The bending stresses corresponding to those elements at the initial design are 6.91psi., 7.44psi., 6.97psi., and 5.74psi., respectively. Thus, there are 10 natural frequency constraints and 4 dynamic stress constraints in the optimization problem.

Figure 10 shows the optimization iteration history for the objective and the constraint functions. In Figure 10, the element numbers are designated by the grid point number of their end points. For example, the element connecting grid points 21 and 22 is designated element 2122.The structural weight remains almost the same in the various iterations. Recall that the weight decreased in the presence of frequency constraints in the first problem and the weight increased in the presence of dynamic stress constraints in the second problem. In the first few

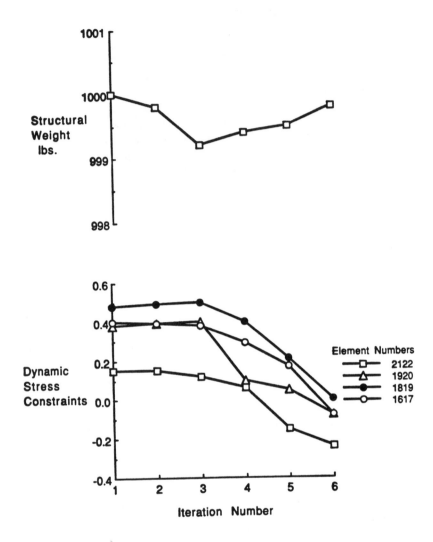

Figure 10. Optimization History for the Preliminary Design Model (Third Problem)

iterations, the dynamic stresses in the elements exceeded their constraint limits, however, the stresses were reduced below their limiting values in the subsequent iterations. The natural frequency constraints (not shown) remained within the feasible region during the iterations.

E.   Optimization Using the Detailed Design Model

A different problem was formulated to illustrate the optimization of the AH-1G helicopter airframe using the detailed design model (Fig. 4) . In this problem, the forced response displacement at selected locations in the airframe was used to formulate both the objective function as well as the constraint functions. The finite element model shown in Figure 2c was used for the analyses. The forced response displacements at various locations in the airframe were computed for a force of 1000 lb. at a frequency of 10.8 Hz (2/rev) acting vertically at the top of the main rotor shaft. The objective function was the forced response displacement at the pilot seat location X_p (which location in the built-up model of Figure 2c approximately corresponds to grid point 8 in the stick model of Figure 2b). Constraints were imposed on the forced response displacements at the nose (grid point 2), gunner (grid point 4), engine (grid point 60), tail boom (grid point 24) and fin (grid point 30) locations. These constraints are all of the form $g_i = X_i/X_a-1$. The constraint limits X_a at the nose, gunner, and the engine locations were 0.0025in. and those at the tail boom and fin locations were 0.005in.

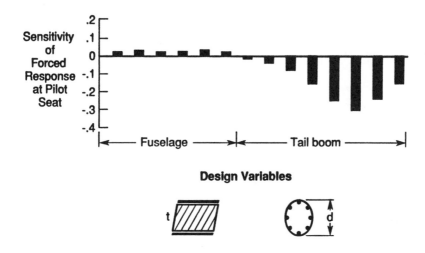

Design Variables

Figure 11. Sensitivity of Forced Response Displacement Constraint

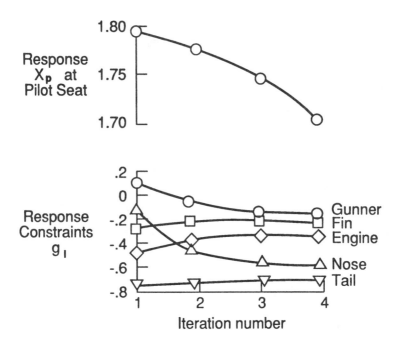

Figure 12. Optimization History for the Detailed Design Model

Figure 11 shows the distribution of the sensitivity coefficients for the forced response displacement at the pilot seat with respect to the fuselage and tail-boom design variables. A comparison of the magnitudes of the sensitivity coefficients in the figure indicates that the response is an order of magnitude more sensitive to changes in the design variables in the tail boom portion of the airframe than with respect to changes in the design variables in the fuselage portion of the airframe. This comparison indicates that the design variables in the tail boom would be more effective in reducing the response at the pilot seat location.

The optimization history of the objective and constraint functions in the various iterations are shown in Figure 12. The objective function was reduced, indicating a trend of decreasing vibration response at the pilot seat. As shown in the figure, all constraints are satisfied throughout the optimization process except for iteration one at the gunner location. During the iterations, the constraints at the gunner and nose location were reduced, indicating a reduction of the vibration response at those locations; however, the constraint values at the fin, tail and engine locations increased gradually.

VIII. SUMMARY AND CONCLUDING REMARKS

The paper has summarized the experiences and results from the investigation on the use of formal, nonlinear programming-based numerical optimization technique for helicopter airframe vibration reduction. The objective and constraint functions and the sensitivity expressions used in the formulation of airframe vibration optimization problems were presented and discussed. Implementation of a new computational procedure based on MSC/NASTRAN and CONMIN in a computer program system called DYNOPT for optimization of airframes subject to strength, frequency, dynamic response and dynamic stress constraints was described. An optimization methodology was proposed which is thought to provide a new way of applying formal optimization techniques during the various phases of the airframe design process. Numerical results obtained from the application of the DYNOPT optimization code to the Bell AH-1G helicopter airframe were discussed.

Specifically, this study has:

(1) Provided significant insight into how the practical problem of airframe vibration reduction can be posed as an optimization problem by examining the various considerations needed in the formulation of optimization problems for both new and existing airframes.

(2) Demonstrated the use of the DYNOPT optimization code for tuning airframe natural frequencies and for reducing weight, dynamic stresses, and vibration amplitudes in the airframe structure by performing numerical computations on both preliminary and detailed design models of the Bell AH-1G helicopter airframe.

(3) Defined an optimization methodology which simplifies the complex task of airframe optimization by dividing the optimization computations into several independent and sequential tasks organized according to the various phases of the airframe design process.

The subject optimization studies have provided considerable practical experience in the systematic application of formal optimization techniques to the problem of vibration reduction in helicopter airframe structures. The results of these applications are quite encouraging. However, more application experience on other airframe models is needed both to asses the proposed optimization methodology and to establish clearly the exact role which optimization can play in the airframe structural design process.

ACKNOWLEDGEMENT

This research study was performed under contract NAS1-19000 from NASA Langley Research Center, Hampton, Virginia. The author gratefully acknowledges Dr. Raymond G. Kvaternik for providing many valuable comments and suggestions in this study and also in preparing the manuscript.

REFERENCES

1. Kvaternik, R. G., Bartlett, F. D., Jr., and Cline, J. H., "A Summary of Recent NASA/Army Contributions to Rotorcraft Vibrations and Structural Dynamics Technology", Proceeding of the 1987 NASA/Army Rotorcraft Technology Conference, NASA CP 2495, pp. 71-179.

2. Kvaternik, R. G., "The NASA/Industry Design Analysis Methods for Vibrations (DAMVIBS) Program - Accomplishments and Contributions", Proceedings of the AHS National Technical Specialists Meeting on Rotorcraft Structures, Williamsburg, VA, October 29-31, 1991.

3. Gabel, R., "The Future: When Helicopter Smooth Surpasses Jet Smooth", Keynote Address, AHS National Specialists' Meeting on Helicopter Vibration, Hartford, CT, Nov. 2-4, 1981.

4. Taylor, R .B., "Helicopter Vibration Reduction by Rotor Blade Modal Shaping," Presented at the 38th Annual Forum of the American Helicopter Society, Washington, D.C., May 1982.

5. Wei, F. S., and Jones, R., "Blade Vibration Reduction Using Minimized Rotor Hub Force Approach,", 2nd International Conference on Basic Rotorcraft Research, College Park, Maryland, Feb. 1988.

6. Adelman, H. M., and Mantey, W. R., "Integrated Multidisciplinary Optimization of Rotorcraft", Journal of Aircraft, Volume 28, Number 1, January 1991, pp. 22-28.

7. Taylor, R. B., and Teare, P. A., "Helicopter Vibration Reduction with Pendulum Absorbers," Journal of the American Helicopter Society, 1975, 20(3).

8. Desjardins, R. A, and Hooper W. E., "Antiresonant Isolation for Vibration Reduction," Presented at the 38th Annual Forum of the American Helicopter Society, Washington, D.C., May 1982.

9. Shipman, D. P., White, J.A., and Cronkhite, J.D., "Fuselage Nodalization," 28th Annual Forum of the American Helicopter Society, Paper No. 611, 1972.

10. Welge, R. T., "Application of Boron/Epoxy Reinforced Aluminum Stringers for the CH-54B Helicopter Tail Cone", NASA CR 111929, July 1971.

11. Sciarra, J. J., "Use of the Finite Element Damped Forced Response Strain Energy Distribution for Vibration Reduction", Boeing-Vertol Company, Report D210-10819-1, U.S. Army Research Office - Durham, Durham, N.C., July 1974.

12. Done, G. T. S., and Hughes, A. D., "Reducing Vibrations by Structural Modification", Vertica, 1976, Vol. 1, pp. 31-38.

13. Sobey, A. J., "Improved Helicopter Airframe Response Through Structural Change", Ninth European Rotorcraft Forum, Paper No. 59, Sept. 13-15, 1983, Stresa, Italy.

14. Hanson, H. W., "Investigation of Vibration Reduction Through Structural Optimization", USAAVRADCOM Report No. TR-80-D-13, July, 1980.

15. King, S. P., "The Modal Approach to Structural Modification", Journal of the American Helicopter Society, Vol. 28, No. 2, April 1983.

16. Done, G. T. S., and Rangacharyulu, M. A. V., " Use of Optimization in Helicopter Vibration Control by Structural Modification", Journal of Sound and Vibration, 1981, 74(4), pp. 507-518.

17. Miura, H., and Chargin, M., "Automated Tuning of Airframe Vibration by Structural Optimization", American Helicopter Society, 42nd Annual Forum, Washington D.C., 1986.

18. Banerjee, D., and Shanthakumaran, P., " Application of Numerical Optimization Methods in Helicopter Industry", Vertica, Vol. 13, No. 1, pp. 17-42, 1989.

19. Smith, M., Rangacharyulu, M. A. V., Wang, B. P, and Chang, Y. K., " Application of Optimization Techniques to Helicopter Structural Dynamics", AIAA 32nd SDM Conference, pp. 227-237, 1991.

20. Sreekanta Murthy, T., "Design Sensitivity Analysis of Rotorcraft Structures for Vibration Reduction", NASA/VPISU Symposium on Sensitivity Analysis In Engineering, NASA CP 2457, September, 1986.

21. Kvaternik, R. G., and Sreekanta Murthy, T., "Airframe Structural Dynamic Considerations in Rotor Design Optimization", NASA TM 101646, August 1989.

22. Sreekanta Murthy, T., "Application of DYNOPT Optimization Program for Tuning Frequencies of Helicopter Airframe Structures", Proceedings of the Army Research Office/Duke University Workshop on Dynamics and Aeroelastic Modeling of Rotorcraft Systems, March 1990.

23. Sreekanta Murthy, T., "Optimization of Helicopter Airframe Structures for Vibration Reduction - Considerations, Formulations, and Applications", Journal of Aircraft, Volume 28, No. 1, Jan. 1991.

24. Sareen, A. K., Schrage, D. P., and Sreekanta Murthy, T., "Rotorcraft Airframe Structural Optimization for Combined Vibration and Fatigue Constraints", Proceedings of the American Helicopter Society 47th Annual Forum, Phoenix, AZ, May 1991.

25. "MSC/NASTRAN, User's Manual: Volume I - II", The MacNeal-Schwendler Corporation, November 1985.

26. Schmit, L. A., "Structural Optimization - Some Key Ideas and Insights", New Directions in Optimum Structural Design, Edited by Atrek, E., et. al, John Wiely & Sons, 1984.

27. Woo, T. H., " Space Frame Optimization Subject to Frequency Constraints", AIAA Journal, Vol. 25, No. 10., Oct. 1987.

28. Vanderplatts, G. N., " CONMIN - A Fortran Program for Constrained Function Minimization, User's Manual", NASA TMX-62282, August 1973.

29. Cronkhite, J. D., Berry, V. L., and Brunken, J. E., "A NASTRAN Vibration Model of the AH-1G Helicopter Airframe", Volume I, U.S. Army Armament Command Report No. R-TR-74-045, June 1974.

30. Sobieszczanski-Sobieski, J., James, B., and Dovi, A., "Structural Optimization by Multi-Level Decomposition", AIAA Journal, Vol. 23, Nov. 1983, pp. 1775-1782.

APPENDIX

CONSIDERATIONS IN FORMULATING OPTIMIZATION PROBLEMS

A. A Proposed Methodology for Optimization

The work involved in the various phases of an airframe design encompasses many disciplinary areas. In such a multi-disciplinary and multi-phase design environment, the formulation of a single optimization problem applicable to all phases of design would be an extremely complex and difficult task. Therefore, a simpler approach to both the formulation and solution of the vibration optimization problem is needed. In an attempt to meet this need, an optimization methodology was identified which is thought to provide a new way of applying formal optimization techniques during the various phases of the airframe design process.

Basically, the methodology involves the formulation and solution of separate optimization problems, one corresponding to each of the phases of the airframe design process: conceptual design, preliminary design, detail design, and ground and flight test, as depicted in Figure 13. In the methodology, the necessary optimization analyses required for the different phases are sequentially organized as depicted in the figure. The optimization tasks, such as the formulation of the problem, structural analysis, sensitivity analysis, and design change computations, are independently performed in each phase as indicated by the flow diagram in each block of the figure. The optimization formulation in each block includes a set of design variables, an objective function, and constraints that are appropriate to that particular phase. The formulation differs from one design phase to another. In a given design phase, necessary analyses are performed to evaluate the objective function, the constraints and the sensitivity derivatives required for the solution of the optimization problem at that particular stage of design. These results are used in a nonlinear programming algorithm for determining the design changes necessary to solve the optimization problem. The optimum design solution determined in one phase is used as an "initial" design in the subsequent phase. Additional and/or redefined design variables and constraints are included as required in each subsequent phase. In this process, the airframe design obtained from the last phase is the "best" or optimum one. The assumption implied here (which is probably reasonable in a practical design situation) is that the design obtained from each phase is successively improved in subsequent phases to obtain an optimum design in the final phase. The sequential and independent organization of optimization computations in this approach avoids the need for complex numerical procedures to link the computations from the different phases. It should be noted that the optimization methodology which is being proposed here falls under what might be termed a

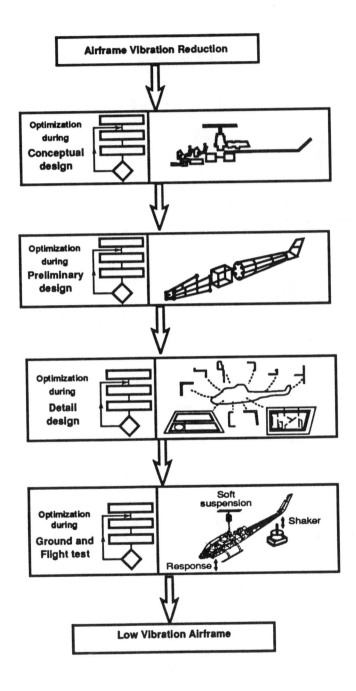

Figure 13.   Optimization Methodology

multi-phase approach to optimization rather than the multi-level decomposition approach which has also been proposed (see, for example, [30]).

B.   Considerations for New Airframes

Consideration required in formulating optimization problems in the conceptual, preliminary, detailed, and ground and flight test phases of a new airframe design are identified and discussed below.

1.   Conceptual Design

Although vibrations considerations are not generally addressed in the conceptual design phase, the proposed optimization methodology includes it because there appears to be some potential for optimization to influence the airframe design for vibrations in this phase. Design variables that can be used in the formulation of the optimization problem in the conceptual design phase are depicted in Figure 14. An initial rough estimate of the vibration characteristics of an airframe can be made in the conceptual design phase based on the knowledge of the past vibration history of a similar class of helicopters. Then configuration details such as the number of blades, rotor speed, flight loads/speed, gross weight, airframe shape and dimensions (Figs. 14a-b), layout of large non-structural masses such as the engine, transmission, fuel, and payload (Figs. 14c-d) can be used to further estimate vibration characteristics. It is expected that even with a basic attention to vibration characteristics, some potential vibration problems could be identified and reduced. Considerations in formulating an optimization problem in this phase should, therefore, be based on those configuration aspects that directly or indirectly influence airframe vibrations. The use of configuration design variables for vibration minimization necessitates consideration of the multi-disciplinary aspects of airframe design involving aerodynamics, layout of components, airframe shape and dimensions, weight, and stability. The modification of a configuration has a direct influence on airframe vibrations because the configuration is directly associated with the distribution of airframe structural stiffness and mass which affects the vibration characteristics. Vibration characteristics will also change significantly with any changes in the location of large mass components in the airframe.

2.   Preliminary Design

In the preliminary design phase, the primary load paths in the airframe are determined, the arrangement of major load carrying members are established, and the materials are selected. Candidate design variables in the preliminary design phase (Fig.15) could include the following: the layout of major structural members such as bulkheads, beams, longerons and stringers (s, l); material

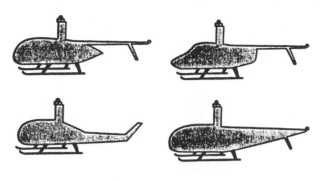

(a) Airframe Shape

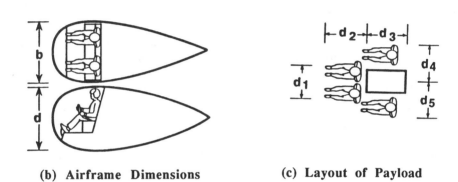

(b) Airframe Dimensions   (c) Layout of Payload

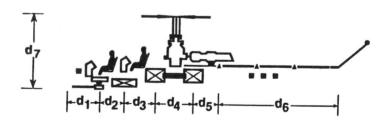

(d) Layout of Large Mass Components

Figure 14. Design Variables in   Conceptual Design Phase

properties of the primary structural members (E); cross-sectional area (A) and moment-of-inertia (I) of major structural members; and overall cross-sectional geometry of the primary structure defined by the distribution of the breadth (b) and depth (d) of the built-up-structural members which carry the major loads of the helicopter. Simple 'elastic-line' or 'stick' models of the airframe (such as shown in Fig. 2b) are usually developed for vibration analysis based on approximate distributions of stiffness and mass of the airframe. Airframe vibration characteristics obtained from such simplified models are much better than those estimated during the conceptual phase of design. Therefore, it is possible to include more detailed vibration considerations in the formulation of an optimization problem in this phase of airframe design. It would appear that the use of optimization in this phase of design has considerable potential to influence the airframe design for minimum vibrations.

3. Detailed Design
The formulation of an optimization problem in the detailed design phase allows for the consideration of constraints evaluated from more detailed discipline-oriented analyses of vibrations, strength, and weight of the airframe.

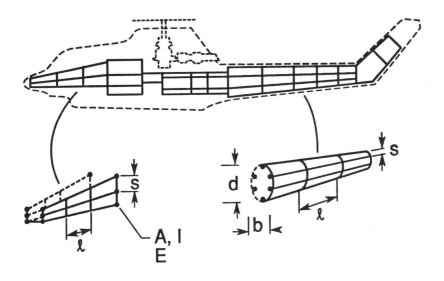

Figure 15. Design Variables in   Preliminary Design Phase

Candidate design variables in this phase are the cross-sectional dimensions of structural members such as the width and the thickness of stringer sections, the depth of beams, and the thickness of panels (Fig. 16). In this phase, the details of the thousands of structural members comprising the airframe and their layout are available. Complete built-up finite element models (such as shown in Fig. 2d) having better representation of the structural, material and geometric properties of an airframe can be developed and used to compute much improved estimates of the structural strength, vibration responses, and weight of the airframe. The use of optimization in this phase is thought to have good potential for influencing airframe design for minimum vibration.

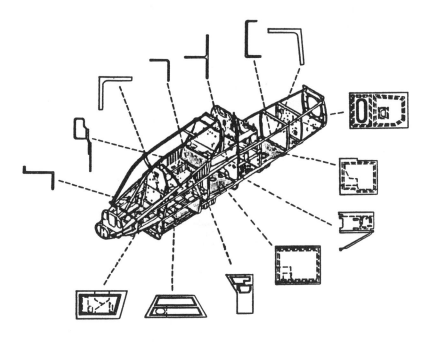

Figure 16. Design Variables in   Detailed Design Phase

4.   Ground and Flight Test

In practice, any serious attempt to address airframe vibrations usually begins late in the helicopter design process - after actual vibration problems are identified during ground and flight test. Severe vibrations in specific areas of the helicopter such as the tail boom, landing gear, and engine supports could possibly be identified in the ground and flight test phase. Vibration alleviation in such areas can be addressed through local structural modifications of the airframe. Considerations to formulate the optimization problem for the ground and flight test phase should be based on the specific vibration problems identified in this phase. Because of the limited choice of design variables, and also because of the narrow bounds which would be placed on the allowable changes in the design variables, it may be difficult to find a feasible optimization solution to the vibration-minimization problem. Therefore, the use of optimization in this phase of design would probably have only a limited potential to influence the design for vibration reduction.

C.   Considerations for Existing Airframes

Considerations needed in the formulation of an optimization problem for an existing airframe were found to be different from those needed in the design of a new airframe. In the case of an existing airframe structure, it is important to note that major modifications to the airframe structure are generally not permissible and that only small modifications to a few structural members can probably be made. In the formulation of the problem, this restriction in the allowable structural modifications needs to be considered by imposing narrow bounds on the allowable changes in the structural design variables. This restriction could severely limit the structural changes that could be made to reduce the vibration to the desired level.

SIZE-REDUCTION TECHNIQUES FOR THE DETERMINATION OF EFFICIENT AEROSERVOELASTIC MODELS

Mordechay Karpel

Faculty of Aerospace Engineering
Technion - Israel Institute of Technology
Haifa 32000, Israel

I. INTRODUCTION

Aeroservoelasticity deals with stability and dynamic response of control augmented aeroelastic systems. Flight vehicles are subjected to dynamic loads, such as those caused by atmospheric gusts, which excite the aeroelastic system and cause structural vibrations. The control system senses the vibrations and commands the motion of the control surfaces according to a control transfer function. This closes the aeroservoelastic loop and affects the airframe stability and response characteristics.

The common approach for formulating the equations of motion of an aeroelastic system starts with normal modes analysis of the structural system [1,2]. A realistic, continuous structural system has an infinite number of vibration modes. However, many flutter, structural dynamic response and aircraft performance issues may be adequately analyzed with a limited set of low-frequency vibration modes (including rigid-body modes). Control-surface deflection modes may be added for aeroservoelastic analysis. Complex gust velocity modes may also be added to analyze the response of the structural and control systems to continuous gust.

The unsteady aerodynamic force coefficients are defined with respect to the structural, control-surface and gust modes which serve as generalized coordinates. The most commonly used unsteady aerodynamic codes, like those based on the Doublet Lattice Method [3], assume that the structure oscillates harmonically. Transcendental unsteady aerodynamic matrices are calculated for various reduced frequency values. Second-order formulations [4] can be used for iterative stability solutions, frequency response, and frequency-domain control synthesis.

The application of various modern control design techniques [5,6], simulation [7] and optimization [8] procedures require the aeroservoelastic equations of motion to be transformed into a first-order, time-domain (state-space) form. This transformation requires the aerodynamic matrices to be approximated by rational functions (ratio of polynomials) in the Laplace domain. The order of the resulting state-space model is a function of the number of selected modes, the number of aerodynamic approximation roots, and the approximation formula. The main considerations in constructing the model are its size (which affects the efficiency of subsequent analyses), its accuracy, and the model construction efforts.

Tiffany and Adams [9] reviewed the most common aerodynamic approximation methods, extended their constraint capabilities and applied them with optimization of the approximation roots. Most methods in [9] are variations of Roger's classic approximation [10] which is based on term-by-term least-square fits. An exception is the Minimum-State (MS) method of Karpel [11, 12] which is based on computationally heavier, iterative, nonlinear least-square solutions, but yields a significant reduction in model size per desired accuracy. Later developments and applications of the MS method [13-16] improved its computational efficiency and introduced a physical weighting algorithm which reduced the number of aerodynamic states required to obtain good accuracy in typical realistic cases to about one fifth of the number of structural states.

A key question that the analyst faces is how many structural modes should be taken into account, and in what manner. The analysis may start with a relatively large number of modes with which the basic aeroservoelastic behavior is studied, such as in [17], and the frequency range of considerable aeroelastic activity is defined. In most applications the analyst selects a number of low-frequency modes to be fully included in the model, and truncates all the modes with higher natural frequency. In many applications, such as control system design, time simulation, parametric studies and structural or control system optimization, it is desired to define a group of modes which are not dominant but may have some effects that should be taken into account. Nissim [18] retained the structural states associated with the less-dominant modes but approximated their aerodynamics with quasi-steady approximations (while the dominant modes are modeled with Roger's approximation), which reduced the number of aerodynamic states. Other applications eliminated the less-dominant modes but used static residualization techniques, such as in [19], to take into account their steady effects without increasing the model size. This process, which yields accurate results in analyzing the rigid-body dynamics, may not be accurate enough in higher frequency flutter mechanisms. Karpel [20] developed a dynamic residualization method in which unsteady effects of the eliminated modes are added to those of the static residualization without increasing the model size.

The purpose of this chapter is to present various size-reduction techniques for the determination of efficient state-space aeroservoelastic models. The frequency-domain equations of motion, which form the basis for the state-space formulation, are reviewed in Section II. Section III develops the aeroservoelastic state-space formulation in general terms. Various aerodynamic approximation techniques are formulated in Section IV using common notations in a way which signifies their differences. The approximation constraints and the least-square solutions are discussed in Section V. The physical weighting algorithm and its use for aerodynamic approximations and for mode selection is described in Section VI. Section VII diescribes the size-reduction techniques which are based on the elimination of structural states. A numerical application which demonstrates and compares the various techniques is given in Section VIII.

II. FREQUENCY-DOMAIN EQUATIONS OF MOTION

The Laplace transform of the open-loop aeroelastic equation of motion in generalized coordinates, excited by control surface motion and atmospheric gusts, is

$$[C_s(s)]\{\xi(s)\} = -([M_c]s^2 + q[Q_c(s)])\{\delta(s)\} - \frac{q}{V}[Q_g(s)]\{w_g(s)\} \quad (1)$$

where

$$[C_s(s)] = [M_s]s^2 + [B_s]s + [K_s] + q[Q_s(s)]$$

where $[M_s]$, $[B_s]$, and $[K_s]$ are the generalized structural mass, damping and stiffness matrices respectively, $[M_c]$ is the coupling mass matrix between the control and the structural modes, $\{\xi\}$ is the vector of generalized structural displacements, $\{\delta\}$ is the vector of control surface commanded deflections, namely the actuator outputs, $\{w_g\}$ is the gust velocity vector, $[Q_s]$, $[Q_c]$ and $[Q_g]$ are the generalized unsteady aerodynamic force coefficient (AFC) matrices associated with the structural, control and gust modes, q is the dynamic pressure and V is the true air speed.

The main difficulty in constructing and solving Eq. (1) is that the AFC matrices are normally not available as an explicit function of s. Common unsteady aerodynamic routines assume that the structure undergoes harmonic oscillations and generate the complex AFC matrices (at a given Mach number) for various values of reduced frequency $k = \omega b/V$ where b is a reference semichord. For open-loop flutter analysis by the $p - k$ method [21], s is replaced in Eq. (1) by pV/b and the roots p_i of $[C_s(p)]$, for which the determinant of $[C_s(p)]$ equals zero, are found iteratively. The AFC matrices $[Q_s(p_i)]$ are approximated by $[Q_s(ik_i)]$, where k_i is the imaginary part of p_i, and are interpolated in each iteration from the tabulated AFC matrices. Flutter conditions are those for which the real part of one of the roots becomes positive (which corresponds to negative damping). Open-loop frequency response to sinusoidal control surface or gust excitation can be calculated by replacing s of Eq. (1) by $i\omega$.

An active control system relates the Laplace transform of the actuator outputs to that of the structural displacements by

$$\{\delta(s)\} = [T(s)][\psi_m]\{\xi(s)\} \tag{2}$$

where $[T(s)]$ is a matrix of transfer functions relating actuator outputs to sensor inputs, and $[\psi_m]$ is the matrix of modal deflections at sensor inputs. With the assumptions that the pilot does not change the steady control commands and that the actuators are irreversible, the substitution of Eq. (2) into Eq. (1) yields the Laplace transform of the closed-loop equation of motion

$$([C_s(s)] + [[M_c]s^2 + q[Q_c(s)]]\,[T(s)][\psi_m])\,\{\xi(s)\} = -\frac{q}{V}[Q_g(s)]\{w_g(s)\} \tag{3}$$

Closed-loop stability and response analyses [4] can be performed similarly to the open-loop analyses, but with $[C_s(s)]$ replaced by the left-side multiplier of $\{\xi(s)\}$ in Eq. (3).

III. STATE-SPACE EQUATIONS OF MOTION

In order to transform Eq. (1) into time-domain constant coefficient equation, the AFC matrices have to be described as rational functions of s. Karpel [11] showed that any rational function approximation of $[Q(s)] = [Q_s\,Q_c\,Q_g]$ that leads to a state-space aeroelastic model can be cast in the form

$$\left[\tilde{Q}(p)\right] = [A_0] + [A_1]p + [A_2]p^2 + [D]\,([I]p - [R])^{-1}\,[E]p \tag{4}$$

where p is the nondimensional complex Laplace variable $p = sb/V$ and all the matrix coefficients are real valued. The $[A_i]$ and $[E]$ matrices in Eq. (4) are column partitioned as

$$[A_i] = [A_{s_i}\,A_{c_i}\,A_{g_i}] \quad (i = 0, 1, 2), \qquad [E] = [E_s\,E_c\,E_g] \tag{5}$$

The gust related columns of $[A_2]$ in Eq. (4) are usually set to $[A_{g_2}] = 0$ to avoid the unnecessary $\ddot{w}_g$ terms in the time-domain model. To facilitate state-space formulation, an augmenting aerodynamic state vector is defined by its Laplace transform as

$$\{x_a(s)\} = \left([I]s - \frac{V}{b}[R]\right)^{-1} \left([E_s]\{\xi(s)\} + [E_c]\{\delta(s)\} + \frac{1}{V}[E_g]\{w_g\}\right) s$$

(6)

The resulting state-space open-loop plant equation of motion is

$$\begin{bmatrix} I & 0 & 0 \\ 0 & -\bar{M}_s & 0 \\ 0 & 0 & I \end{bmatrix} \left\{\begin{array}{c} \dot{\xi} \\ \ddot{\xi} \\ \dot{x}_a \end{array}\right\} = \begin{bmatrix} 0 & I & 0 \\ \bar{K}_s & \bar{B}_s & \bar{D} \\ 0 & E_s & \bar{R} \end{bmatrix} \left\{\begin{array}{c} \xi \\ \dot{\xi} \\ x_a \end{array}\right\} +$$

$$+ \begin{bmatrix} 0 & 0 & 0 \\ \bar{K}_c & \bar{B}_c & \bar{M}_c \\ 0 & E_c & 0 \end{bmatrix} \left\{\begin{array}{c} \delta \\ \dot{\delta} \\ \ddot{\delta} \end{array}\right\} + \begin{bmatrix} 0 & 0 \\ \bar{K}_g & \bar{B}_g \\ 0 & \bar{E}_g \end{bmatrix} \left\{\begin{array}{c} w_g \\ \dot{w}_g \end{array}\right\}$$

(7)

where

$$[\bar{M}_s] = [M_s] + \frac{qb^2}{V^2}[A_{s_2}], \quad [\bar{K}_s] = [K_s] + q[A_{s_0}], \quad [\bar{B}_s] = [B_s] + \frac{qb}{V}[A_{s_1}],$$

$$[\bar{M}_c] = [M_c] + \frac{qb^2}{V^2}[A_{c_2}], \quad [\bar{K}_c] = q[A_{c_0}], \quad [\bar{B}_c] = \frac{qb}{V}[A_{c_1}], \quad [\bar{D}] = q[D],$$

$$[\bar{R}] = \frac{V}{b}[R], \quad [\bar{K}_g] = q[A_{g_0}], \quad [\bar{B}_g] = \frac{qb}{V}[A_{g_1}], \quad [\bar{E}_g] = \frac{1}{V}[E_g]$$

The resulting number of aerodynamic augmenting states in $\{x_a\}$ is the order (n_a) of the aerodynamic lag matrix $[R]$.

It is assumed in this work that the control system, including the actuator dynamics, can be expressed in a state-space form as

$$\{\dot{x}_c\} = [A_c]\{x_c\} + [B_c]\{u\}$$

$$\left\{\begin{array}{c} \delta \\ \dot{\delta} \\ \ddot{\delta} \end{array}\right\} = \begin{bmatrix} C_{c_0} \\ C_{c_1} \\ C_{c_2} \end{bmatrix} \{x_c\}$$

(8)

where $\{x_c\}$ is the control state vector and $\{u\}$ is the control system input vector. It is also assumed that a gust filter can be defined in the state form

$$\{\dot{x}_g\} = [A_g]\{x_g\} + \{B_g\}w$$

$$\left\{\begin{array}{c} w_g \\ \dot{w}_g \end{array}\right\} = \begin{bmatrix} C_{g_0} \\ C_{g_1} \end{bmatrix} \{x_g\}$$

(9)

such that the power spectral density functions of the gust velocities in $\{w_g\}$ are obtained when w represents a white-noise process. It should be noted that, unlike in Ref. [20], the output equation in (9) does not contain a w term in its right side. This may require additional gust filter states, but it allows the acceleration measurements below to be formulated without a white-noise term.

Equations (7-9) combine for the open-loop state-space aeroservoelastic equation of motion

$$
\begin{bmatrix}
I & 0 & 0 & 0 & 0 \\
0 & -\bar{M}_s & 0 & 0 & 0 \\
0 & 0 & I & 0 & 0 \\
0 & 0 & 0 & I & 0 \\
0 & 0 & 0 & 0 & I
\end{bmatrix}
\begin{Bmatrix}
\dot{\xi} \\
\ddot{\xi} \\
\dot{x}_a \\
\dot{x}_c \\
\dot{x}_g
\end{Bmatrix}
=
$$

$$
\begin{bmatrix}
0 & I & 0 & 0 & 0 \\
\bar{K}_s & \bar{B}_s & \bar{D} & \bar{F}_c & \bar{F}_g \\
0 & E_s & \bar{R} & \hat{E}_c & \hat{E}_g \\
0 & 0 & 0 & A_c & 0 \\
0 & 0 & 0 & 0 & A_g
\end{bmatrix}
\begin{Bmatrix}
\xi \\
\dot{\xi} \\
x_a \\
x_c \\
x_g
\end{Bmatrix}
+
\begin{bmatrix}
0 \\
0 \\
0 \\
B_c \\
0
\end{bmatrix}
\{u\}
+
\begin{Bmatrix}
0 \\
0 \\
0 \\
0 \\
B_g
\end{Bmatrix} w
\qquad (10)
$$

where

$$
\begin{aligned}
\left[\hat{E}_c\right] &= [E_c][C_{c_1}] \\
\left[\hat{E}_g\right] &= [\bar{E}_g][C_{g_1}] \\
[\bar{F}_c] &= [\bar{K}_c][C_{c_0}] + [\bar{B}_c][C_{c_1}] + [\bar{M}_c][C_{c_2}] \\
[\bar{F}_g] &= [\bar{K}_g][C_{g_0}] + [\bar{B}_g][C_{g_1}]
\end{aligned}
$$

or

$$
[Z]\{\dot{x}\} = [A]\{x\} + [B]\{u\} + \{B_w\}w \qquad (11)
$$

where the matrices and vectors correspond to the respective ones in Eq. (10). Outputs of the aeroservoelastic system are expressed in the form

$$
\{y\} = [C]\{x\} \qquad (12)
$$

Structural outputs are based on the matrix $[\psi_m]$ of modal deflections and rotations. Displacements are expressed with $[C] = [\psi_m\ 0\ 0\ 0\ 0]$, velocities with $[C] = [0\ \psi_m\ 0\ 0\ 0]$ and accelerations with

$$
[C] = -[\psi_m][\bar{M}_s]^{-1}\begin{bmatrix} \bar{K}_s & \bar{B}_s & \bar{D} & \bar{F}_c & \bar{F}_g \end{bmatrix} \qquad (13)
$$

Equation (12) can be used to augment Eq. (11) by sensor models. The system outputs in this case would be based on the sensor states. Whether sensor dynamics are included or ignored, the aeroservoelastic loop is closed by relating the control inputs to system outputs by

$$\{u\} = [G_c]\{y\} \tag{14}$$

where $[G_c]$ is the control gain matrix. The substitution of Eq. (14) into Eq. (10) yields the closed-loop equation of motion

$$[Z]\{\dot{x}\} = [\bar{A}]\{x\} + \{B_w\}w \tag{15}$$

where

$$[\bar{A}] = [A] + [B][G_c][C]$$

It can be noticed from Eq. (10) that only the fourth row of $[\bar{A}]$ is different than that of $[A]$. A $\{u\}$ term can be left in the right side of Eq. (15) to reflect pilot commands.

IV. RATIONAL FUNCTION APPROXIMATIONS

The most commonly used rational function approximations of the AFC matrices are reviewed in this section with emphasis on the resulting number of aerodynamic augmenting states. The approximations are formulated here in the matrix form of Eq. (4), from which their effects on the subsequent state-space model, Eq. (10), can be deduced. All the approximation processes described herein start with the definition of a rational function of the imaginary part (ik) of the nondimensional Laplace variable p. Least-square procedures are then used to calculate the approximation coefficients that best fit the tabulated $[Q(ik_\ell)]$ matrices. The approximation is then expanded to the entire Laplace domain by replacing ik with sb/V. The least-square solution procedures are discussed in Section V.

The Minimum-State (MS) method [11,12] uses the most general expression of Eq. (4) which, with p replaced by ik, is

$$\left[\tilde{Q}(ik)\right] = [A_0] + ik[A_1] - k^2[A_2] + ik[D]\left(ik[I] - [R]\right)^{-1}[E] \tag{16}$$

The MS procedure assumes that $[R]$ is diagonal with n_L distinct negative values. The number of the resulting aerodynamic states in Eq. (10) is $n_a = n_L$. By allowing all the other terms in the coefficient matrices of Eq. (16) to be variables in the fit process, the MS method involves a relatively complicated least-square solution, but it yields a minimal number of aerodynamic states per desired accuracy.

Roger [10] approximated each aerodynamic term separately but with common aerodynamic lags,

$$\left[\tilde{Q}(ik)\right] = [A_0] + ik[A_1] - k^2[A_2] + \sum_{l=1}^{n_L} \frac{ik}{ik + \gamma_l}[A_{l+2}] \tag{17}$$

which can be cast in the form of Eq. (16) with

$$[D] = \begin{bmatrix} I & I & \cdots \end{bmatrix}, \quad [R] = - \begin{bmatrix} \gamma_1 I & & \\ & \gamma_2 I & \\ & & \ddots \end{bmatrix}, \quad [E] = \begin{bmatrix} A_3 \\ A_4 \\ \vdots \end{bmatrix} \quad (18)$$

The resulting number of aerodynamic states is $n_a = n_L \times n_m$ where n_m is the number of vibration modes.

The Modified Matrix Padé (MMP) method [9] applies Eq. (17) to each column separately

$$\left\{ \tilde{Q}(ik) \right\}_j = \{A_0\}_j + ik\{A_1\}_j - k^2\{A_2\}_j + \sum_{l=1}^{n_{L_j}} \frac{ik}{ik + \gamma_{l_j}} \{A_{l+2}\}_j \quad (19)$$

where the number of lags and their values can be different for different columns, which can lead to a lower number of states. When the number of lags (but not their values) are equal, Eq. (19) can be cast in the form of Eq. (16) with

$$[D] = \begin{bmatrix} A_3 & A_4 & \cdots \end{bmatrix}, \quad [R] = \begin{bmatrix} R_1 & & \\ & R_2 & \\ & & \ddots \end{bmatrix}, \quad [E] = \begin{bmatrix} I \\ I \\ \vdots \end{bmatrix} \quad (20)$$

where

$$[R]_l = - \begin{bmatrix} \gamma_{l_1} & & \\ & \gamma_{l_2} & \\ & & \ddots \end{bmatrix}$$

which is also an alternative representation of Roger's approximation but it yields a larger number of aerodynamic states, $n_a = (n_m + n_c + n_g) \times n_L$ where n_c and n_g are the number of control and gust modes respectively. However, for each $n_{L_j} < n_{l_j} \leq n_L$, the $j + (n_m + n_c + n_g) \times (l-1)$ column in $[D]$ and $[R]$, and the $j + (n_m + n_c + n_g) \times (l-1)$ row in $[E]$ and $[R]$ are eliminated. The resulting number of aerodynamic states is $n_a = \sum n_{L_j}$. It should be noticed that some n_{L_j} values can be zero, which corresponds to quasi-steady aerodynamics and does not add aerodynamic states.

Several studies [9,11-16] compared the aeroelastic models resulting from the MS, Roger and MMP approximations. The number of aerodynamic states, n_a, in the MS models were about 75% smaller than those of the Roger's models with similar accuracy, and $40 - 70\%$ smaller than those of the MMP models. On the other hand, the consideration of both $[D]$ and $[E]$ as free-coefficient matrices in the MS method makes the MS solution more complicated than the others.

Nissim [18] suggested that, with some insight, many aeroelastic problems can be modeled with quasi-steady approximation for most vibration modes. Nissim partitioned the AFC matrices and approximated $[Q_{s_{11}}]$, $[Q_{c_1}]$ and $[Q_{g_1}]$ with Roger's approximation of Eq.(17), and $[Q_{s_{12}}]$, $[Q_{s_{21}}]$, $[Q_{s_{22}}]$, $[Q_{c_2}]$ and $[Q_{g_2}]$ with no lag terms, where subscript 1 relates to structural modes with significant influences on flutter (group one), and 2 relates to the other modes (group two). This approximation can be first cast in the form of Eq. (16) with $[D]$, $[E]$ and $[R]$ of Eq. (18), and then reduced by eliminating the columns in $[D]$ and $[R]$ and the rows in $[R]$ and $[E]$ which are associated with group two. The resulting number of aerodynamic states is $n_a = n_L \times n_{m_1}$, which is demonstrated in [18] to be significantly lower than that of the full Roger's approximation, with a minor loss of accuracy in flutter results. It may be noticed that after the row elimination, the columns of $[E]$ in Eq. (18) associated with group two are zero, which indicates that the $[Q_{s_{12}}]$ partition could be approximated as well by Roger's formula without causing an increase of n_a.

Eversman and Tewari [22] extended Roger's formulation to allow multiple aerodynamic poles. Their baseline distinct-pole approximation is different than Eq. (17) by having one instead of ik in the numerators of the lag terms. Since they both have the same order of numerator polynomials when brought to common denominators, it can be shown that the two approximations are equivalent. The disadvantage in the baseline approximation in [22] is that its $[A_0]$ term does not represent steady aerodynamics as in all the other formulations above. The extended multiple-pole approximation, however, may improve the lag optimization process discussed in the next section.

V. APPROXIMATION PROCEDURES

To facilitate real-valued algebra, the complex approximation expression of Eq. (16) is separated into real and imaginary parts,

$$Re[\tilde{Q}(ik)] \equiv [\tilde{F}(k)] = [A_0] - k^2[A_2] + k^2[D]\left(k^2[I] + [R]^2\right)^{-1}[E] \quad (21)$$

and

$$Im[\tilde{Q}(ik)] \equiv [\tilde{G}(k)] = k[A_1] - k[D]\left(k^2[I] + [R]^2\right)^{-1}[R][E] \quad (22)$$

The comparison of Eqs. (21) and (22) with the real and imaginary parts, $[F(k_\ell)]$ and $[G(k_\ell)]$, of the tabulated AFC matrices $[Q(ik_\ell)]$ provides an overdetermined set of approximate equations. The task is to find the free approximation coefficients that minimize, under some constraints, the total least-square approximation error

$$\varepsilon_t = \sqrt{\sum_{i,j,\ell} |\tilde{Q}_{ij}(ik_\ell) - Q_{ij}(ik_\ell)|^2 W_{ij\ell}^2} \qquad (23)$$

where $W_{ij\ell}$ is the weight assigned to the ijth term of the ℓth tabulated AFC matrix. All the procedures associated with the methods presented in Section IV start with an initial set of aerodynamic lags which define $[R]$. The denominator coefficients are then calculated by a sequence of least-square solutions. Some procedures [9, 11, 16, 22] then apply an optimization scheme to modify the lags and repeat the solution for a better fit, until convergence occurs.

The sequence of least-square solutions in each method depends on the formulation and the way constraints and weighting functions are applied. Each individual solution in the sequence is based on the general approximate equation

$$[W^*]_\ell [A^*]_\ell \{x^*\} \approx [W^*]_\ell \{b^*\}_\ell \qquad for\ \ell = 1, n_k \qquad (24)$$

where n_k is the number of tabulated AFC matrices, $\{b^*\}_\ell$ is based on the tabulated data, $[W^*]_\ell$ is a diagonal matrix of weights associated with the terms of $\{b^*\}_\ell$, $\{x^*\}$ is a subset of unknown coefficients which are uncoupled with others, and $[A^*]_\ell$ is a function of k_ℓ, the aerodynamic lag values and the constraints associated with the unknowns. The weighted least-square solution for $\{x^*\}$ is obtained by solving

$$\left(\sum_\ell [A^*]_\ell^T [W^*]_\ell^2 [A^*]_\ell \right) \{x^*\} = \sum_\ell [A^*]_\ell^T [W^*]_\ell^2 \{b^*\}_\ell \qquad (25)$$

The contents of $[A^*]_\ell$, $\{x^*\}$ and $\{b^*\}_\ell$ in the various approximation methods, and the number of least-square solutions in each case, are given in the remaining of this section. For the sake of simplicity, a uniform weighting, $[W^*]_\ell = [I]$, is assumed. Weighting methods are discussed in Section VI. It should be noted that the least-square solutions are not necessarily solved one by one. The solutions for different $\{x^*\}$ vectors that have the same coefficient matrix, in the right side of Eq. (25), can be grouped together to reduce the computational cost.

A. The Unconstrained Problem

In all the approximation methods of Section VI, except the MS method, either $[D]$ or $[E]$ of Eqs. (21) and (22) are constant, which yields a linear least-square problem for a given set of aerodynamic lags. Furthermore, the structure of Eqs. (17) and (19) indicates that the solution in these cases can be performed for each aerodynamic term separately by performing $n_m \times (n_m + n_c + n_g)$ solutions of Eq. (25) with

$$[A^*]_\ell = \begin{bmatrix} 1 & 0 & -k_\ell^2 & \frac{k_\ell^2}{k_\ell^2+\gamma_{1_j}^2} & \frac{k_\ell^2}{k_\ell^2+\gamma_{2_j}^2} & \cdots \\ 0 & k_\ell & 0 & \frac{k_\ell\gamma_{1_j}}{k_\ell^2+\gamma_{1_j}^2} & \frac{k_\ell\gamma_{2_j}}{k_\ell^2+\gamma_{2_j}^2} & \cdots \end{bmatrix},$$

$$\{x^*\} = \left\{ \begin{array}{c} A_{0_{ij}} \\ A_{1_{ij}} \\ \vdots \end{array} \right\}, \qquad \{b^*\}_\ell = \left\{ \begin{array}{c} F_{ij}(k_\ell) \\ G_{ij}(k_\ell) \end{array} \right\} \qquad (26)$$

where the number of unknowns in each solution is $n_{L_j} + 3$ in the MMP approximation, Eq. (19), $n_L + 3$ in Roger's approximation, Eq.(17), and either $n_L + 3$ or 3 in Nissim's approximation.

With both $[D]$ and $[E]$ in Eqs. (21) and (22) being unknown, the MS problem for a given $[R]$ is nonlinear. The unconstrained problem is solved iteratively by starting with an initial guess of $[D]$ in which at least one term in each row and each column is nonzero. The unknown matrices $[A_0]$, $[A_1]$, $[A_2]$ and $[E]$ are then calculated by performing $n_m + n_c + n_g$ column-by-column solutions of Eq.(25) with

$$[A^*]_\ell = \begin{bmatrix} I & 0 & -k_\ell^2 I & k_\ell^2 [D] \left(k_\ell^2[I] + [R]^2\right)^{-1} \\ 0 & k_\ell I & 0 & -k_\ell [D] \left(k_\ell^2[I] + [R]^2\right)^{-1} [R] \end{bmatrix},$$

$$\{x^*\} = \left\{ \begin{array}{c} A_{0_j} \\ A_{1_j} \\ A_{2_j} \\ E_j \end{array} \right\}, \qquad \{b^*\}_\ell = \left\{ \begin{array}{c} F_j(k_\ell) \\ G_j(k_\ell) \end{array} \right\} \qquad (27)$$

where the j indices relate to the jth columns of the respective matrices and where the number of unknowns in each solution is $3n_m + n_L$. The calculated $[E]$ is then used to recalculate $[A_0]$, $[A_1]$, $[A_2]$ and $[D]$ by performing n_m row-by-row solutions of Eq.(25) with

$$[A^*]_\ell = \begin{bmatrix} I & 0 & -k_\ell^2 I & k_\ell^2 [E]^T \left(k_\ell^2[I] + [R]^2\right)^{-1} \\ 0 & k_\ell I & 0 & -k_\ell [E]^T \left(k_\ell^2[I] + [R]^2\right)^{-1} [R] \end{bmatrix},$$

$$\{x^*\} = \left\{ \begin{array}{c} A_{0_i}^T \\ A_{1_i}^T \\ A_{2_i}^T \\ D_i^T \end{array} \right\}, \qquad \{b^*\}_\ell = \left\{ \begin{array}{c} F_i^T(k_\ell) \\ G_i^T(k_\ell) \end{array} \right\} \qquad (28)$$

where the i indices relate to the ith rows of the respective matrices and where the number of unknowns in each solution is $3(n_m + n_c + n_g) + n_L$.

Ref. [9] solves the unconstrained problem by repeating the $[D] \to [E] \to [D]$
iterations of Eqs. (27) and (28) until convergence is obtained or until the
specified maximum number of iterations is reached. Ref. [13] demonstrated
that the convergence process does not strongly depend on the initial guess
of $[D]$. Theoretically, each iteration reduces the total approximation error defined in Eq. (23). It should be noted, however, that numerical ill-
conditioning problems may occur [13,15] when too many aerodynamic lag
terms are used. It is therefore recommended that the number of structural
modes (n_m) is the upper limit on the number of lags (n_L). This upper
limit (which yields the same number of aerodynamic states as Roger's approximation with one lag) is usually more than sufficient to obtain a good
aerodynamic fit [11-16, 23].

The iterative nature of the MS procedure and the relatively large number of unknowns solved for simultaneously in each iteration of the unconstrained problem, Eqs. (27) and (28), require a considerably larger computation time than that of the other methods, especially when optimization
of the aerodynamic lag values is performed [9]. A major reduction in the
MS computation efforts may be obtained by applying approximation constraints.

B. Approximation constraints

It is often desired to apply approximation constraints in order to obtain
exact fits at specified reduced frequencies or to null out some coefficients,
even though the constraints increase the total fit errors. The most frequently used constraint is a match of the steady-aerodynamics data ($k = 0$).
An imaginary-part data-match constraint at a k close to 0 yields the tabulated low-frequency quasi-steady aerodynamic damping. Data-match constraints at higher k values are sometimes desired to increase the accuracy
of anticipated flutter mechanisms. The structural columns, $[A_{s_2}]$ of the
apparent mass matrix, $[A_2]$, are sometimes set to 0 to avoid repetitive inversions of $[\bar{M}_s]$ of Eq. (7). The gust columns of $[A_2]$ are often set to
$[A_{g_2}] = 0$, as discussed above Eq. (7). The option of nulling out terms in
the aerodynamic damping matrix $[A_1]$ is also available but is usually not
recommended.

Up to three linear equality constraints for each aerodynamic term can
be used to explicitly determine $[A_0]$, $[A_1]$ and $[A_2]$, which reduces the approximation problem size. The three constraints are:

1. Steady aerodynamics match which yields

$$A_{0_{ij}} = F_{ij}(0) \tag{29}$$

2. A real-part match constraint at a nonzero $k = k_{f_{ij}}$ which yields

$$A_{2_{ij}} = \frac{1}{k_{f_{ij}}^2}\left(A_{0_{ij}} - F_{ij}(k_{f_{ij}})\right) + [D_i]\left(k_{f_{ij}}^2[I] + [R]^2\right)^{-1}\{E_j\} \quad (30)$$

which can be replaced by the $A_{2_{ij}} = 0$ constraint [13, 14], and

3. An imaginary-part match constraint at a nonzero $k = k_{g_{ij}}$ which yields

$$A_{1_{ij}} = \frac{1}{k_{g_{ij}}}G_{ij}(k_{g_{ij}}) + [D_i]\left(k_{g_{ij}}^2[I] + [R]^2\right)^{-1}[R]\{E_j\} \quad (31)$$

which can be replaced by the $A_{1_{ij}} = 0$ constraint.

The most drastic effect of using the constraints to reduce the approximation problem size is in the MS method. When the three data-match constraints of Eqs. (29)-(31) are applied simultaneously, and when all $k_{f_{ij}} = k_f$ and all $k_{g_{ij}} = k_g$, the MS $[D] \to [E] \to [D]$ least-square matrices, Eqs. (27) and (28), are replaced by

$$[A^*]_\ell = \begin{bmatrix} k_\ell^2[D]\left[\left(k_\ell^2[I] + [R]^2\right)^{-1} - \left(k_f^2[I] + [R]^2\right)^{-1}\right] \\ -k_\ell[D]\left[\left(k_\ell^2[I] + [R]^2\right)^{-1} - \left(k_g^2[I] + [R]^2\right)^{-1}\right][R] \end{bmatrix},$$

$$\{x^*\} = \{E_j\}, \qquad \{b^*\}_\ell = \left\{ \begin{array}{c} \bar{F}_j(k_\ell) \\ \bar{G}_j(k_\ell) \end{array} \right\} \quad (32)$$

and

$$[A^*]_\ell = \begin{bmatrix} k_\ell^2[E]^T\left[\left(k_\ell^2[I] + [R]^2\right)^{-1} - \left(k_f^2[I] + [R]^2\right)^{-1}\right] \\ -k_\ell[E]^T\left[\left(k_\ell^2[I] + [R]^2\right)^{-1} - \left(k_g^2[I] + [R]^2\right)^{-1}\right][R] \end{bmatrix},$$

$$\{x^*\} = \{D_i^T\}, \qquad \{b^*\}_\ell = \left\{ \begin{array}{c} \bar{F}_i^T(k_\ell) \\ \bar{G}_i^T(k_\ell) \end{array} \right\} \quad (33)$$

where the terms of $[\bar{F}]$ and $[\bar{G}]$ are

$$\bar{F}_{ij}(k_\ell) = F_{ij}(k_\ell) - F_{ij}(0) - (F_{ij}(k_f) - F_{ij}(0))\frac{k_\ell^2}{k_f^2}$$

$$\bar{G}_{ij}(k_\ell) = G_{ij}(k_\ell) - G_{ij}(k_g)\frac{k_\ell}{k_g} \quad (34)$$

The application of different constraint sets to different terms requires each row of $[A^*]_\ell$ and $\{b^*\}_\ell$ of Eqs. (32) and (33) to be calculated with a different

k_f or k_g. The applications of the $A_{1_{ij}} = 0$ and $A_{2_{ij}} = 0$ constraint options are performed by deleting the terms which include k_g and k_f respectively. With Eqs. (32) and (33), the number of unknowns in each application of Eq. (25) is reduced to n_L, which is typically about one seventh of that of the unconstrained problem.

A different way to enforce data match at $k \neq 0$ was presented in [9]. Instead of defining $[A_1]$ and $[A_2]$ explicitly, as in Eqs. (31) and (30), the data match was enforced by introducing Lagrange variables which increase the problem size. While this approach allowed a more flexible number of constraints, it made the computation time of an optimized MS application to a seven mode problem with one constraint to be more than 100 times that of the application of Roger's method. Since the number of aerodynamic lags, n_L, in the MS method is larger than in the other methods of similar accuracy, optimization of their values is more time consuming but not as important. A typical comparison between the applications of the MS and Roger's methods is given in [23] for a 10 mode case. An 8-lag MS approximation (which yields 8 aerodynamic states) with three constraints and without optimization was of the same level of fit accuracy as a 3-lag optimized Roger's approximation (which yielded 30 aerodynamic states) with one constraint (at $k = 0$). Based on the size and the number of least-square solutions in [23], it was argued that both applications required similar computational efforts.

VI. DATA WEIGHTING

The percentage deviations of uniformly weighted least-square curve fits from large data values are generally smaller than from small data values. This raises two problems in the context of aerodynamic approximations. One problem is that the approximation might be affected by the way the modes are normalized, even though the accurate solution is not. The other problem is that small aerodynamic data values are not necessarily less important than larger ones. The first problem can be resolved by a weighting that has the effect of data normalization (Subsection A). The physical weighting suggested in Subsection B resolves both problems.

A. Data Normalization

A way to avoid the effect of the manner in which structural modes are normalized is by defining the terms of the weight matrix $[W]_\ell$ as

$$W_{ij\ell} = \frac{\epsilon}{\max_\ell\{|Q_{ij}(ik_\ell)|, \epsilon\}} \tag{35}$$

where ϵ is a user-defined small positive parameter. The resulting absolute value of a weighted aerodynamic term is

$$\bar{Q}_{ij}(k_\ell) = W_{ij\ell}|Q_{ij}(ik_\ell)| \tag{36}$$

The effect of this weighting is renormalization of the input data such that
the maximum $|\bar{Q}(k_\ell)|$ of each ijth term is ϵ, with the exception that terms
with maximum $|Q(ik_\ell)|$ of less than ϵ are not normalized. With $\epsilon = 1$,
ε_t of Eq. (23) is consistent with the "common measure of approximation
performance" in [9]. It should be noticed that the data-normalization $[W]_\ell$
matrices are the same for all ℓ values. Consequently, they have to be applied
in the least-square solutions only when the unknowns appearing in a $\{x^*\}$
of Eq. (24) relate to different aerodynamic terms. Among the methods
discussed above, this is the case only with the MS method. In the other
methods, where the inner-loop approximations are performed term by term,
these weights should be applied only to calculate ε_t by Eq. (23) for the
evaluation of the approximation performance or for the optimization of the
aerodynamic lag values.

B. Physical Weighting

The physical-weighting algorithm developed in [13-15] was designed to
weight each term of the tabulated data such that the magnitude of the
weighted term, Eq. (36), indicates its "aeroelastic importance". The idea
is that the weight assigned to a data term should be proportional to the
estimated effect of a unit approximation error on a representative aeroelastic property. Different representative properties are selected below for the
structural, control and gust-related partitions of the AFC matrices. The
error effects are estimated by the differentiation of the selected aeroelastic
properties with respect to the aerodynamic terms. The weight calculations
are performed with the frequency-domain coefficients of Eqs. (1) and (2)
(with s replaced by $ik_\ell V/b$) which are already known at this stage of the
modeling process for a nominal aeroservoelastic configuration. The weighting dynamic pressure is $q = q_d$ at which the open-loop system is stable.

The weights assigned to the terms of a structure-related AFC matrix
$[Q_s(ik_\ell)]$ are based on their effect on the determinant of the system matrix
$[C_s(ik_\ell)]$ of Eq. (1). The absolute values of the partial derivatives of this
determinant of with respect to $Q_{s_{ij}}(ik_\ell)$, divided by the determinant itself,
is shown in [13] to be the ijth term of the weight matrix

$$[\bar{W}_s]_\ell = q_d \left|[C_s(ik_\ell)]^{-1}\right|^T \tag{37}$$

The weights assigned to the terms in the jth column of a control-related
AFC matrix $[Q_c(ik_\ell)]$ is based on the open-loop frequency response of the
jth actuator to excitation by the jth control surface, derived from Eqs.
(1) and (2). The magnitude of the partial derivative of this Nyquist signal

with respect to $Q_{c_{ij}}(ik_\ell)$ is shown in [13] to be the ijth term of the weight matrix

$$[\bar{W}_c]_\ell = q_d \left| [T(ik_\ell)][\psi_m][C_s(ik_\ell)]^{-1} \right|^T \tag{38}$$

The physical-weighting transfer functions in $[T(ik_\ell)]$ should represent a basic control system. Structural, narrow-band filters with high sensitivity to parametric changes should not be included as it may result in the assignment of low weights to important aerodynamic data terms.

The weights assigned to the terms of the jth column of a gust-related AFC matrix $[Q_g(ik_\ell)]$ is based on the power spectral density (PSD) of the open-loop response of a selected structural acceleration to continuous gust, derived from Eq. (1),

$$\Phi_{z_j}(k_\ell) = \left| \frac{k_\ell^2 q_d V}{b^2} [\psi_{z_j}][C_s(ik_\ell)]^{-1}\{Q_{g_j}(ik_\ell)\} \right|^2 \Phi_{w_j}(k_\ell) \tag{39}$$

where $[\psi_{z_j}]$ is a row vector of modal displacements at the selected response point, and $\Phi_{w_j}(k)$ is the PSD function of the associated gust velocity. The partial derivative of $\sqrt{\Phi_{z_j}(k_\ell)}$ with respect to $Q_{g_{ij}}(ik_\ell)$ is the ijth term of the weight matrix

$$[\bar{W}_g]_\ell = \frac{k_\ell^2 q_d V}{b^2} \left| [\psi_z][C_s(ik_\ell)]^{-1} \right|^T [\bar{\Phi}_w]_\ell \tag{40}$$

where $[\bar{\Phi}_w]_\ell$ in an $n_g \times n_g$ diagonal matrix whose elements are $\sqrt{\Phi_{w_j}(k_\ell)}$.

The variations of terms in the weight groups $[\bar{W}_s]_\ell$, Eq. (37), $[\bar{W}_c]_\ell$, Eq. (38), and $[\bar{W}_g]_\ell$, Eq. (40), with k may have very sharp peaks. In addition, the peak values of many terms may be several orders-of-magnitude smaller than other peaks. The extreme weight variations have the effect of neglecting much of the data, which may cause numerical ill-conditioning problems and unrealistic curve fits. To ensure realistic interpolation between the tabulated k values, and to facilitate the application of the resulting aeroelastic model to a variety of flow conditions, structural modifications, and control parameters, it may be desirable to moderate the weight variations. This is performed by widening the weight peaks and by scaling up the extremely low weights. The peak widening is performed in n_{wd} cycles where, in each cycle, $\bar{W}_{ij}(k_\ell)$ is changed to $\max\{\bar{W}_{ij}(k_{\ell-1}), \bar{W}_{ij}(k_\ell), \bar{W}_{ij}(k_{\ell+1})\}$ of the previous cycle. The weight matrices are then normalized and combined to the final weight matrix

$$[W]_\ell = \begin{bmatrix} [W_s]_\ell & [W_c]_\ell & [W_g]_\ell \end{bmatrix} \tag{41}$$

where a term in $[W_s]_\ell$ is

$$W_{s_{ij\ell}} = \left(\max\left\{ \frac{1}{\max_{i,j}\{\tilde{W}_{s_{ij}}\}}, \frac{W_{cut}}{\tilde{W}_{s_{ij}}} \right\} \right) \bar{W}_{s_{ij\ell}} \tag{42}$$

where

$$\tilde{W}_{s_{ij}} = \max_{\ell} \left\{ |Q_{s_{ij}}(ik_\ell)| \bar{W}_{s_{ij\ell}} \right\}$$

and the terms of $[W_c]_\ell$ and $[W_g]_\ell$ are calculated similarly (but separately). The upscale parameter W_{cut} is defined by the analyst. The resulting magnitudes of the weighted terms, $\bar{Q}_{ij}(k_\ell)$ of Eq. (36), fall between W_{cut} and 1.0 when the value of 1.0 typically appears only once in each group. The modified physical weighting is actually a compromise between the the unmodified one (with n_{wd} and W_{cut} equal zero) and the data-normalization weighting, Eq. (35). With $n_{wd} = n_k$ and $W_{cut} = 1.0$, all the physical-weighting effects are suppressed and the weighting becomes a data-normalization one. Recommended parameters in typical cases [14] are $n_{wd} = 2$ and $W_{cut} = 0.01$. Various applications demonstrated that the resulting aeroservoelastic models were adequate for analyses with large variations of dynamic pressures [13-16], control gains [8], and structural parameters [24].

C. Modal Measures of Aeroelastic Importance

The physical weights can be used to rate the vibration modes according to their relative aeroelastic importance. Based on the magnitudes of the weighted aerodynamic data terms, Eq. (36), calculated with n_{wd} and W_{cut} equal zero, three modal measures of aeroelastic importance are defined for each structural vibration mode by

$$
\begin{aligned}
Q_{s_i}^* &= \max_{j,\ell} \left\{ |\bar{Q}_{s_{ij}}(ik_\ell)| \right\}, & Q_{c_i}^* &= \max_{j,\ell} \left\{ |\bar{Q}_{c_{ij}}(ik_\ell)| \right\}, \\
Q_{g_i}^* &= \max_{j,\ell} \left\{ |\bar{Q}_{g_{ij}}(ik_\ell)| \right\}
\end{aligned}
\tag{43}
$$

These measures can be interpreted as indicators of the aeroelastic activity of the ith vibration mode, on a scale of 0 to 1, in three categories: a) influence on the open-loop system roots (Q_s^*); b) role in the aeroservoelastic loop (Q_c^*); and c) contribution to gust response (Q_g^*). Being based on a limited analysis, these measures should be used with caution. Their main usage is in supplying physical insight and in pointing out the structural modes that can be eliminated from the model without causing significant errors.

VII. RESIDUALIZATION OF STRUCTURAL STATES

A. General

It is assumed at this point that an aeroservoelastic model, Eqs. (11) and (12), has already been established with an initial set of vibration modes and that it is now desired to reduce the model size by eliminating structural states which have a negligible effect on the model accuracy. The modes to be eliminated are identified by either performing a preliminary analysis or by inspecting the modal measures of aeroelastic importance discussed in Section VI-C above. The state vector $\{x\}$ is partitioned into two subsets, $\{x_r\}$ and $\{x_e\}$, where $\{x_e\}$ is to be constrained and eliminated from the state vector, and $\{x_r\}$ is to be retained. In our case, $\{x_e\}$ includes only the structural states $\{\xi_e\}$ and $\{\dot{\xi}_e\}$ which represent the eliminated vibration modes. The formulation in this section follows that of [20]. The main differences stem from the fact that the gust filter in this work, Eq. (9), does not include a noise term in its output.

The partitioning of Eqs. (11) and (12) into $\{x_r\}$ and $\{x_e\}$ related partitions yields

$$
\begin{bmatrix} Z_{rr} & Z_{re} \\ Z_{er} & Z_{ee} \end{bmatrix} \begin{Bmatrix} \dot{x}_r \\ \dot{x}_e \end{Bmatrix} = \begin{bmatrix} A_{rr} & A_{re} \\ A_{er} & A_{ee} \end{bmatrix} \begin{Bmatrix} x_r \\ x_e \end{Bmatrix} + \begin{bmatrix} B_r \\ 0 \end{bmatrix} \{u\}
$$
$$
+ \begin{Bmatrix} B_{w_r} \\ 0 \end{Bmatrix} w \qquad (44)
$$

and

$$
\{y\} = \begin{bmatrix} C_r & C_e \end{bmatrix} \begin{Bmatrix} x_r \\ x_e \end{Bmatrix} \qquad (45)
$$

With the assumption that the effects of $\{\dot{x}_e\}$ on the top partition of Eq. (44) may be neglected, the most general constraints which allow model size reduction are

$$
\{x_e\} = [F]\{x_r\} + [G]\{u\} + \{G_w\}w + [H]\{\dot{x}_r\} \qquad (46)
$$

The coefficient matrices in Eq. (46) are defined in the following subsections for the various size-reduction methods. It should be mentioned that Eq. (46) does not necessarily define the motion of the eliminated modes. It only defines the portion of this motion that affects the remaining model. The substitution of Eq. (46) into the top partition of Eq. (44) with $[Z_{re}]\{\dot{x}_e\} = 0$ yields the residualized equation of motion

$$
\left[\tilde{Z}\right]\{\dot{x}_r\} = [\tilde{A}]\{x_r\} + [\tilde{B}]\{u\} + \{\tilde{B}_w\}w \qquad (47)
$$

and the associated output equation

$$
\{y\} = [\tilde{C}]\{x_r\} + [\tilde{D}]\{u\} + \{\tilde{D}_w\}w \qquad (48)
$$

where the matrix coefficients are

$$
\left[\tilde{Z}\right] = [Z_{rr}] - [A_{re}][H], \qquad \left[\tilde{A}\right] = [A_{rr}] + [A_{re}][F],
$$

$$\left[\tilde{B}\right] = [B_r] + [A_{re}][G], \qquad \left\{\tilde{B}_w\right\} = \{B_{w_r}\} + [A_{re}]\{g_w\},$$

$$\left[\tilde{C}\right] = [C_r] + [C_e]\left([F] + [H][\tilde{Z}]^{-1}[\tilde{A}]\right),$$

$$\left[\tilde{D}\right] = [C_e]\left([G] + [H][\tilde{Z}]^{-1}[\tilde{B}]\right),$$

$$\left\{\tilde{D}_w\right\} = [C_e]\left(\{G_w\} + [H][\tilde{Z}]^{-1}\{\tilde{B}_w\}\right) \qquad (49)$$

B. Mode Truncation

The simplest and most commonly used size-reduction technique is based on the assumption that the eliminated states have no effects on the dynamics and on the output of the system, namely the coefficient matrices of Eq. (46) are all zero. The resulting model is that of Eqs. (44) and (45) where all the rows and columns associated with $\{x_e\}$ are truncated.

C. Static Residualization

Static residualization is based on the principles of static aeroelastic analysis using vibration modes (such as in [19]) where the aerodynamics associated with the deflections of the elastic modes have a major impact on the rigid-body aerodynamics. In the present application it is assumed that the $\{\xi_e\}$ effects are important but the $\{\dot{\xi}_e\}$ and $\{\ddot{\xi}_e\}$ effects may be neglected. With these assumptions, the bottom row partition of Eq. (44) and the matrix definitions of Eq. (10) yield

$$\left[\bar{K}_{s_{ee}}\right]\{\xi_e\} = [F_1]\{x_r\} - [\bar{M}_{s_{er}}]\{\ddot{\xi}_r\} \qquad (50)$$

where

$$[F_1] = -\left[\begin{array}{ccccc} \bar{K}_{s_{er}} & \bar{B}_{s_{er}} & \bar{D}_e & \bar{F}_{c_e} & \bar{F}_{g_e} \end{array}\right]$$

Equation (50) and $\{\dot{\xi}_e\} = 0$ yield the constraint matrices of Eq. (46) for the case of static residualization

$$[F] = \left[\begin{array}{c} \bar{K}_{s_{ee}}^{-1} \\ 0 \end{array}\right][F_1], \qquad [G] = [0], \qquad \{G_w\} = \{0\},$$

$$[H] = -\left[\begin{array}{c} \bar{K}_{s_{ee}}^{-1} \\ 0 \end{array}\right]\left[\begin{array}{ccccc} 0 & \bar{M}_{s_{er}} & 0 & 0 & 0 \end{array}\right] \qquad (51)$$

The statically residualized model of Eqs. (47) and (48) can now be constructed using Eqs. (49) and (51). It can be shown from Eqs. (10), (49), and (51) that the matrix coefficients of the resulting state-space equation have the same topology (non-zero partitions) as those of the full-size equation (10), and that the output equation remains with no control and noise terms, namely $[\tilde{D}] = 0$ and $\{\tilde{D}_w\} = 0$.

D. Dynamic Residualization

The dynamic residualization suggested in [20] does not neglect apriori the effects of $\{\dot{\xi}_e\}$. The extension of Eq. (50), using Eq. (10), to include these effects reads

$$[\bar{K}_{s_{ee}}] \{\xi_e\} = [F_1]\{x_r\} - [\bar{M}_{s_{er}}]\{\ddot{\xi}_r\} - [\bar{B}_{s_{ee}}]\{\dot{\xi}_e\} \qquad (52)$$

where the first term on the right side is the primary contributor to the generalized forces acting on the eliminated modes and the other terms have secondary effects. The differentiation of Eq. (52) with respect to time, while neglecting the derivatives of the secondary effects and using the definition of $[F_1]$ in Eq. (50), yields an equation for $\{\dot{\xi}_e\}$:

$$[\bar{K}_{s_{ee}}] \{\dot{\xi}_e\} = -[\bar{K}_{s_{er}}]\{\dot{\xi}_r\} - [\bar{B}_{s_{er}}]\{\ddot{\xi}_r\} - [\bar{D}_e]\{\dot{x}_a\} - [\bar{F}_{c_e}]\{\dot{x}_c\} - [\bar{F}_{g_e}]\{\dot{x}_g\} \qquad (53)$$

which, by using Eq. (10) for $\{\dot{x}_a\}$, $\{\dot{x}_c\}$, and $\{\dot{x}_g\}$, becomes

$$[X_{22}] \{\dot{\xi}_e\} = [F_2]\{x_r\} - [\bar{F}_{c_e}][B_c]\{u\} - [\bar{F}_{g_e}]\{B_g\}w - [\bar{B}_{s_{er}}]\{\ddot{\xi}_r\} \qquad (54)$$

where

$$[X_{22}] = [\bar{K}_{s_{ee}}] + [\bar{D}_e][E_{s_e}]$$
$$[F_2] = -\left[\, 0 \quad \bar{K}_{s_{er}} + \bar{D}_e E_{s_r} \quad \bar{D}_e \bar{R} \quad \bar{D}_e \hat{E}_c + \bar{F}_{c_e} A_c \quad \bar{D}_e \hat{E}_g + \bar{F}_{g_e} A_g \,\right]$$

Equations (52) and (54) yield the constraint matrices of Eq. (46) for the case of dynamic residualization

$$[F] = [X]^{-1}\begin{bmatrix} F_1 \\ F_2 \end{bmatrix}, \qquad\qquad [G] = -[X]^{-1}\begin{bmatrix} 0 \\ \bar{F}_{c_e}B_c \end{bmatrix},$$

$$\{G_w\} = -[X]^{-1}\begin{Bmatrix} 0 \\ \bar{F}_{g_e}B_g \end{Bmatrix},$$

$$[H] = -[X]^{-1}\begin{bmatrix} 0 & \bar{M}_{s_{er}} & 0 & 0 & 0 \\ & \bar{B}_{s_{er}} & & & \end{bmatrix} \qquad (55)$$

where

$$[X]^{-1} = \begin{bmatrix} \bar{K}_{s_{ee}}^{-1} & -\bar{K}_{s_{ee}}^{-1}\bar{B}_{s_{ee}}X_{22}^{-1} \\ 0 & X_{22}^{-1} \end{bmatrix}$$

The dynamically residualized model of Eqs. (47) and (48) can now be constructed using Eqs. (49) and (55). It can be shown from Eqs. (10), (49) and (55) that $[\tilde{Z}]$ of Eq. (47) still has the same topology as $[Z_{rr}]$. However, unlike the static residualization, the topology of $[\tilde{A}]$, $[\tilde{B}]$, and $\{\tilde{B}_w\}$ resulting from the dynamic residualization does not keep the original topology. In addition, $[\tilde{D}]$ and $\{\tilde{D}_w\}$ are not equal zero. It should be noted that setting $\{G_w\} = 0$ instead of $\{G_w\}$ of Eq. (55), which is usually an adequate assumption, would null out $\{\tilde{D}_w\}$ such that there would not be a noise term in the output equation. The added terms in Eq. (48) modify the closed-loop equation, Eq. (15), which becomes

$$\left[\tilde{Z}\right]\{\dot{x}_r\} = [\hat{A}]\{x_r\} + \{\hat{B}_w\}w \tag{56}$$

where

$$\left[\hat{A}\right] = [\tilde{A}] + [\tilde{B}]\left([I] - [G_c][\tilde{D}]\right)^{-1}[G_c][\tilde{C}]$$

$$\left\{\hat{B}_w\right\} = \{\tilde{B}_w\} + [\tilde{B}]\left([I] - [G_c][\tilde{D}]\right)^{-1}[G_c]\{\tilde{D}_w\}$$

The columns of $[A_2]$ of Eq. (4) which relate to the eliminated modes are neglected in all the size-reduction techniques discussed in this section. Constraining them to be zero in the aerodynamic approximation process may improve the accuracy of the reduced-size model. The use of dynamically residualized models in aeroservoelastic optimization is discussed in [25].

VIII. NUMERICAL EXAMPLE

A. The Mathematical Model

The numerical application deals with the mathematical model of NASA's Drone for Aerodynamic and Structural Testing - Aerodynamic Research Wing 1 (DAST-ASW1) [5]. A top view of the model geometry is given in Fig. 1. The same model was also used by Nissim [18] to demonstrate his aerodynamic approximation method in comparison with Roger's method. The main results of [18] are compared in this section to flutter characteristics derived from aeroservoelastic models which are based on the Minimum-State (MS) approximation method.

The model consists on ten symmetric natural vibration modes and one trailing-edge control-surface mode. The oscillatory AFC matrices were calculated using doublet lattice aerodynamics [3] at Mach 0.9. The control surface is driven by a third-order actuator whose transfer function is

$$\frac{\delta}{\delta_c} = \frac{1.915 \times 10^7}{(s + 214)(s^2 + 179.4s + 89450)} \tag{57}$$

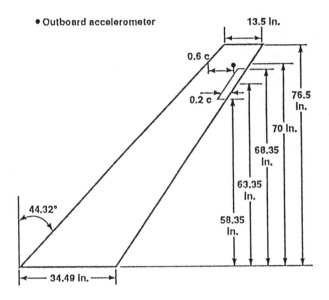

Fig. 1. Model geometry of DAST-ARW1.

Closed-loop results will be shown for two control laws which relate the actuator input, δ_c, to the acceleration, $\ddot{h}_{outb}$, measured by an outboard accelerometer (Fig. 1). The transfer function of the first control law (CL1), developed in [5], is

$$\frac{\delta_c}{\ddot{h}_{outb}} = \frac{0.1}{(s+10)} \frac{s}{(s+1)} \frac{628.3^2}{(s+628.3)^2} \left[\frac{s^2+30.79s+14692}{s^2+572.6s+88578}\right]$$
$$\times \left[\frac{s^2+47.37s+72436}{s^2+568.6s+86972}\right] \frac{rad}{in/sec^2} \qquad (58)$$

The transfer function of the second control law (CL2), developed in [26], is

$$\frac{\delta_c}{\ddot{h}_{outb}} = 5.745 \times 10^{-5} \left[\frac{s+185.8}{s+630}\right] \left[\frac{s^2+155s+29658}{s^2+13.53s+272.25}\right] \frac{rad}{in/sec^2} \qquad (59)$$

The dynamics of the accelerometer are neglected. The total control transfer function is obtained by multiplying CL1 or CL2 by the actuator transfer function, Eq. (57). The realization of the control system in state-space form, Eq. (8), yields 11 control states for CL1 and 6 for CL2.

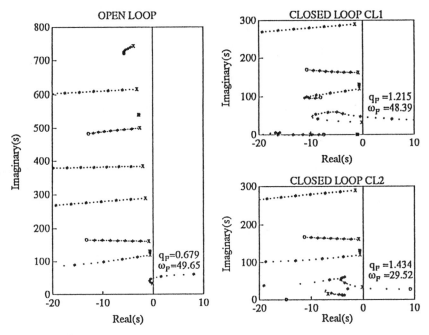

Fig. 2. Open and closed-loop root loci of DAST-ARW1.

B. Flutter Results with Various Aerodynamic Approximations

The results shown in this subsection are based on the same aerodynamic data used in [18]. This data, which will be referred to as Data 1, consists on 10 AFC matrices calculated at $k_l = 0.$, 0.05, 0.1, 0.2, 0.3, 0.4, 0.5, 0.6, 0.7, and 0.8. The reference "true" results are those obtained with Roger's method with 4 aerodynamic lag terms, $\gamma_l = 0.1, 0.2, 0.3,$ and 0.4. These lags were used in all the analyses of this subsection. When the number of lags is less than four, the first ones are taken. The generalized structural damping matrix, $[B_s]$ of Eq. (1), is assumed to be

$$[B_s] = 2\zeta[M_s][\omega_n] \tag{60}$$

where $[\omega_n]$ is a diagonal matrix of natural frequencies, and ζ is the modal damping coefficient, 0.01 in this work. The open- and closed-loop flutter conditions were found by root-locus analysis of the respective system matrix with variable dynamic pressure (q). Typical root-loci plots, where q varies in constant increments from 0 to 1.6 psi (marked by X and O respectively) are shown if Fig. 2. It can be observed that all the flutter mechanisms are of low frequency but there are significant differences between the flutter conditions of the three cases. The high-frequency closed-loop branches (not shown) are almost identical to the open-loop ones.

The MS modeling method was applied to Data 1 with 4 aerodynamic lags and with three data-match constraints, one at $k = 0$ and two at $k_f = 0.8$, assigned to each aerodynamic term. The least-square solution was performed in 50 $[D] \rightarrow [E] \rightarrow [D]$ iterations with the physical weighting of Section VI-B, performed at $q_d = 0.5$ psi and modified with $n_{wd}=2$ and $W_{cut}=0.01$. These typical parameters are based on the experience gained in previous applications of the MS method [13-16]. The trunster function used to calculate the control-column weights, Eq. (38), was that of the actuator only. The open and closed-loop flutter results are compared in Table 1 to some of those given in [18] for Roger's and Nissim's methods.

Method	n_L	Modes with lags	n_a	q_f (psi)	ω_f rad/s	ε_q (%)	ε_ω (%)
Open-loop							
Roger	4	All	40	0.677	49.6	0.0	0.0
Roger	2	All	20			-0.3	-0.2
Roger	1	All	10			0.2	-0.2
Nissim	1	1,2	2			0.1	-0.2
MS	4	All	4	0.679	49.7	0.3	0.2
Closed-loop CL1							
Roger	4	All	40	1.215	48.4	0.0	0.0
Roger	2	All	20			-0.1	-0.2
Roger	1	All	10			2.6	0.6
Nissim	1	1,2	2			-3.9	5.1
Nissim	1	1,2,4,5	4			3.7	-0.4
Nissim	2	1,2	4			-5.9	5.1
Nissim	2	1,2,4,5	8			-0.3	-1.0
MS	4	All	4	1.197	49.7	-1.5	2.6
Closed-loop CL2							
Roger	4	All	40	1.426	29.4	0.0	0.0
Roger	2	All	20			-1.0	2.7
Roger	1	All	10			1.6	-3.7
Nissim	1	1,2	2			0.7	3.7
MS	4	All	4	1.435	29.8	0.6	1.4

Table 1. Comparison of flutter results with various modeling methods.

Due to a slight difference between the definitions of structural damping, the true flutter results in this work, q_F and ω_F, are slightly different than those of [18]. It is assumed, however, that the differences has no effect on the flutter percentage errors, ε_q and ε_ω. It can be observed that, in this example, the MS and Nissim's methods yield similar numbers of aerodynamic states (n_a) per desired accuracy. It should be noted, however, that while the presented Nissim's results were obtained after a careful selection of the best lag values and the modes to which they should be applied, the MS modeling did not require such an investigation.

C. Effects of Data Points and Approximation Constrains

Since flutter is more likely to involve the low-frequency modes than the high-frequency ones, the common practice is to include in the tabulate data more AFC matrices in the low-frequency range than in the higher one. To conform with this approach, and to investigate the effects of various data-match constraints, a new aerodynamic data set, refered to as Data 2, was generated. The new data consists on 11 AFC matrices at $k_\ell = 0.$, 0.02, 0.0544, 0.07, 0.0896, 0.11, 0.15, 0.2, 0.3, 0.5 and 0.8. The 0.0544 and 0.0896 valued were chosen because they are the true flutter reduced frequency values, k_F, of the closed-loop cases CL2 and CL1 respectively. The open-loop true k_F is 0.0919.

Data 2 was used to construct MS models with the same physical weighting parameters as above, but with three different constraint sets, one with all the aerodynamic terms constrained to match the data at $k_f = 0.8$ (as in the Data 1 case), one with $k_f = 0.0896$, and one with $k_f = 0.0544$. In addition, all the approximations were constrained to match the data at $k = 0$. A comparison of the resulting flutter errors is given in Table 2. The comparison between the first two cases shows that the larger amount of low-frequency data in Data 2 resulted in smaller errors than with Data 1. A further error reduction is obtained when the data-match constraints are at k values which are closer to k_F.

Data	k_f	Open-loop		Closed-loop CL1		Closed-loop CL2	
		ε_q	ε_ω	ε_q	ε_ω	ε_q	ε_ω
1	0.8	0.28	0.16	-1.48	2.60	0.63	1.36
2	0.8	0.31	0.22	-0.99	1.94	0.56	0.78
2	0.0896	0.03	0.02	0.00	0.00	0.35	-0.61
2	0.0544	-0.18	-0.06	-0.66	-0.27	0.00	0.00

Table 2. Effects of data points and constraints on percentage flutter errors.

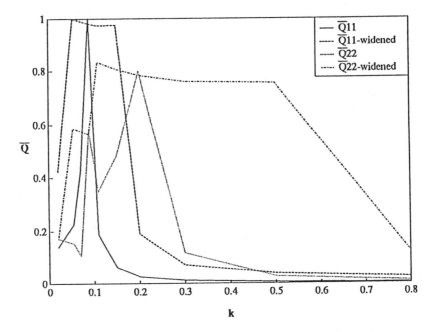

Fig. 3. Magnitudes of weighted aerodynamic terms versus reduced frequency.

It should be remembered that a major reason for applying the constraints when using the MS method is that they reduce the problem size.
In our numerical example, in each of the 50 least-square iterations there
are two inversions of 4x4 matrices. Without the constraints, the inversions
in each iteration would have been one of a 34x34 matrix and one of a 37x37
matrix. Even though the application of data-match constraints at the highest k_ℓ value yielded the largest flutter error, it had the least effect on the
total approximation error (not shown).

D. The Physical Weights

When the physical weighting is performed with $n_{wd} = 0$ and $W_{cut} = 0$,
the magnitudes of the aerodynamic data terms, $\bar{Q}_{ij}(k_\ell)$ of Eq. (36), indicate
their aeroelastic importance. Typical variations of $\bar{Q}_{ij}$ values of important
terms, calculated with Data 2, versus k are shown in Fig. 3. It can be
observed that $\bar{Q}_{11}$ has a sharp peak which may cause the approximation
to deteriorate at off-peak values. The variations of the same terms, but
with the peaks widened in two widening cycles ($n_{wd} = 2$), are also shown.
With the modified weights, good least-square fits are expected over a wider

range of reduced frequencies, which makes the accuracy of the resulting model less sensitive to parametric changes.

The largest $\bar{Q}$ values, calculated separately for each aerodynamic term and rounded to the third decimal place, are given in Table 3. The rows and columns associated with modes 3 and 8 are not shown because all their values are rounded to zero. The maximal value in each row among the ten structural columns, and the associated value at the control column are the modal measures of aeroelastic importance, $Q^*_{s_i}$ and $Q^*_{c_i}$ of Eq. (43). Table 3 indicates that modes 1, 2, 4 and 5 are the most important (in decreasing order of importance). This is in agreement with the modes to which aerodynamic lags are applied in the Nissim's method cases of Table 1, and with the aeroelastic behavior of the modes in Fig. 2. Eighty percent of the structural $\bar{Q}_{max}$ values of Table 3, and fifty percent of the control $\bar{Q}_{max}$ values, are less that 0.01. With W_{cut} =0.01, the weights assigned to these terms are upscaled such that their $\bar{Q}_{max}$ =0.01.

Minimum-State approximations of Data 2 were performed with the number of aerodynamic lags, n_L, varying from 2 to 10. The n_L diagonal values of $[R]$ where arbitrarily chosen between -0.05 and -1.5 with more values close to -0.05 than to -1.5. Two models were constructed for each n_L case, one with the physical weighting used to construct the model of the second case in Table 2, and one with data-normalization weighting. The RMS values of the resulting errors in flutter dynamic pressure and flutter frequency of the three open and closed-loop cases together are shown in Fig. 4. These results are in agreement with those of [13-16] that showed that the number of aerodynamic states required for a certain level of model accuracy when physical weighting is used is about 50% of that required when physical weighting is not used.

	j=1	j=2	j=4	j=5	j=6	j=7	j=9	j=10	j=11
i=1	1.000	.595	.038	.014	.000	.001	.000	.000	1.000
i=2	.593	.800	.049	.025	.001	.002	.000	.000	.355
i=4	.047	.051	.344	.021	.002	.000	.000	.000	.081
i=5	.029	.016	.020	.321	.005	.001	.000	.000	.039
i=6	.001	.001	.002	.005	.048	.001	.000	.000	.013
i=7	.001	.001	.000	.001	.001	.029	.000	.000	.002
i=9	.000	.000	.000	.000	.000	.000	.011	.000	.000
i=10	.000	.000	.000	.000	.000	.000	.000	.003	.000

Table 3. Maximum magnitudes of weighted aerodynamic terms.

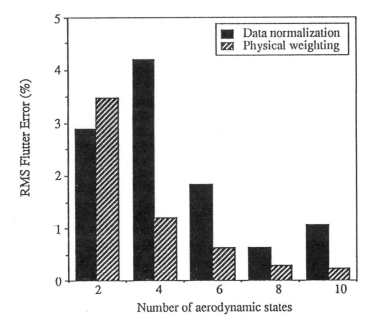

Fig. 4. Minimum-State flutter errors versus number of aerodynamic states.

E. Convergence in the Minimum-State Procedure

All the MS approximations of Fig. 4 were performed with 50 $[D] \rightarrow [E] \rightarrow [D]$ least-square iterations. The variations of the total approximation error, ε_t of Eq. (23), with the number of iterations are shown in Fig. 5. Because the data-normalization (N) weights, Eq. (35) with $\epsilon = 1$, are very different than the physical weights (PW), the errors of the two cases are not comparable. The relative effects of adding an approximation root in each of the two cases are similar. Since most of the weights in the PW cases are very small, ther is effectively less data to fit than in the N cases. Consequently, the PW cases converge faster and more consistently. Most error reductions are in the first few iterations. While the error in each case is reduced monotonously, the convergence rate may not, which may cause numerical difficulties when a convergence criteria is applied or when optimization of the lag values is attempted.

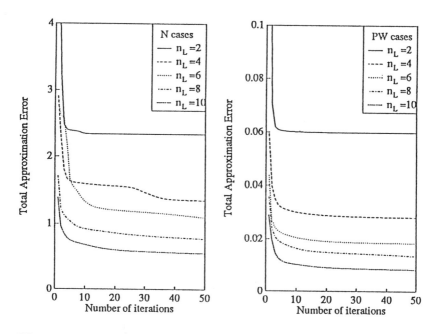

Fig. 5. Approximation errors versus number of Minimum-State iterations.

F. Elimination of Structural States

As shown above, high-accuracy aeroservoelastic models can be constructed for our application with only four aerodynamic states. A "full-size" 10-mode model has 20 structural states, the largest among the structural, aerodynamic and control state groups. Hence, a major improvement of the model efficiency can be achieved when some of the structural states are eliminated.

The third case of Table 2 was used as a baseline for the investigation of the effect of eliminating structural states. As mentioned before, the measures of aeroelastic importance given in Table 3 indicate that modes 1 and 2 are the most important and that modes 3 and 8 are the least important. Figure 6 shows the effects of mode truncation, and static and dynamic residualizations on the RMS errors in q_F and ω_F, calculated separately for the open-loop and the two closed-loop cases. Modes 3 and 8 were truncated in all the cases. Other modes were then truncated one by one, in decreasing frequency order. Figure 6 shows that up to 4 modes (8 states) can be truncated without any visible impact on the accuracy. The errors are still less

than 1% with the truncation of 6 modes, but with the truncation of more
modes, the errors grow to unacceptable levels where the largest errors are
in the open-loop flutter dynamic pressure, more than 100%. With static
and dynamic residualizations, 8 modes can be eliminated with RMS error
levels of less than 3% and 1% respectively. The open-loop model in this
case has only 8 states (4 structural and 4 aerodynamic). The CL1 and CL2
models have additional states, 11 and 6 respectively.

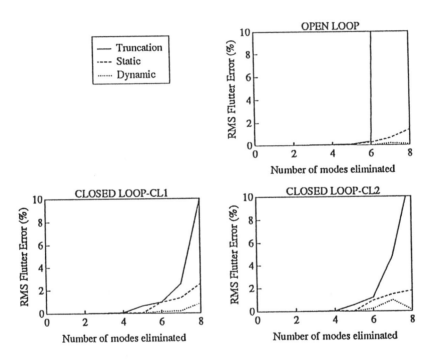

Fig. 6. Percentage RMS flutter errors versus number of eliminated modes.

IX. CONCLUSIONS

Size-reduction techniques for the determination of efficient time-domain,
state-space aeroservoelastic models were presented. Various rational function approximation methods of the unsteady aerodynamic force coefficients
were brought to a common notation, which emphasizes their differences.
Among those, the classic Roger's method is the easiest to apply but its
resulting number of aerodynamic states is typically equal to or larger than

the number of structural states. On the other side of the spectrum is the Minimum-State method which typically reduces the number of aerodynamic states by 70% or more, but requires the solution of an iterative, nonlinear least-square solution. The Minimum-State computational efforts are reduced significantly when three approximation constraints are applied. These constrained can be used to improve the fit at a desired reduced frequency. A selective application of Roger's formula, where aerodynamic lags are assigned only to a small number of modes, may also result in small-size models, but only in cases where the aeroelastic behavior is dominated by a small number of modes, and these modes are known apriori. The presented physical weighting of the aerodynamic data yields a further reduction (about 50%) in the number of states per desired accuracy. The physical weights can also be used to rate the structural modes according to their relative aeroelastic importance. The elimination of structural states associated with the less important modes by performing dynamic residualization was shown to be considerably more accurate than mode truncation and static residualization.

ACKNOWLEDGMENT

This research was supported by NASA Grant NAGW-1708. This support is gratefully acknowledged.

REFERENCES

1. R. L. Bisplinghoff, H. Ashley, and R. L. Halfman, "Aeroelasticity," Addison-Weseley, Cambridge, Mass., 1955.

2. R.L. Bisplinghoff and H. Ashley, "Principles of Aeroelasticity," Wiley, New York, 1962.

3. E. Albano and W. P. Rodden, "A Doublet-Lattice Method for Calculating Lifting Disturbances of Oscillating Surfaces in Subsonic Flow", *AIAA J.* **7**(2), 279-285 (1969).

4. W. M. Adams Jr., S. H. Tiffany, J. R. Newsom and E. L. Peele, "STABCAR - A Program for Finding Characteristic Roots of Systems Having Transcendental Stability Matrices", **NASA TP-2165** (1984).

5. J. R. Newsom, I. Able and H. J. Dunn, "Application of Two Design Methods for Active Flutter Suppression and Wind-Tunnel Test Results", **NASA TP-1653** (1980).

6. V. Mukhopadhyay, J. R. Newsome and I. Abel, "A Method for Obtaining Reduced-Order Control Laws for High-Order Systems Using Optimization Techniques," **NASA TP-1876** (1981).

7. P. D. Arbuckle, C. S. Buttrill, and T. A. Zeiler, "A New Simulation Model Building Process for use in Dynamics Systems Integration Research", *Proc. AIAA Flight Simulation Technologies Conf., Monterey, Ca* (1987).

8. M. Karpel, "Sensitivity Derivatives of Flutter Characteristics and Stability Margins for Aeroservoelastic Design", *J. Aircraft*, **27**(4), 368-375 (1990).

9. S. H. Tiffany and W. M. Adams Jr., "Nonlinear Programming Extensions to Rotational Approximation Methods of Unsteady Aerodynamic Forces", **NASA TP-2776** (1988).

10. K. L. Roger, "Airplane Math Modeling and Active Aeroelastic Control Design," **AGARD-CP-228**, 4.1-4.11 (1977).

11. M. Karpel, "Design for Active and Passive Flutter Suppression and Gust Alleviation", **NASA CR-3482** (1981)

12. M. Karpel, "Design for Active Flutter Suppression and Gust Alleviation Using State-Space Aeroelastic Modeling", *J. Aircraft*, **19**(3), 221-227 (1982).

13. M. Karpel, "Time-Domain Aeroservoelastic Modeling Using Weighted Unsteady Aerodynamic Forces", *J. Guidance, Control, and Dyn.*, **13**(1), 30-37 (1990).

14. M. Karpel, "Extension to the Minimum-State Aeroelastic Modeling Method", *AIAA J.*, **29**(11), 2007-2009 (1991).

15. M. Karpel and S. T. Hoadley, "Physically Weighted Approximations of Unsteady Aerodynamic Forces Using the Minimum-State Method", **NASA TP-3025** (1991).

16. S. T. Hoadley and M. Karpel, "Application of Aeroservoelastic Modeling Using Minimum-State Unsteady Aerodynamic Approximations", *J. Guidance, Control, and Dyn.*, **14**(6), 1267-1276 (1991).

17. E. Nissim and I. Lottati, "An Optimization Method for the Determination of the Important Flutter Modes", *J. Aircraft*, **18**(8), 663-668 (1981).

18. E. Nissim, "Reduction of Aerodynamic Augmented States in Active Flutter Suppression Systems," *J. Aircraft*, **28**(1), 82-93, (1991).

19. Z. Sheena and M. Karpel, "Static Aeroelastic Analysis Using Aircraft Vibration Modes", *Coll. Papers Second Int. Symp. Aeroelasticity and Struct. Dyn.*, Aachen, Germany, 229-232 (1985).

20. M. Karpel, "Reduced-Order Aeroelastic Models via Dynamic residualization", *J. Aircraft*, **27**(5), 449-455 (1990).

21. H. J. Hassig, "An Approximate True Damping Solution of the Flutter Equation by Determinant Iteration," *J. Aircraft*, **8**(11), 885-889 (1971).

22. W. Eversman and A. Tewari, "Consistent Rational-Function Approximation for Unsteady Aerodynamics," *J. Aircraft*, **28**(9), 545-552 (1991).
23. M. Karpel, "Reduced Size First-Order Subsonic and Supersonic Aeroelastic Modeling", *Proc. AIAA/ASME/ASCE/AHS 31st Struct., Str. Dyn. and Materials Conf.*, Long Beach, Ca, 1405-1417 (1990).
24. M. Karpel and C. D. Wieseman "Modal Coordinates for Aeroelastic Analysis with Large Local Structural Variations", *Proc. Int. Forum on Aeroelasticity and Struct. Dyn. 1991*, Aachen, Germany, 364-370 (1991).
25. I. Herszberg and M. Karpel, "Sensitivity Derivatives for Residualized Aeroservoelastic Optimum Design Models", *Proc. Int. Forum on Aeroelasticity and Struct. Dyn. 1991*, Aachen, Germany, 371-378 (1991).
26. E. Nissim, "Design of Control Laws for Flutter Suppression Based on the Aerodynamic Energy Concept and Comparisons With Other Design Methods", **NASA TP-3056** (1990).

SENSITIVITY ANALYSIS OF EIGENDATA
OF AEROELASTIC SYSTEMS

V. R. Murthy
Yi Lu

Department of Mechanical and Aerospace Engineering

Syracuse University

Syracuse, New York 13244

I. INTRODUCTION

In design of engineering systems, sensitivity analyses play a key role in arriving at optimum solutions. Sensitivity analyses call for the sensitivity derivatives, which are defined as the ratios of the variations in the system characteristics to the variations in the design parameters. The sensitivity derivatives with respect to the design variables provide an important information to the designer enabling him to understand the constraints and proceed methodically with the optimization. Also, the sensitivity derivatives are necessary in automated minimization procedures, which are central to the computer based optimizations.

The sensitivity derivatives can be calculated by finite differencing the entire system analysis with respect to the design parameters. This repetitive solution approach is excessively time consuming to arrive at an optimum design. In addition, if the system has a large number of design parameters, safety and performance constraints, this will add to the enormity of the problem. Therefore, several analytical methods are developed to calculate the sensitivity derivatives.

CONTROL AND DYNAMIC SYSTEMS, VOL. 54

297

The sensitivity analyses of systems with stability constraints often involve the calculation of derivatives of eigenvalues and eigenvectors with respect to the design parameters. The direct methods available to calculate these derivatives are reviewed in Refs. [1, 2]. The methods to calculate the derivatives of self-adjoint problems are presented in Refs. [3-5] while the non-self-adjoint problems are treated in Refs. [6-11].

The optimal design of aerospace systems often involve the aeroelastic stability constraints such as flutter. The relations given in Refs. [3-11] are not usually convenient for direct applications to flutter problems. The reasons are

1.   Flutter problems are double eigenvalue problems (velocity V, and frequency ω) in the sense that they are not the real and imaginary parts of any single complex eigenvalue.

2.   In general, the eigenvalues do not appear in a polynomial form in the equations of motion and the dependence of aerodynamic forces on the eigenvalues is complex.

3.   Flutter problem is a nonlinear eigenvalue problem in the sense that the coefficient matrix depends upon the eigenvalue as follows: $[A(\lambda)]\{x\} = \lambda\{x\}$.
Several methods to calculate the derivatives of eigenvalues of aeroelastic systems exist in the literature such as those given in Refs. [12-19]. The linear eigenvalue problems are treated in Refs. [12-16] while the quadratic and nonlinear problems are treated in Refs. [17, 18]. Recently, Lu and Murthy [19] presented a general formulation for a nonlinear aeroelastic eigenvalue problem in the Laplace transform domain and these nonlinear problems can also be classified as s-dependent systems.

Several periodically-time-varying systems exist in the various engineering disciplines. Typically, the helicopter rotor aeroelastic problems in forward flight are governed by periodic equations. The literature dealing with the sensitivity analysis of time-invariant systems is extensive, as can be seen in the review paper by Adelman and Haftka [1] with 148 citations, while it is very scarce for periodic systems. Sensitivity derivatives of rotary-wing aeroelastic problems are calculated usually by either finite difference type methods [20-23] or direct chain rule differentiation methods [24-27]. Recently, Lu and Murthy [28] have developed analytical formulation for the derivatives of eigenvalues and eigenvectors of periodic systems.

In this article, several methods to calculate the derivatives of eigendata are presented. The class of problems included are (1) linearly dependent time-invariant systems, (2) nonlinearly dependent time-invariant systems and (3) periodic systems. Three applications of these methods to rotary-wing aeroelastic prob-lems are presented with a few numerical results. The problems considered are (1) hovering rotor, (2) rotor blade flapping in forward flight and (3) coupled rotor-body problem in hover.

II. TIME-INVARIANT SYSTEMS

In general, two types of time-invariant eigenvalue problems can be identified. In one type, the problem linearly depends upon the eigenvalue, while in the other, the dependence is nonlinear.

A. LINEARLY DEPENDENT SYSTEMS

This class of systems are governed by an equation of the form

$$\{\dot{y}(t)\} = [A] \{y(t)\} \tag{1}$$

where [A] is a constant square matrix of order n. In the Laplace transform domain, Eq. (1) becomes

$$([A] - s[\ I\ ])\ \{x(s)\} = \{0\} \tag{2}$$

where 's' is the Laplace transform variable, $[\ I\ ]$ is a unit matrix and $\{x\}$ is the Laplace transform of $\{y(t)\}$. Therefore, the stability of the system described by Eq. (1) is determined by the eigenvalues λ_j of $[A]$. The eigenvalues and eigenvectors $\{p_j\}$ satisfy

$$([A] - \lambda_j[\ I\ ])\ \{p_j\} = \{0\} \tag{3}$$

where $j = 1, 2, ..., n$. The bi-orthogonal vectors $\{q_i\}$ of $[A]$ safisfy the following adjoint eigenvalue problem:

$$([A]^T - \lambda_i[\ I\ ])\ \{q_i\} = \{0\} \tag{4}$$

The eigenvalues of $[A]$ and $[A]^T$ are same and satisfy the following relation:

$$\lceil\ \lambda\ \rfloor = [Q]^T[A][P] \tag{5}$$

where $[P]$ and $[Q]$ are matrices containing the eigenvectors and bi-orthogonal vectors of $[A]$ as shown below:

$$\left.\begin{array}{l}([P] = [\{p_1\}, \{p_2\}, ..., \{p_j\}, ..., \{p_n\}])\\ ([Q] = [\{q_1\}, \{q_2\}, ..., \{q_i\}, ..., \{q_n\}])\end{array}\right\} \tag{6}$$

Also, the vectors $\{p\}$ and $\{q\}$ are orthogonal, which implies

$$\{q_i\}^T\{p_j\} = 0, \ i \neq j \tag{7}$$

A normalization condition to make the product $[Q]^T[P]$ unique is considered as

$$[Q]^T[P] = [\,I\,] \tag{8}$$

The derivatives of eigendata of matrix [A] can be computed by two methods: (1) adjoint method; and (2) direct method.

1. Adjoint Method

Differentiating Eq. (3) once with respect to a design variable 'd' yields

$$[A]'\{p_j\} + [A]\{p_j\}' = \lambda_j'\{p_j\} + \lambda_j\{p_j\}' \tag{9}$$

Premultiplying Eq. (9) by $\{q_i\}^T$ yields

$$\{q_i\}^T[A]'\{p_j\} + \{q_i\}^T[A]\{p_j\}' = \lambda_j'\{q_i\}^T\{p_j\} + \lambda_j\{q_i\}^T\{p_j\}' \tag{10}$$

Substituting the transpose of Eq. (4) into Eq. (10) yields

$$\{q_i\}^T [A]'\{p_j\} + \lambda_i\{q_i\}^T\{p_j\}' = \lambda_j'\{q_i\}^T\{p_j\} + \lambda_j\{q_i\}^T\{p_j\}' \tag{11}$$

For $i = j$, Eq. (11) by virtue of Eq. (8) reduces to

$$\lambda_j' = \{q_j\}^T [A]'\{p_j\} \tag{12}$$

For a problem with distinct eigenvalues, the associated eigenvectors are linearly independent and hence can be regarded as a basis in the n-dimensional space. Any vector in this space can be expanded in either the set $\{p\}$ or $\{q\}$ and hence the derivative vector of $\{p\}$ can be written as

$$\{p_j\}' = \sum_{i=1}^{n} \varepsilon_{ij} \{p_i\}, \quad j = 1, 2, ..., n. \tag{13a}$$

or

$$[P]' = [P][\varepsilon] \tag{13b}$$

where

$$\varepsilon_{ij} = \{q_i\}^T \{p_j\}' \tag{14}$$

For $i \neq j$, Eq. (11) by virtue of Eq. (8) becomes

$$(\lambda_j - \lambda_i) \{q_i\}^T \{p_j\}' = \{q_i\}^T [A]'\{p_j\} \tag{15}$$

Comparison of Eqs. (14) and (15) yields

$$\varepsilon_{ij} = \{q_i\}^T [A]'\{p_j\}/(\lambda_j - \lambda_i), \quad i \neq j. \tag{16}$$

For $i = j$, Eq. (16) is not valid and ε_{ii} is arbitrary since the eigenvectors $\{p\}$s are not unique. This makes the derivative given in Eq. (13) non-unique to the extent of an additive multiple of $\{p_i\}$. It should be recognized that the normalizing condition given by Eq. (8) does not make the eigenvectors $\{p\}$s unique. Therefore, it is required to normalize the eigenvectors independent of their bi-orthogonal vectors. On such condition, similar to the self-adjoint case, is $\{p_i\}^T \{p_i\} = 1$. However, this condition can lead to numerical difficulties since the elements of the eigenvectors can assume complex values. If the elements are complex, then there exists a possibility of product $\{p_i\}^T \{p_i\}$ being or nearly being zero. Therefore, to alleviate this difficulty, the normalizing condition can be written as

$$p_{ij} = 1, \quad j = 1, 2, ..., n. \tag{17}$$

where i = index of the maximum absolute value of the element of the jth eigenvector. From Eq. (17) p'_{ij} is equal to zero and substituting this result into Eq. (13a) yields

$$\varepsilon_{jj} = -\sum_{\substack{i=1 \\ i \neq j}}^{n} \varepsilon_{ij}\, p_{ji} \tag{18}$$

Similarly, the second-order derivatives of the eigendata are obtained by differentiating Eq. (3) twice and premultiplying the resulting equation by $\{q_i\}^T$:

$$\{q_i\}^T [A]''\{p_j\} + 2\{q_i\}^T [A]'\{p_j\}' + \{q_i\}^T [A]\{p_j\}''$$

$$= \lambda''_j \{q_i\}^T\{p_j\} + 2\lambda'_j \{q_i\}^T\{p_j\}' + \lambda_j\{q_i\}^T\{p_j\}'' \tag{19}$$

Similar to the first-order derivative case, by specializing Eq. (19) for i = j and i ≠ j, the following equations can be obtained for the derivatives of the eigendata:

$$\lambda''_j = \{q_j\}^T [A]''\{p_j\} + 2\{q_j\}^T ([A]' - \lambda'_j[I])\{p_j\}' \tag{20}$$

$$[P]'' = [P][\eta] \tag{21}$$

where

$$\eta_{ij} = (\{q_i\}^T [A]''\{p_j\} + 2\{q_i\}^T [A]'\{p_j\}' - 2\lambda'_j \varepsilon_{ij})/(\lambda_j - \lambda_i)$$

$$i \neq j \tag{22a}$$

$$\eta_{jj} = -\sum_{\substack{i=1 \\ i \neq j}}^{n} \eta_{ij}\, p_{ji} \tag{22b}$$

2. Direct Method

Equation (9) can be arranged into a matrix form as

$$[\{p_j\}, \lambda_j[\,I\,]-[A]\,] \left\{ \begin{array}{c} \lambda'_j \\ \{p_j\}' \end{array} \right\} = [A]'\{p_j\} \qquad (23)$$

$$\text{nx(n+1)} \qquad \text{(n+1)x1} \quad \text{nxn nx1}$$

From Eq. (17), $p'_{ij} = 0$ and substituting this result into the above equation yields

$$[B] \left\{ \begin{array}{c} \lambda'_j \\ \{p_j\}'_i \end{array} \right\} = [A]'\{p_j\} \qquad (24)$$

$$\text{nxn} \quad \text{nx1} \qquad \text{nxn nx1}$$

where $[B] = [\{p_j\}, \lambda_j[\,I\,]-[A]\,]$ with the ith column omitted.

$\{p_j\}'_i = \{p_j\}'$ with the ith column omitted.

Equation (24) yields

$$\left\{ \begin{array}{c} \lambda'_j \\ \{p_j\}'_i \end{array} \right\} = [B]^{-1}[A]'\{p_j\} \qquad (25)$$

The inverse of [B] in the above equation exists if the eigenvalues are non-repetitive. The second derivatives of the eigendata are obtained by differentiating Eq. (24) and solving the resulting equation as shown below:

$$\left\{ \begin{matrix} \lambda''_j \\ \{p_j\}''_i \end{matrix} \right\} = [B]^{-1} \left([A]''\{p_j\} + [A]'\{p_j\}' - [B] \left\{ \begin{matrix} \lambda'_j \\ \{p_j\}'_i \end{matrix} \right\} \right) \qquad (26)$$

In the direct method, the bi-orthogonal vectors are not needed unlike in the case of adjoint method. The relative efficeincy of these methods depends upon the problem parameters [2], such as (i) the size of the matrix, (ii) the number of design parameters, (iii) the number of eigenvalues of interest, (iv) the sparsity of the system matrix and (v) the derivative data of interest, i.e., whether only first order derivatives required or both the first and second order derivatives required.

B. NONLINEARLY DEPENDENT SYSTEMS

The equation of motion of a typical homogeneous aeroelastic system in the Laplace transform domain can be written as

$$[A(s)]\{x(s)\} = \{0\} \qquad (27)$$

where 's' is the Laplace transform variable. The associated eigenvalue problem is written as

$$[A(\lambda_j)]\{p_j\} = \{0\} \qquad (28)$$

These class of problems are sometimes known as the 'nonlinear eigenvalue problems' or 's-dependent systems' since the dependence of the operator A on the eigenvalue, in general, is nonlinear. The eigenvector can be normalized as

$$\{p_j\}^T\{p_j\} = 1 \qquad (29)$$

The variations of Eqs. (28) and (29) are

$$\left(\frac{\partial[A(\lambda)]}{\partial\lambda}\Big|_{\lambda=\lambda_j}\delta\lambda_j + \delta[A(\lambda_j)]\right)\{p_j\} + [A(\lambda_j)]\delta\{p_j\} = \{0\} \tag{30}$$

$$\{p_j\}^T\delta\{p_j\} = 0 \tag{31}$$

Arranging Eqn. (31) and (31) into matrix form yields

$$\begin{bmatrix} \frac{\partial[A(\lambda)]}{\partial\lambda}\Big|_{\lambda=\lambda_j}\{p_j\} & [A(\lambda_j)] \\ 0 & \{p_j\}^T \end{bmatrix} \begin{Bmatrix} \delta\lambda_j \\ \delta\{p_j\} \end{Bmatrix} = - \begin{Bmatrix} \delta[A(\lambda_j)]\{p_j\} \\ 0 \end{Bmatrix} \tag{32}$$

The derivatives of eigendata with respect to a design variable 'd' can be obtained from Eq. (32) as

$$\begin{Bmatrix} \dfrac{\partial\lambda_j}{\partial d} \\ \dfrac{\partial\{p_j\}}{\partial d} \end{Bmatrix} = [B]_j \begin{Bmatrix} \dfrac{\partial[A]}{\partial d}\{p_j\} \\ 0 \end{Bmatrix} \tag{33}$$

where

$$[B]_j = - \begin{bmatrix} \frac{\partial[A(\lambda)]}{\partial\lambda}\Big|_{\lambda=\lambda_j}\{p_j\} & [A(\lambda_j)] \\ 0 & \{p_j\}^T \end{bmatrix}^{-1} \tag{34}$$

The special cases of interest are

   1) Derivatives with respect to an element of the system matrix, then

$$d = a_{mk} = (m, k)\text{th element of } [A]$$

$$\left\{ \begin{array}{c} \dfrac{\partial[A]}{\partial d}\{p_j\} \\[2mm] 0 \end{array} \right\} = \{\ I\ \}_m p_{kj} \qquad (35)$$

where

$\{\ I\ \}_m$ = null column vector except the mth element, which is equal to

unity.

p_{kj} = kth element of $\{p_j\}$.

2) Linearly dependent eigenvalue problem, then the matrix [A] takes the
form

$$\left. \begin{array}{l} [A(\lambda)] = [A] - \lambda[\ I\ ] \\[2mm] \dfrac{\partial[A(\lambda)]}{\partial\lambda} = -\ [\ I\ ] \end{array} \right\} \qquad (36)$$

3) Quadratric eigenvalue problem, this results in

$$\left. \begin{array}{l} [A(\lambda)] = \lambda^2[M] + \lambda[C] + [K] \\[2mm] \dfrac{\partial[A(\lambda)]}{\partial\lambda} = 2\lambda[M] + [C] \end{array} \right\} \qquad (37)$$

III. PERIODIC SYSTEMS

A periodic homogeneous system can be described by

$$\{\dot{x}(t)\} = [A(t)]\{x(t)\} \qquad (38)$$

where the system matrix is of the form

$$[A(t)] = [A(t+T)] \qquad (39)$$

where T = period of the system.

The solution for Eq. (38), whether periodic or not, can be written in terms of the system state transition matrix as

$$\{X(t)\} = [\Phi(t)]\{X(0)\} \tag{40}$$

The transition matrix can be determined by solving

$$\frac{d[\Phi(t)]}{dt} = [A(t)][\Phi(t)] \tag{41}$$

with the initial condition

$$[\Phi(0)] = [\ I\ ] \tag{42}$$

The stability of periodic systems can be determined by the Floquet-Liapunov theory [29], which states that the state transition matrix of a periodic system is of the form

$$[\Phi(t)] = [U(t)]\ e^{[\beta]t} \tag{43}$$

where $\qquad\qquad [U(t) = [U(t+T)]$, periodic matrix $\qquad\qquad$ (44)

$$[\beta] = \text{constant matrix} \tag{45}$$

The transition matrix up to a single period can be written as

$$[\alpha] = [\Phi(T)] = e^{[\beta]T} \tag{46}$$

Equation (46) follows from Eqs. (43), (42) and (44). From the product rule of transition matrices, which states $[\Phi(t_2)] = [\Phi(t_2,t_1)][\Phi(t_1)]$, and from Eqs. (43) and (46), one can write

$$[\Phi(t+mT)] = [\Phi(t)] \, [\alpha]^m, \qquad m = 0, 1, 2, ..., \text{integer.} \qquad (47)$$

If $\lceil \lambda \rfloor$ and $[\mu]$ are the eigenvalue and modal matrices of $[\beta]$, it follows from the matrix theory that

$$[\mu]^{-1}[\beta][\mu] = \lceil \lambda \rfloor \qquad (48)$$

$$[\mu]^{-1}e^{[\beta]t}[\mu] = \lceil e^{\lambda t} \rfloor \qquad (49)$$

Since $[\alpha] = e^{[\beta]T}$ from Eq. (46), it follows from Eq. (49) that $[\mu]$ is also a modal matrix of the Floquet transition matrix $[\alpha]$. The eigenvalues of the Floquet transition matrix are then given by

$$[\mu]^{-1}[\alpha][\mu] = \lceil \Lambda \rfloor \qquad (50)$$

Therefore, it follows from Eqs. (49) and (50)

$$[\mu]^{-1}[\alpha]^m[\mu] = \lceil \Lambda^m \rfloor \qquad (51)$$

$$\lambda = \frac{1}{T} \, Ln\Lambda \qquad (52)$$

From Eqs. (40), (47), (51) and (52), it follows that the solution is unstable if either

$$|\Lambda| > 1 \qquad (53)$$

or

$$Re\lambda > 0 \tag{54}$$

Since the logarithm of a complex variable is a multivalued function, Eq. (52) can be written as a principal part plus integer multiples of $2\pi i/T$ as

$$\lambda = \frac{1}{T}[\ Ln\ |\Lambda| + i\ \angle\Lambda\ ] + \frac{2m\pi i}{T} \tag{55}$$

where   m = any integer.

$$|\Lambda| = \sqrt{(Re\Lambda)^2 + (Im\Lambda)^2}$$

$$\angle\Lambda = \tan^{-1}(\frac{Im\Lambda}{Re\Lambda})$$

Substituting Eqs. (43) and (49) into the transient response given by Eq. (40) yields

$$\{x(t)\} = [U(t)][\mu]\lceil e^{\lambda t}\rfloor[\mu]^{-1}\{x(0)\} \tag{56}$$

Equation (56) can be written as

$$\{x(t)\} = [P(t)]\{y(t)\} \tag{57}$$

where

$$[P(t)] = [U(t)][\mu] \tag{58}$$

$$\{y(t)\} = \lceil e^{\lambda t}\rfloor\{y(0)\} \tag{59}$$

$$\{y(0)\} = [\mu]^{-1}\{x(0)\} \tag{60}$$

The matrix $[P(t)]$ is considered as a periodic modal matrix of the system, and the solution given by Eq. (58) can be written in terms of the eigenvectors as

$$\{x(t)\} = \sum_{j=1}^{n} \{p_j(t)\} \; e^{\lambda_j t} \; y_j(0) \tag{61}$$

Substituting Eqs. (40) and (43) into Eq. (38) yields

$$[\dot{U}(t)] = [A(t)][U(t)] - [U(t)][\beta] \tag{62}$$

Substitution of Eq. (48) into Eq. (62) yields

$$[\dot{U}(t)][\mu] = [A(t)][U(t)][\mu] - [U(t)][\mu][\lceil \lambda \rfloor] \tag{63}$$

Since the modal matrix is defined as $[P(t)] = [U(t)][\mu]$, the differential equation governing the eigenvector $\{p_j(t)\}$ can then be derived from Eq. (63) as

$$\{\dot{p}_j(t)\} = ([A(t)] - \lambda_j[\,I\,])\{p_j(t)\} \tag{64}$$

The homogeneous boundary condition for this eigenvalue problem is the periodic condition given by

$$\{p_j(t)\} = \{p_j(t+T)\} \tag{65}$$

The first order variation of Eq. (64) is given by

$$\delta\{\dot{p}_j(t)\} = (\delta[A(t)] - \delta\lambda_j[\,I\,])\{p_j(t)\} + ([A(t)] - \lambda_j[\,I\,])\delta\{p_j(t)\} \tag{66}$$

Premultiplying Eq. (66) by the transpose of the bi-orthogonal vector $\{q_j(t)\}$ of $\{p_j(t)\}$ yields

$$\{q_j(t)\}^T \delta\{\dot{p}_j(t)\} = \{q_j(t)\}^T (\delta[A(t)] - \delta\lambda_j[\,I\,])\{p_j(t)\}$$

$$+ \{q_j(t)\}^T ([A(t)] - \lambda_j[\,I\,])\delta\{\dot{p}_j(t)\} \qquad (67)$$

The bi-orthogonal matrix $[Q(t)]$ of $[P(t)]$ satisfies

$$[Q(t)]^T[P(t)] = [\,I\,] \qquad (68)$$

The variation of the eigenvector can be expanded as

$$\delta\{p_j(t)\} = \sum_{i=1}^{n} \delta\varepsilon_{ij}(t)\{p_i(t)\}, \quad j = 1, 2, ..., n. \qquad (69)$$

where $\delta\varepsilon_{ij}(t)$ is a periodic function of 't' since both $\delta\{p_j(t)\}$ and $\{p_j(t)\}$ are periodic functions. Substituting Eq. (69) into Eq. (67), after simplification, yields

$$\delta\lambda_j = -\,\delta\dot{\varepsilon}_{jj}(t) + \{q_j(t)\}^T \delta[A(t)]\{p_j(t)\} \qquad (70)$$

or

$$\frac{\partial\lambda_j}{\partial d} = -\frac{\partial\dot{\varepsilon}_{jj}(t)}{\partial d} + \{q_j(t)\}^T \frac{\partial[A(t)]}{\partial d}\{p_j(t)\} \qquad (71)$$

The derivative of the eigenvalue with respect to a design variable 'd' can be defined as

$$\lambda_j' = \frac{1}{T} \int_0^T \frac{\partial \lambda_j}{\partial d} \, d\psi \tag{72}$$

Substituting Eq. (71) into Eq. (72) yields

$$\lambda_j' = \frac{1}{T} \int_0^T \{q_j(t)\}^T \frac{\partial [A(t)]}{\partial d} \{p_j(t)\} \, dt \tag{73}$$

Premultiplying Eq. (66) by the transpose of the bi-orthogonal vector $\{q_i(t)\}$, $i \neq j$, yields

$$\{q_i\}^T \delta\{\dot{p}_j\} = \{q_i\}^T (\delta[A] - \delta\lambda_j[\,I\,])\{p_j\} + \{q_i\}^T ([A] - \lambda_j[\,I\,])\delta\{p_j\} \tag{74}$$

Substituting Eq. (69) into Eq. (74), and simplifying the resulting equation, yields

$$\delta\dot{\varepsilon}_{ij}(t) + (\lambda_j - \lambda_i)\delta\varepsilon_{ij}(t) = \{q_i(t)\}^T \delta[A(t)]\{p_j(t)\} \tag{75}$$

$$i \neq j.$$

The derivative of Eq. (75) with respect to a design variable is given by

$$\frac{d}{dt}\varepsilon_{ij}' + (\lambda_j - \lambda_i)\varepsilon_{ij}' = \{q_i(t)\}^T [A(t)]'\{p_j(t)\} \tag{76}$$

$$i, j = 1, 2, ..., n; \quad i \neq j.$$

The boundary condition for this differential equation (76) is the periodicity condition given by

$$\varepsilon_{ij}'(t) = \varepsilon_{ij}'(t+T), \qquad i \neq j. \qquad (77)$$

The ε_{ii}' can be determined by using the normalizing condition of the eigenvector shown as

$$\int_0^T \{p_i(t)\}^T \{p_i(t)\} \, dt = 1 \qquad (78)$$

The first order variation of Eq. (78) is given by

$$\int_0^T \{p_i(t)\}^T \delta\{p_i(t)\} \, dt = 0 \qquad (79)$$

Substituting Eq. (69) into Eq. (79) yields

$$\varepsilon_{ii}' = - \sum_{\substack{j=1 \\ i \neq j}}^n \int_0^T \{q_i(t)\}^T \varepsilon_{ij}' \{p_j(t)\} \, dt \qquad (80)$$

Finally, the derivative of the modal matrix follows from Eq. (69)

$$[p]' = [P][\varepsilon]' \qquad (81)$$

where the matrix $[\varepsilon]'$ can be computed from Eqs. (76) and (80).

IV.   APPLICATIONS

The sensitivity methods presented in parts II and III are applied to three rotary-wing aeroelastic problems. The problems considered are (1) hovering rotor, (2) rotor blade flapping in forward flight and (3) coupled rotor-body in hover. The first problem is a time-invariant problem while the second gives rise to a periodic system. The third problem leads to a periodic system in the individual blade coordinates while it is a constant system in the multiblade coordinates.

A. HOVERING ROTOR

The hovering rotor equations are time-invariant, and the sensitivity of the aeroelastic stability can conveniently be analyzed by the method developed for the nonlinear eigenvalue systems in Part II.

1. Blade Model

A nonlinear elastic blade model is used in the hovering analysis and the corresponding equations of motion are derived by Hodges and Dowell [30]. The equations of motion by assuming $e_t = B_1^* = B_2^* = C_1^* = 0$, for algebraic simplicity, are

$$-[(GJ + S_x k_A^2)\phi']' + (EI_z - EI_y)[(w''^2 - v''^2)\cos\theta\sin\theta + v''w''\cos2\theta]$$

$$+ m\, k_m^2\ddot{\phi} + m\Omega^2\phi(k_{m2}^2 - k_{m1}^2)\cos2\theta + my_c[\Omega^2 x(w'\cos\theta - v'\sin\theta)$$

$$- (\ddot{v} - \Omega^2 v)\sin\theta + \ddot{w}\cos\theta] + m\Omega^2(k_{m2}^2 - k_{m1}^2)\cos\theta\sin\theta$$

$$+ xmy_c\Omega^2\beta_p\cos\theta = (M_x)^A \qquad (82)$$

$$- (S_x v')' + \{[EI_z \cos^2(\theta + \phi) + EI_y \sin^2(\theta + \phi)]v''$$

$$+ (EI_z - EI_y)w'' \cos(\theta + \phi)\sin(\theta + \phi)\}'' + m\ddot{v} - \ddot{\phi}my_c \sin\theta$$

$$- 2my_c \Omega(\dot{v}'\cos\theta + \dot{w}\sin\theta) - m\Omega^2[v + y_c \cos(\theta + \phi)]$$

$$+ 2(\dot{u} - \beta_p \dot{w})\Omega m - \{my_c[x\Omega^2 \cos(\theta + \phi) + 2\Omega\dot{v}\cos\theta]\}'$$

$$= (F_y)^A \tag{83}$$

$$- (S_x w')' + \{[EI_z \sin^2(\theta + \phi) + EI_y \cos^2(\theta + \phi)]w''$$

$$+ v''(EI_z - EI_y)\cos(\theta + \phi)\sin(\theta + \phi)\}'' + m\ddot{w} + \ddot{\phi}my_c \cos\theta$$

$$+ 2m\Omega\beta_p\dot{v} + m\Omega^2 x\beta_p - \{my_c[x\Omega^2 \sin(\theta + \phi) + 2\Omega\dot{v}\sin\theta]\}'$$

$$= (F_z)^A \tag{84}$$

where

$$S_x = \frac{1}{2}m\Omega^2(R^2 - x^2) + 2m\Omega \int_x^R \dot{v}\, dx$$

$$\tag{85}$$

and

$$\dot{u} = - \int_x^R (v'\dot{v} + w'\dot{w})\, dx$$

If two dimensional quasi-steady aerodynamic theory is employed [31], then

$$(F_y)^A = k_2\{v_i^2 - \Omega^2 x^2 C_{d_a} - \Omega x v_i(\theta + \phi) - [2\Omega x k_3 + v_i(\theta + \phi)]\dot{v}$$

$$+ [2v_i - \Omega x(\theta + \phi)]\dot{w} - k_1 v_i(\dot{v} + w' + \beta_p)$$

$$- \Omega k_1 \dot{w}(w' + \beta_p)\} \tag{86}$$

$$(F_z)^A = k_2\{\Omega^2 x^2(\theta + \phi + v) - \Omega x v_i - \Omega^2 x v(\beta_p + w')$$

$$+ \Omega^2 x k_1(\beta_p + w') + [2\Omega x(\theta + \phi) - v_i - \Omega k_1(w' + \beta_p)]\dot{v}$$

$$- x\Omega(\dot{w} + k_1\dot{\phi})\} \tag{87}$$

$$(M_x)^A = e_a(F_z)^A - k_2 c^2 [x\Omega\dot\phi + \Omega^2 x(\beta_p + w') + \Omega\dot{v}(w' + \beta_p)]/16$$

$$(88)$$

where

$$k_1 \equiv \frac{1}{2}c - e_a = b - e_a, \qquad k_2 = \rho ac/2, \qquad k_3 = C_d/a$$

$$v_i = \pi\sigma\Omega R(\sqrt{1+12(\theta + \phi)/\pi\sigma} - 1)$$

$$v = \int_0^x v'w'' \, dx$$

w = Flapwise deflection;   v = Chordwise deflection

ϕ = Torsional deflection; θ = Blade pretwist + Collective

u = Axial deflection;   b = Semi-chord

c = Chord;   R = Rotor radius

σ = Solidity $Nc/\pi R$;   N = Number of blades

β_p = Precone;   m = Mass of blade per unit length

C_d = Drag coefficient;   a = Lift-curve slope

e_a = Chordwise offset between aerodynamic center (A.C.) and elastic
     axis (or feathering axis) of blade, positive for A.C. forward

y_c = Chordwise offset between center of gravity (C.G.) and elastic
     axis (or feathering axis) of blade, positive for C.G. forward

x = Blade spanwise independent variable

I_y, I_z = Blade area moments of inertia for flapwise and chordwise
     bendings, respectively

k_A = Polar radius of gyration of blade cross-section about the elastic
     axis

k_{m1}, k_{m2} = Mass polar radii of gyration about major neutral axis and about an axis perpendicular to the chord through the elastic axis, respectively

2. Stability Model

The first task in the stability analysis is the calculation of the trim state. Then the blade equations are linearized for small perturbations about this trim state. The blade motions can be expressed in terms of the trim quantities (v_0, w_0, ϕ_0) and the perturbation variables (Δv, Δw, $\Delta \phi$) as

$$\left. \begin{array}{l} v(t,x) = v_0 + \Delta v \\ w(t,x) = w_0 + \Delta w \\ \phi(t,x) = \phi_0 + \Delta \phi \end{array} \right\} \tag{89}$$

Substituting Eq. (89) into Eqs.(82)-(84) and neglecting the nonlinear perturbation terms yield

$$-[(GJ + S_x k_A^2)\phi']' + (EI_z - EI_y)[(w_0'')\sin2\theta_t + v_0''\cos2\theta)w''$$

$$+ (w_0'')\cos2\theta + v_0''\sin2\theta)v''] + m k_m^2 \ddot{\phi} + m\Omega^2\phi(k_{m2}^2 - k_{m1}^2)\cos2\theta$$

$$+ my_c[\Omega^2x(w'\cos\theta - v'\sin\theta) - (\ddot{v} - \Omega^2v)\sin\theta + \ddot{w}\cos\theta] = (M_x)^A \tag{90}$$

$$- (S_x v')' + \{[EI_z\cos^2(\theta + \phi_0) + EI_y\sin^2(\theta + \phi_0)]v''$$

$$+ [(EI_z - EI_y)v_0''\sin^2(\theta + \phi_0) + (EI_z - EI_y)w_0''\cos2(\theta + \phi_0)]\phi$$

$$+w''(EI_z - EI_y)\cos(\theta + \phi_0)\sin2(\theta + \phi_0)\}'' - m\Omega^2[v - \phi y_c\cos(\theta + \phi)]$$

$$+ my_c[\phi x\Omega^2\sin(\theta + \phi_0)]' = (F_y)^A - m\ddot{v} - \ddot{\phi}my_c\sin\theta$$

$$- 2my_c\Omega(\dot{v}'\cos\theta + \dot{w}'\sin\theta) + 2\beta_p m\Omega \, \dot{w} + 2\Omega(y_c \dot{v}\cos\theta)'$$

$$+ 2m \left(\int_0^x (v_0'\dot{v} + w_0'\dot{w}) \, dx + \Omega v_0' \int_x^R \dot{v} \, dx \right)' \tag{91}$$

$$- (S_x w')' + \{[EI_z\sin^2(\theta + \phi_0) + EI_y\cos^2(\theta + \phi_0)]w''$$

$$+ (EI_z - EI_y)v''\cos(\theta + \phi_0)\sin(\theta + \phi_0) + (EI_z - EI_y)[v_0''\cos2(\theta + \phi_0)$$

$$+ w_0''\sin2(\theta + \phi_0)]\phi\}'' + m\ddot{w} + \ddot{\phi}my_c\cos\theta + 2m\Omega\beta_p\dot{v}$$

$$+ m\Omega^2 x\beta_p - \{my_c[x\phi\Omega^2\cos(\theta + \phi_0) + 2\Omega\dot{v}\sin\theta]\}'$$

$$= (F_z)^A + 2m\Omega(w_0' \int_x^R \dot{v} \, dx)' \tag{92}$$

where the perturbation quasi-steady aerodynamic forces are given by

$$(M_x)^A = e_a (\Delta F_z)^A - k_2 c^2[x\Omega\dot{\phi} + \Omega^2 xw' + \Omega\dot{v}(w_0' + \beta_p)]/16$$

$$(F_y)^A = k_2\{ - \Omega xv_i\phi - [2\Omega xk_3 + v_i(\theta + \phi_0)]\dot{v}$$

$$+ [2v_i - \Omega x(\theta + \phi_0)]\dot{w} - k_1 v_i (\dot{\phi} - w') - \Omega k_1 \dot{w}(w_0' + \beta_p)\}$$

$$(F_z)^A = k_2\{\Omega^2 x^2\phi + v - \Omega^2 xv(\theta + w_0') - (\Omega^2 xv_0 - \Omega^2 xk_1)w'$$

$$+ [2\Omega x(\theta + \phi_0) - v_i - \Omega k_1(w_0' + \beta_p)]\dot{v} - x\Omega(\dot{w} - k_1\dot{\phi}) \}$$

$$\tag{93}$$

where $\quad v = \int_0^x v_0'w'' \, dx + \int_0^x w_0'' v' \, dx$

The perturbation unsteady aerodynamic forces are given by [31]

$$(M_x)^A = C^* e_a (F_z)^A \big|_q - \frac{1}{8} k_2 c^2 \{ (\frac{1}{2} - \frac{e_a}{b}) [\ddot{w} - \ddot{v}(\theta + \phi_0)] + (\frac{e_a}{b} - (\frac{e_a}{b})^2 $$
$$ - \frac{3}{8} c) \ddot{\phi} b \} - \frac{1}{16} k_4 c^2 [x\Omega \dot{\phi} + x\Omega^2 w' + \Omega(w'_0 + \theta)\dot{v}] $$
$$ - \frac{1}{8} k_2 c^2 [\frac{e_a}{b} \Omega(w'_0 + \theta)\dot{v} + x(\frac{1}{2} - \frac{e_a}{b})\Omega\dot{\phi} - \frac{1}{2}\Omega(w'_0 + \theta)\dot{v}] $$

$$(F_y)^A = C^* (F_y)^A \big|_q + \frac{c}{4} k_2 (\theta + \phi_0) [\ddot{w} - (\frac{b}{2} - e_a)\ddot{\phi}] - \frac{1}{2} xk_2 \Omega \dot{\phi}(\theta + \phi) $$

$$(F_z)^A = C^* (F_y)^A \big|_q - \frac{c}{4} k_2 [\ddot{w} - (\theta + \phi_0) \ddot{v} $$
$$ - (\frac{b}{2} - e_a)\ddot{\phi} + \Omega(\theta + \phi_0)\dot{v} - x\Omega\dot{\phi}] \qquad (94) $$

where $(F_y)^A \big|_q$ and $(F_z)^A \big|_q$ are quasi-steady aerodynamic forces given by Eq. (93) and C^* is the generalized Theodorsen function. In Eqs. (90)-(94), all Δ notations in front of the perturbation variables and forces are dropped for the sake of simplicity.

Equations (90)-(94) are linear partial differentio-integral equations, and they are reduced to a linear ordinary differential form by a modal method. This procedure starts with

$$\{X(t, x)\} = [P(x)]\{\eta(t)\} \qquad (95)$$

where

$$\{X(t,x)\} \equiv \left\{ \begin{matrix} \phi(t,x) \\ v(t,x) \\ w(t,x) \end{matrix} \right\} \qquad (96)$$

The matrix $[P(x)]$ is a modal matrix of the undamped free vibration, and is obtained by solving Eqs. (90)-(92) after the first order time derivative and the aerodynamic force terms are dropped. These free vibration equations are of the form

$$[L_K]\{X(t,x)\} = -[L_M]\{\ddot{X}(t,x)\} \tag{97}$$

where $[L_K]$ and $[L_M]$ are matrix operators. The orthogonality between the modal functions is

$$\int_0^R [P(x)]^T [L_M][P(x)]dx = [M] \tag{98}$$

where

$\qquad [M] = $ constant diagonal matrix

$\qquad [P(x)] \equiv [\{p_1(x)\}, \{p_2(x)\}, ..., \{p_n(x)\}]$

$\qquad n = $ number of modes used

$$\{p_i(x)\} \equiv \begin{Bmatrix} \phi_i(t,x) \\ v_i(t,x) \\ w_i(t,x) \end{Bmatrix}$$

Substituting Eqs. (93)-(98) into Eqs. (90)-(92) yields

$$[M]\{\ddot{\eta}(t)\} + [C]\{\dot{\eta}(t)\} + [K]\{\eta(t)\} = \{0\} \tag{99}$$

The matrices $[M]$, $[C]$ and $[K]$ in Eq. (99) are given in Appendix A. The eigenvalues of Eq. (99), which provide the stability information, satisfy

$$[A(\lambda)]\{\eta\} = \{0\} \tag{100}$$

where

$$[A(\lambda)] = \lambda^2[M] + \lambda[C] + [K]$$

3. Sensitivity

The stability of the helicopter rotor in hovering flight, which is described by Eqs. (82)-(88), is examined by solving the eigenvalue problem given by Eq. (100). The stability equation, Eq. (100), is of the form given by Eq. (28) and hence its sensitivity can be determined from Eq. (33). The blade properties for numerical calculations are taken from Ref. [32] and presented in Table I. The first five vibration mode shapes are used in the stability and sensitivity analyses. These five modes include three flapping modes, one chordwise bending mode and one torsional mode. The trim state and the natural frequencies and mode shapes about the initial and trim states are computed by using the transfer matrix method and the results are presented in Table II. The stability eigenvalues and their derivatives computed by the present approach are presented in Table III. The following trend is evident from the results presented in this table.

a)   The precone angle (β_p) affects mainly the first flap and the lead-lag motion. Since the real part of $\partial\lambda_2/\partial\beta_p \approx 3.8$ sec.$^{-1}$, the precone has a strong destabilizing effect on the lead-lag mode.

b)   An increase in the pretwist (θ_{tw}) improves the lead-lag mode stability, since $Re(\partial\lambda_2/\partial\theta_{tw}) \approx -0.65$ sec.$^{-1}$, but degrades the stability of the first flapping slightly.

c)   The chordwise offset of the aerodynamic center from the elastic axis has a significant effect on the torsional mode and an instability will occur in this mode with an increase of this offset.

d)   An increase in the rotor speed improves the stability of the system.

Table I. Properties of Hovering Rotor

Parameter	Value
Rotor radius	40 in.
Collective pitch (θ_0)	0.3 rad.
Blade pretwist (θ_{tw})	0.0
Blade precone ((β_p)	0.0
Chord (c)	π in.
Mass per unit length (m)	1.9737 lb-sec.^{2}/in.^{2}
Flapwise bending stiffness (EI_y)	4.6736×10^6 lb-in.^{2}
Chordwise bending stiffness (EI_z)	5.3427×10^6 lb-in.^{2}
Torsional stiffness (GJ)	1.8154×10^6 lb-in.^{2}
Chordwise offset between aerodynamic center and elastic axis (e_a)	0.0
Chordwise offset between C. G. and elastic axis (y_c)	0.0
Lift-curve slope (a)	2π rad^{-1}
Drag coefficient (C_d)	0.01
Rotational speed (Ω)	$25/\pi$ sec-1
Polar Radius of gyration squared of the blade cross-section about the elastic axis (k_A^2)	1.5 in.^{2}
Mass polar moment of inertia (mk_m^2)	1.9737 lb-sec.^{2}
(mk_{m1}^2)	0.0

e)   A decrease in the Lock number degrades the stability of all modes [Lock number $\gamma = \rho a c R^4 / I_b$].

f)   The unsteady aerodynamic parameter, C^*, improves the system stability due to inertial and lag effects of the unsteady aerodynamics.

Table II. Trim State and Natural Frequencies

Mode	Natural Frequencies (rad./sec.)	
	Initial State	Trim State[*]
I Flap (ω_1)	9.15760	8.50100
I Lag (ω_2)	11.9370	12.0340
II Flap (ω_3)	29.3043	28.9906
I Torsion (ω_4)	39.7896	40.3551
III Flap (ω_5)	68.1363	67.5876

[*] Trim states:

$(\phi_0)_{tip} = -0.04143$, $(v_0/R)_{tip} = -0.005048$, $(w_0/R)_{tip} = 0.01153$.

B. ROTOR BLADE FLAPPING

The dynamics of an isolated rotor blade flapping in forward flight is governed by a periodic equation. The equation of motion is given by [33]

$$\ddot{\beta}(\psi) + \frac{\gamma}{8}(1 + \frac{4}{3}\mu\sin\psi)\dot{\beta}(\psi) + (v_\beta^2 + \frac{\gamma}{8}\mu(\frac{4}{3}\cos\psi + \mu\sin2\psi))\beta(\psi) = 0$$

(101)

Equation (101) in a state vector form is given by

$$\{\dot{x}(\psi)\} = [A(\psi)] \{x(\psi)\} \qquad (102)$$

where

$$\{x(\psi)\} \equiv \lfloor \beta(\psi), \dot{\beta}(\psi) \rfloor^T$$

Table III. Eigenvalues and Derivatives

λ_i	-2.28 + 7.693i		-0.4468 + 12.5757i		-2.1164 + 28.925i		-2.815 + 40.223i		-1.8538 + 67.5757i	
Parameter	λ'_1		λ'_2		λ'_3		λ'_4		λ'_5	
	Re	Im	Re	Im	Re	Im	Re	Im	Re	Im
β_p	-3.4523	-3.3030	3.81910	2.82520	-.29243	-.17744	-.07745	-.05158	0.00152	-.05361
θ	0.26858	0.52444	-.65843	-.66643	-.51417	-.09343	0.01650	0.00344	0.51043	-.20430
e_a	-.30144	-.17577	-.00744	0.03241	-.01744	-.01857	0.88852	-2.0375	-.00645	-.00115
Ω	-.22525	-.08443	-.09855	0.10734	-.24936	0.00329	-.18541	-.03044	-.22154	0.00456
γ	-.38818	-.09042	-.14743	-.00404	-.41402	-.02244	-.29244	-.01642	-.36754	-.00653
ω_1	-.01352	1.22310	-.00845	-.09810	0.00005	0.00003	0.00004	0.00032	0.00004	0.00001
ω_2	0.01444	-.69452	0.01010	1.01880	0.00000	0.00002	0.00152	0.00051	0.00029	0.00001
ω_3	0.00024	0.00001	0.00001	0.00028	-.00100	1.00260	0.00035	0.00003	0.00000	0.00008
ω_4	0.00001	0.00007	-.00140	0.00000	0.00000	0.00002	0.00155	1.00260	0.00003	0.00014
ω_5	0.00000	0.00000	0.00010	0.00002	0.00000	0.00001	0.00021	0.00005	0.00029	0.99987
EI_y	-.08001	6.87040	-.04147	-.53845	0.04446	85.6010	0.03651	0.01346	-.01851	253.510
EI_z	0.06144	-2.9725	0.04150	4.36010	0.00044	0.00148	0.00022	0.00000	0.00000	0.00000
GJ	0.07540	-.18655	-.39744	0.02544	0.14043	-.05146	0.59045	408.580	0.00017	0.00004
C^*	-1.9062	-.59615	-.77301	0.06811	-2.0712	-.12619	-.02103	-.03122	-1.8402	-.04013

$$[A(\psi)] = \begin{bmatrix} 0 & 1 \\ -a_1(\psi) & -a_2(\psi) \end{bmatrix}$$

$$a_1(\psi) \equiv v_\beta^2 + \frac{\gamma}{8} \mu(\frac{4}{3} \cos\psi + \mu\sin2\psi)$$

$$a_2(\psi) \equiv \frac{\gamma}{8} (1 + \frac{4}{3} \mu \, \sin\psi)$$

Equation (102) is of the form of Eq. (38) and its stability characteristics can be examined by the eigenvalues of its Floquet transition matrix. The stability sensitivity can be computed from Eqs. (73) and (81).

The eigenvalues of a time-invariant system are unique, and its modal matrix will be unique if a normalization condition is specified. It is to be noted that the eigenvalues of a periodic system are nonunique, but can differ from one another by integer multiples of $2\pi i/T$ as shown in Eq. (55). This does not make any difference in the unique determination of the solution of the problem. From Eq. (61), it can be seen that the actual response of the system depends upon the product $\{p(t)\}e^{\lambda t}$. If an integer multiple of the fundamental frequency, $2\pi i/T$, is added to the eigenvalue, λ, then to keep the response product $\{p(t)\}e^{\lambda t}$ unchanged, the vector $\{p(t)\}$ should be multiplied by the periodic function $e^{-2\pi ni/T}$. One can see, from Eq. (65), that the Floquet theory requires only the eigenvector be periodic, but never restricts the allotment of periodicity between the eigenvalue and its eigenvector. The expansion of $\{p(t)\}e^{\lambda t}$ as a Fouries series yields

$$\{p(t)\}e^{\lambda t} = \sum_{n=-\infty}^{n=\infty} \{p_n\}e^{\lambda+2\pi ni/T} \qquad (103)$$

It follows, from Eq. (103), that the natural vibration of periodic system contains the principal eigenvalue plus or minus all integer multiples of the fundamental frequency, ω_0 (= $2\pi/T$), as shown below:

$$\lambda_{j,a} = \lambda_{j,p} + \omega_0 ki, \qquad j = 1, 2, ..., n. \qquad (104)$$
$$k = \pm 1, \ \pm 2, \ ..., \ \pm\infty.$$

The Fourier coefficients $\{p_n\}$ in Eq. (103) give the information as to how much of each harmonic will occur in the total natural vibration. The dominant frequency in the natural vibration is usually identified by the frequency content of the eigenvector. The principal eigenvalue, λ_p, and the associated principal eigenvector can be determined uniquely, and the harmonic of the largest magnitude in this eigenvector gives the dominant frequency. In helicopter problems, however, the dominant frequency can usually be determined also from the requirement that the roots be continuous as the periodicity drops out in the limit of zero advance ratio. The following observations can be made from the results of the sensitivity formulation.

1. The eigenvalue derivatives of periodic systems, as given by Eq. (73), are unique despite the non-uniqueness of $\{p_k(t)\}$, since the non-uniqueness in the eigenvector and its bi-orthogonal negate each other.

2. Different normalization conditions, such as the one given by Eq. (78), yield different eigenvector derivatives, which all will be periodic.

3. It is to be noted that the eigenvalues are continuous functions of the system parameters and therefore their derivatives always exist.

4. For periodic systems, three types of principal eigenvalues can result: a complex conjugate pair, or a single root, or a root with a 1/2 per revolution imaginary part. It is to be noted that the integer multiples of the fundamental frequencies can always be added to the principal eigenvalues. This can give rise to a frequency-locked instability at a frequency

that is an integer multiple of one-half of the fundamental frequency. The derivative of the eigenvalue developed in the present formulation, as given by Eq. (73), is valid in all the regions including the frequency-locked. However, the regular perturbation expansion of the eigenvector variation, as given by Eq. (69), breaks down when the frequency lock occurs.

C. COUPLED ROTOR-BODY IN HOVER

A simple coupled rotor-body problem in hover, as shown in Fig.1, is considered here. This problem provides an excellent means to validate the formulation since it yields a periodic system in the individual blade coordinate system while yielding a constant system in the multiblade coordinates. The helicopter body is assumed to be rigid with pitch and roll degrees-of-freedom. A three-bladed rotor with flapping motion restrained by linear elastic spring, and quasi-steady aerodynamics, is used in the derivation of equations of motion. The resulting equations are

$$[M(\psi)] \{\ddot{y}(\psi)\} + [C(\psi)] \{\dot{y}(\psi)\} + [K(\psi)] \{y(\psi)\} = \{0\} \qquad (105)$$

Here, the flapping motion is described in individual blade coordinates which makes some terms in the coefficient matrices periodic. This can be seen in the coefficient matrices given in Appendix B. Equation (105) in a state vector form becomes

$$\{\dot{x}(\psi)\} = [A(\psi)] \{x(\psi)\} \qquad (106)$$

where

$$\{x(\psi)\} \equiv \lfloor \{y(\psi)\}, \{\dot{y}(\psi)\} \rfloor^T$$

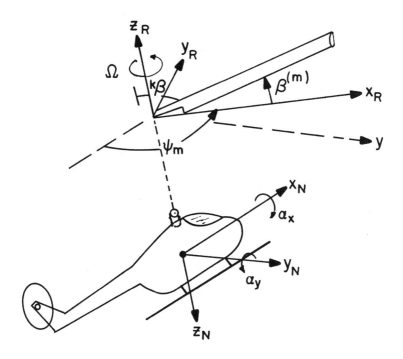

Fig. 1   Coupled rotor body model.

$$[A(\psi)] \equiv \begin{bmatrix} [0] & [\ I\ ] \\ -[M]^{-1}[K] & -[M]^{-1}[C] \end{bmatrix}$$

The periodicity in Eq. (106) can be eliminated by employing multiblade coordinates [34]. The resulting time-invariant equation in the multiblade coordinates is given by

$$[M^*]\ \{\ddot{y}^*(\psi)\} + [C^*]\ \{\dot{y}^*(\psi)\} + [K^*]\ \{y^*(\psi)\} = \{0\} \qquad (107)$$

where $$\{y^*(\psi)\} = [T(\psi)]\ \{y(\psi)\}$$

Table IV. Nondimensional Stability Eigenvalues of the Coupled
Rotor-Body Problem. $\gamma = 5$; $\nu_\beta = 1.1$; $I_p = 3$; $I_q = 4$.

Periodic System		Equivalent Constant System	
Re	Im	Re	Im
-.31247	-1.0547	-.31250	-1.0547
-.31247	1.0547	-.31250	01.0547
-.30576	-2.0548	-.30571	-2.0549
-.30576	2.0548	-.30571	2.0549
-.16683	-.21722	-.16737	-.22074
-.16683	0.21722	-.16737	0.22074
-.15235	-.29353	-.15192	-.29417
-.15235	0.29353	-.15192	0.29417

In the state vector form, Eq. (107) becomes

$$\{\dot{x}^*(\psi)\} = [A^*]\ \{x^*(\psi)\} \qquad (108)$$

where

$$\{x^*(\psi)\} \equiv \lfloor\{y^*(\psi)\}, \{\dot{y}^*(\psi)\}\rfloor^T$$

$$[A^*] \equiv \begin{bmatrix} [0] & [I] \\ -[M^*]^{-1}[K^*] & -[M^*]^{-1}[C^*] \end{bmatrix}$$

The matrices $[T(\psi)]$, $[M^*]$, $[C^*]$ and $[K^*]$ are also given in Appendix B.

The eigenvalues computed from the periodic description as well as its equivalent constant system are presented in Table IV. The derivatives of eigenvalues with respect to several system parameters are presented in Table V. The results are compared with those obtained by using a finite difference method. The results presented in this table clearly validate the present formulation.

V. SUMMARY

The derivatives of eigendata play a key role in the design of systems with stability constraints. Methods to determine these derivatives for time-invariant as well as periodic systems are presented in this article. The methods are applied to determine the sensitivity derivatives of the stability eigenvalues of three rotary-wing aeroelastic problems. The first example is selected to demonstrate the application of the nonlinear eigenvalue derivative formulation to a complex aeroelastic system. The second example dealing with the isolated rotor-blade flapping in forward flight is considered to bring out the important features of the formulation for periodic systems. Finally, the coupled rotor-body problem is selected to demonstrate the validation of the formulation.

Table V. Derivatives of Eigenvalues

	$\dfrac{\partial \lambda_k}{\partial \gamma}$		$\dfrac{\partial \lambda_k}{\partial \nu_\beta}$		$\dfrac{\partial \lambda_k}{\partial I_p}$		$\dfrac{\partial \lambda_k}{\partial I_q}$	
	Re	Im	Re	Im	Re	Im	Re	Im
	-.62500E-01	0.18519E-01	0.11269E-17	-.10430E+01	-.22308E-20	0.27105E-19	0.25451E-20	0.54210E-19
	-.62500E-01	-.18519E-01	0.12319E-18	0.10430E+01	0.84863E-21	0.11796E-31	0.35452E-20	-.67763E-19
	-.61087E-01	0.19302E-01	0.64842E-01	-.10671E+01	-.13293E-02	0.36493E-04	-.70222E-03	0.21427E-04
PDM	-.61087E-01	-.19302E-01	0.64842E-01	0.10671E+01	-.13293E-02	-.36493E-04	-.70222E-03	-.21427E-04
	-.36670E-01	0.10860E-01	-.28623E-02	-.13169E+01	-.12874E-02	0.95663E-02	0.51299E-02	0.34971E-01
	-.36670E-01	-.10860E-01	-.28623E-02	0.13169E+01	-.12874E-02	-.95663E-02	0.51299E-02	-.34971E-01
	-.27244E-01	0.24368E-01	-.61980E-01	-.22608E+01	0.26167E-02	0.51521E-01	-.44277E-02	0.57736E-02
	-.27244E-01	-.24368E-01	-.61980E-01	0.22608E+01	0.26167E-02	-.51521E-01	-.44277E-02	-.57736E-02
	-.62497E-01	0.18513E-01	0.10502E-03	-.10429E+01	0.20817E-13	0.27756E-13	0.69389E-14	0.00000E+00
	-.62497E-01	-.18513E-01	0.10502E-03	0.10429E+01	0.20817E-13	-.27756E-13	0.69389E-14	0.00000E+00
	-.61085E-01	0.19298E-01	0.64874E-01	-.10672E+01	-.13280E-02	0.37163E-04	-.70160E-03	0.21756E-04
FDM	-.61085E-01	-.19298E-01	0.64874E-01	0.10672E+01	-.13280E-02	-.37163E-04	-.70160E-03	-.21756E-04
	-.36669E-01	0.10861E-01	-.23014E-02	-.13107E+01	-.12859E-02	0.95709E-02	0.51275E-02	0.34968E-01
	-.36669E-01	-.10861E-01	-.23014E-02	0.13107E+01	-.12859E-02	-.95709E-02	0.51275E-02	-.34968E-01
	-.27240E-01	0.24364E-01	-.62338E-01	-.22553E+01	0.26135E-02	0.51499E-01	-.44261E-02	0.57683E-02
	-.27240E-01	-.24364E-01	-.62338E-01	0.22553E+01	0.26135E-02	-.51499E-01	-.44261E-02	-.57683E-02
	-.62500E-01	0.18519E-01	0.13878E-16	-.10430E+01	0.46954E-36	-.28522E-34	0.36833E-35	-.24482E-34
	-.62500E-01	-.18519E-01	-.68090E-33	0.10430E+01	0.46954E-36	0.28522E-34	0.36833E-35	0.24482E-34
	-.61072E-01	0.19311E-01	0.65536E-01	-.10674E+01	-.13292E-02	0.36896E-04	-.73837E-03	0.22613E-04
ECS	-.61072E-01	-.19311E-01	0.65536E-01	0.10674E+01	-.13292E-02	-.36896E-04	-.73837E-03	-.22613E-04
	-.37011E-01	0.10186E-01	0.73059E-03	-.13134E+01	-.14021E-02	0.10577E-01	0.55215E-02	0.35361E-01
	-.37011E-01	-.10186E-01	0.73059E-03	0.13134E+01	-.14021E-02	-.10577E-01	0.55215E-02	-.35361E-01
	-.26918E-01	0.24618E-01	-.66266E-01	-.22747E+01	0.27313E-02	0.50526E-01	-.47831E-02	0.66749E-02
	-.26918E-01	-.24618E-01	-.66266E-01	0.22747E+01	0.27313E-02	-.50526E-01	-.47831E-02	-.66749E-02

PDM = Present direct method; FDM = Finite difference method; ECS = Equivalent constant system.

APPENDIX A.   EQUATIONS OF MOTION OF HOVERING ROTORS

The equation of motion governing the hovering rotor considered is given by Eq. (99) as

$$[M]\{\ddot{\eta}(t)\} + [C]\{\dot{\eta}(t)\} + [K]\{\eta(t)\} = \{0\}$$

where $[M] = \int_0^R [P(x)]^T [L_M] [P(x)] \, dx$

$$[L_M] = m \begin{bmatrix} k_m^2 & -y_c\sin\theta & y_c\cos\theta \\ -y_c\sin\theta & 1 & 0 \\ y_c\sin\theta & 0 & 1 \end{bmatrix}$$

$$[C] = [C_1] + [C_2] + [C_3] + [C_4] + [C_5]$$

$$[K] = [M][\omega^2] + [K_1] + [K_2] + [K_3]$$

where

$$[K_1] = -\int_0^R [P(x)]^T[D_1][P(x)] \, dx \,, \quad [K_2] = -\int_0^R [P(x)]^T[D_3][P(x)] \, dx$$

$$[K_3] = -\int_0^R [P(x)]^T[F_1(x)] \, dx \,, \quad [C_1] = -\int_0^R [P(x)]^T[D_2(x)][P(x)] \, dx$$

$$[C_2] = -\int_0^R [P(x)]^T[D_6(x)][P(x)]' \, dx, \quad [C_3] = -\int_0^R [P(x)]^T[D_7(x)][P(x)] \, dx$$

$$[C_4] = - \int_0^R [P(x)]^T [F_2(x)] \, dx$$

$$[C_5] = - 2m\Omega \int_0^R [P(x)]^T [D_9(x)]([D_{10}(x)] - [D_{11}(x)]) \, dx$$

$$[F_1(x)] = - e_a\Omega^2 \int_0^x x_1([D_4(x_1)][P(x_1)]' + [D_5(x_1)][P(x_1)]) \, dx_1$$

$$[F_2] = \int_0^x [D_8(x_1)][P(x_1)]' \, dx_1$$

$$[\omega^2] = \begin{bmatrix} \omega_1^2 & 0 & 0 & 0 & 0 \\ 0 & \omega_2^2 & 0 & 0 & 0 \\ 0 & 0 & \omega_3^2 & 0 & 0 \\ 0 & 0 & 0 & \omega_4^2 & 0 \\ 0 & 0 & 0 & 0 & \omega_5^2 \end{bmatrix}$$

$$[D_1] = \begin{bmatrix} e_a k_2 \Omega^2 x & -k_3 k_2 \Omega^2 x(w_0' + \beta_p) & 0 \\ -k_2 \Omega^2 x v_i & 0 & 0 \\ k_2 \Omega^2 x & -k_2 \Omega^2 x(w_0' + \beta_p) & 0 \end{bmatrix}$$

$$[D_2] = \begin{bmatrix} d_{11} & d_{12} & d_{13} \\ d_{21} & d_{22} & d_{23} \\ d_{31} & d_{32} & d_{33} \end{bmatrix}$$

$$d_{11} = e_a k_2 \Omega x - k_2 \Omega x c^2 / 16$$

$$d_{12} = -e_a k_2 [2\Omega x(\theta + \phi_0) - v_i + \Omega k_1(w_0' + \beta_p)] - c^2 k_2 \Omega (w_0' + \beta_p)/16$$

$$d_{13} = -e_a k_2 \Omega x, \qquad\qquad d_{21} = -k_1 k_4 v_i$$

$$d_{22} = -k_2[2\Omega x k_3 + (\theta + \phi_0)v_i]$$

$$d_{23} = -k_2[2v_i\Omega x(\theta + \phi_0) - v_i - k_1(w_0' + \beta_p)], \qquad d_{31} = k_1 k_2 \Omega x$$

$$d_{32} = k_2[2\Omega x(\theta + \phi_0) - v_i + \Omega k_1(w_0' + \beta_p)], \qquad d_{33} = -k_2\Omega x$$

$$[D_3] = \begin{bmatrix} 0 & 0 & e_a k_2\Omega^2 x(k_1-v_0)-k_2\Omega^2 x^2 c/16 \\ 0 & 0 & -k_1 k_2\Omega v_i \\ 0 & 0 & k_2\Omega^2 x(k_1-v_0) \end{bmatrix}, \qquad [D_4] = \begin{bmatrix} 0 & e_a w_0'' & 0 \\ 0 & 0 & 0 \\ 0 & w_0'' & 0 \end{bmatrix}$$

$$[D_5] = \begin{bmatrix} 0 & 0 & e_a v_0' \\ 0 & 0 & 0 \\ 0 & 0 & v_0' \end{bmatrix}, \qquad\qquad [D_6] = m e_a \Omega \begin{bmatrix} 0 & 0 & 0 \\ 0 & 4\cos\theta & 2\sin\theta \\ 0 & -2\sin\theta & 0 \end{bmatrix}$$

$$[D_7] = 2m\beta_p\Omega \begin{bmatrix} 0 & 0 & 0 \\ 0 & 0 & 1 \\ 0 & -1 & 0 \end{bmatrix}, \qquad\qquad [D_8] = \begin{bmatrix} 0 & 0 & 0 \\ 0 & v_0' & w_0' \\ 0 & 0 & 0 \end{bmatrix}$$

$$[D_9] = \begin{bmatrix} 0 & 0 & 0 \\ 0 & v_0' & 0 \\ 0 & w_0' & 0 \end{bmatrix}, \quad [D_{10}] = \int_0^R [P(x)]\, dx, \quad [D_{11}] = \int_0^x [P(x_1)]\, dx_1$$

APPENDIX B.   DEFINITION OF MATRICES IN THE EQUATIONS OF MOTION OF THE COUPLED ROTOR-BODY PROBLEM

$$[M(\psi)] = \begin{bmatrix} 1 & 0 & 0 & \sin\psi_1 & \cos\psi_1 \\ 0 & 1 & 0 & \sin\psi_2 & \cos\psi_2 \\ 0 & 0 & 1 & \sin\psi_3 & \cos\psi_3 \\ 0 & 0 & 0 & 1 & 0 \\ 0 & 0 & 0 & 0 & 1 \end{bmatrix}$$

$$[C(\psi)] = \begin{bmatrix} \gamma/8 & 0 & 0 & 2\cos\psi_1+\gamma\sin\psi_1/8 & 2\sin\psi_1-\gamma\cos\psi_1/8 \\ 0 & \gamma/8 & 0 & 2\cos\psi_2+\gamma\sin\psi_2/8 & 2\sin\psi_2-\gamma\cos\psi_2/8 \\ 0 & 0 & \gamma/8 & 2\cos\psi_3+\gamma\sin\psi_3/8 & 2\sin\psi_3-\gamma\cos\psi_3/8 \\ 0 & 0 & 0 & 0 & 0 \\ 0 & 0 & 0 & 0 & 0 \end{bmatrix}$$

$$[K(\psi)] = \begin{bmatrix} \nu_\beta^2 & 0 & 0 & 0 & 0 \\ 0 & \nu_\beta^2 & 0 & 0 & 0 \\ 0 & 0 & \nu_\beta^2 & 0 & 0 \\ \dfrac{(1-\nu_\beta^2)}{I_p}\sin\psi_1 & \dfrac{(1-\nu_\beta^2)}{I_p}\sin\psi_2 & \dfrac{(1-\nu_\beta^2)}{I_p}\sin\psi_3 & 0 & 0 \\ \dfrac{(\nu_\beta^2-1)}{I_q}\sin\psi_1 & \dfrac{(\nu_\beta^2-1)}{I_q}\sin\psi_2 & \dfrac{(\nu_\beta^2-1)}{I_q}\sin\psi_3 & 0 & 0 \end{bmatrix}$$

$$[T(\psi)] = \begin{bmatrix} 1 & \cos\psi_1 & \sin\psi_1 & 0 & 0 \\ 1 & \cos\psi_2 & \sin\psi_2 & 0 & 0 \\ 1 & \cos\psi_3 & \sin\psi_3 & 0 & 0 \\ 0 & 0 & 0 & 1 & 0 \\ 0 & 0 & 0 & 0 & 1 \end{bmatrix}$$

$$[M^*] = \begin{bmatrix} 3 & 0 & 0 & 0 & 0 \\ 0 & 3/2 & 0 & 0 & -3/2 \\ 0 & 0 & 3/2 & 3/2 & 0 \\ 0 & 0 & 0 & 1 & 0 \\ 0 & 0 & 0 & 0 & 1 \end{bmatrix}$$

$$[C^*] = \begin{bmatrix} 3\gamma/8 & 0 & 0 & 0 & 0 \\ 0 & 3\gamma/16 & 3 & 3 & -3\gamma/16 \\ 0 & -3 & 3\gamma/16 & 3\gamma/16 & 3 \\ 0 & 0 & 0 & 0 & 0 \\ 0 & 0 & 0 & 0 & 0 \end{bmatrix}$$

$$[K^*] = \begin{bmatrix} 3v_\beta^2 & 0 & 0 & 0 & 0 \\ 0 & 3(v_\beta^2-1)/2 & 3\gamma/16 & 0 & 0 \\ 0 & -3\gamma/16 & 3(v_\beta^2-1)/2 & 0 & 0 \\ 0 & 0 & \dfrac{3(1-v_\beta^2)}{2I_p} & 0 & 0 \\ 0 & \dfrac{3(v_\beta^2-1)}{2I_q} & 0 & 0 & 0 \end{bmatrix}$$

where

$$\psi_m \equiv 2m\pi/3, \qquad m = 1, 2, 3.$$

I_p = Rolling moment of inertia of body / I_b

I_q = Pitching moment of inertia of body / I_b

I_b = Blade flapping moment of inertia

v_β = Flapping rotating natural frequency of the blade

REFERENCES

1. H. M. Adelman and R. T. Haftka, "Sensitivity Analysis of Discrete Structural Systems", AIAA Journal 24, 823-832 (1986).

2. D. V. Murthy and R.T . Haftka, "Survey of Methods for Calculating Sensitivity of General Eigenproblems", NASA Symposium on Sensitivity Analysis in Engineering, Hampton, Virginia, 177-196 (1987).

3. W. H. Wittrick, "Rates of Change of Eigenvalues with Reference to Buckling and Vibration Problems," Journal of the Royal Aeronautical society 66, 590-591 (1962).

4. R. L. Fox and M. P. Kapoor, "Rates of Change of Eigenvalues and Eigenvectors", AIAA Journal 6, 2426-2429 (1968).

5. M. S. Zarghamee, "Minimum Weight Design with Stability Constraint", Journal of the Structural Division of ASCE 96, 1697-1710 (1970).

6. L. C. Rogers, "Derivatives of Eigenvalues and Eigenvectors", AIAA Journal 8, 943-944 (1970).

7. R. H. Plaut and K. Huseyin, "Derivatives of Eigenvalues and Eigenvectors in Non-Self-Adjoint Systems", AIAA Journal 11, 250-251 (1973).

8. S. Garg, "Derivatives of Eigensolutions for a General Matrix", AIAA Journal 11, 1191-1194 (1973).

9. C. S. Rudisill, "Derivatives of Eigenvalues and Eigenvectors of a General Matrix," AIAA Journal 12, 721-722 (1974).

10. R. B. Nelson, "Simplified Calculation of Eigenvector Derivatives", AIAA Journal 14, 1201-1205 (1976).

11. C. S. Rudisill and Y. Y. Chu, "Numerical Methods for Evaluating the Derivatives of Eigenvalues and Eigenvectors", AIAA Journal 13, 834-837 (1975).

12. C. S. Rudisill and K. G. Bhatia, "Optimization of Complex Structures to Satisfy Flutter Requirements", AIAA Journal 9, 1487-1491 (1971).

13. C. S. Rudisill and K. G. Bhatia, "Second Derivatives of the Flutter Velocity and the Optimization of Aircraft Structures", AIAA Journal 10, 1596-1572 (1972).

14. K. G. Bhatia, "An Automated Method for Determining the Flutter Velocity and the Matched Point", Journal of Aircraft 11, 21-27 (1974).

15. S. S. Rao, "Rates of Change of Flutter Mach Number and Flutter Frequency", AIAA Journal 10, 1526-1528 (1972).

16. A. P. Seyranian, "Sensitivity Analysis and Optimization of Aeroelastic Stability", International Journal of Solids and Structures 18, 791-807 (1982).

17. C. Cardani and P. Mantegazza, "Calculation of Eigenvalue and Eigenvector Derivatives for Algebraic Flutter and Divergence Eigen-Problems," AIAA Journal 17, 408-412 (1979).

18. P. Bindolino and P. Mantegazza, "Aeroelastic Derivatives as a Sensitivity Analysis of Nonlinear Equations," AIAA Journal 25, 1145-1146 (1987).

19. Y. Lu and V. R. Murthy, "Stability Sensititvity Studies for Synthesis of Aeroelastic Systems", Journal of Aircraft 27, 849-850 (1990).

20. R. B. Taylor, "Helicopter Vibration Reduction by Rotor Blade Modal Shaping", Proceedings of the 38th American Helicopter Society Annual Forum (1982).

21. R. H. Blackwell, "Blade Design for Reduced Helicopter Vibration", Journal of the American Helicopter Society 28, 33-41 (1983).

22. P. P. Friedmann and P. Shanthakumaran, "Optimum Design of Rotor Blades for Vibration Reduction in Forward Flight", Journal of the American Helicopter Society 29, 77-80 (1984).

23. D. A. Peters, M. P. Rossow, A. Korn and T. Ko, "Design of Helicopter Blade for Optimum Dynamic Characteristics", Computers and Mathematics with Applications 12, 85-109 (1986).

24. J. W. Lim and I. Chopra, "Aeroslastic Optimization of a Helicopter Rotor", Journal of the American Helicopter Society 34, 52-62 (1989).

25. J. W. Lim and I. Chopra, "Response and Hub Loads Sensitivity Analysis of a Helicopter Rotor", AIAA Journal 28, 75-82(1990).

26. J. W. Lim and I. Chopra, "Stability Sensitivity Analysis of a Helicopter Rotor", AIAA Journal 28, 1089-1097 (1990)

27. J. W. Lim and I. Chopra, "Aeroelastic Optimization of Helicopter Rotor Using an Efficient Sensitivity Analysis", Journal of Aircraft 28, 29-37 (1991).

28. Y. Lu and V. R. Murthy, "Sensitivity Analysis of Discrete Periodic Systems with Applications to Rotor Dynamics", Proceedings of the AIAA /ASME/ASCE/AHS/ASC 32nd Structures, Structural Dynamics and Materials Conference, 348-393 (1991).

29. N. J. Pullman, " Matrix Theory and Its Applications-Selected Topics", Marcel Dekker, New York, N. Y. (1976).

30. D. H. Hodges, and E. H. Dowell, "Nonlinear Equations of Motion for the Elastic Bending and Torsion of Twisted Nonuniform Rotor Blades," NASA TN D-7818 (1974).

31. L. A. Shultz, "Dynamic Analysis of Multiple-Load-Path Blades by the Transfer Matrix Method", Ph.D. Dissertation, To be submitted to the Department of Mechanical and Aerospace Engineering, Syracuse University, Syracuse, N. Y. (1992).

32. W. B. Stephens, D. H. Hodges, J. H. Avila and R. M. Kung, "Stability of Nonuniform Blades in Hover Using a Mixed Formulation", Vertica 6, 97-109 (1982).

33. A. R. S. Bramwell, "Helicopter Dynamics", Edward Arnold, London (1976).

34. K. H. Hohenemser and S. K .Yin, "Some Applications of the Method of Multiblade Coordinates", Journal of the American Helicopter Society 17, 3-12 (1972).

A SIMPLIFIED GENERAL SOLUTION METHODOLOGY FOR TRANSIENT STRUCTURAL DYNAMIC PROBLEMS WITH LOCAL NONLINEARITIES

EDWIN E. HENKEL
RENE' HEWLETT
RAYMOND MAR

Space Systems Division
Rockwell International Corporation
12214 Lakewood Boulevard
P. O. Box 7009
Downey, California 90241-7009

I. Introduction

The linear transient solutions for large structural problems are usually performed in modal coordinates which are obtained from an eigensolution. The primary advantages include uncoupling the equations of motion, resulting in a direct and efficient solution, and reducing the problem size by enabling the analyst to discard the mode shapes that are negligible to the solution, usually the higher order modes.

Interest in dynamic solutions involving local nonlinearities was generated by the National Space Transportation System (NSTS) payload community. The use of sliding trunnions to couple the payloads to the Space Shuttle allows for friction induced loads, with its resulting sticking and sliding. Such systems are very large, usually ~1,000 Degrees of Freedom (DOF). A trunnion in the static friction state (i.e., stuck or locked) constitutes an internal load path which alters the NSTS/payload nonfriction system modal properties. Thus these problems became quite complex and numerically intensive. Sliding friction forces are treated as externally applied forces, their magnitudes dependent upon the respective Orbiter/payload interface forces and their directions based upon the respective trunnion's relative motion. The static friction forces occur for locked trunnions, resulting in a change in the system modal properties. This force acts as an internal load until the trunnion transitions back into the sliding state. Any practical solution technique would need to use an invariant set of eigenvalues and eigenvectors, while at the same time accurately account for the static friction induced changes in the structural system's modal characteristics. Ref. 1 presented such a methodology . From this work, it was learned how to treat a large class of local nonlinearities.

Reference 2 addressed the dynamic liftoff problem of a booster/spacecraft system separating from its launch pad. This type of problem also involves a local nonlinear phenomena in that the constraints between the booster and pad will vary as the booster's engines reach full thrust and lifts the system away from the pad. The approach is much the same as that presented in Ref. 1.

As pointed out in Ref. 3, the Space Station program has generated considerable interest in the solution of time variant problems, e.g., a telerobot moving along flexible tracks or on the Space Station's truss structure. Two sample problems were presented in Ref. 4 and solved by several methods using time dependent modal properties. Not only were time dependent eigensolutions utilized, but the first and second time derivatives of the eigenvectors were calculated by use of a cubic spline interpolation. Such an approach to the transient friction solution of Ref. 1 is impossible since the transition in modal states happens instantaneously. Reference 4 demonstrated that treatment of structural time varying parameters as local nonlinear forces could greatly simplify at least some of these type problems and accurately solve them with only one set of invariant modal properties.

II. Modal Transient Solution

The linear modal equation of motion for the ith mode is given by the following:

$$m_i \ddot{\xi}_i + b_i \dot{\xi}_i + k_i \xi_i = P_i \tag{1}$$

This equation is more conveniently rewritten as follows:

$$\ddot{\xi}_i + 2\beta \dot{\xi}_i + \omega_0^2 \xi_i = \frac{P_i}{m_i} \tag{2}$$

This second order differential modal equation of motion can be solved by numerous solution techniques. A very large number of finite difference approaches exist. Ref. 1 employed the modified Newmark-Beta method. However, Refs. 2 and 4 employed a more computationally efficient method by utilizing a closed form solution. The closed form approach is possible because the modal equations of motion are uncoupled, i.e., each mode acts as an independent single DOF oscillator. Such an approach requires that the time history of the externally applied forces be defined as a function of time. This is seldom the case. Most often, the forces are defined for specific time slices. The solution technique most often used, assumes that the applied forces vary linearly between solution time steps. With this assumption, the solution is derived in a closed form method. The incremental solution, for time step n+1, is given by the following three equations:

$$\xi_{i,\,n+1} = F\,\xi_{i,\,n} + G\,\dot{\xi}_{i,\,n} + A\,P_{i,\,n} + B\,P_{i,\,n+1} \qquad (3)$$

$$\dot{\xi}_{i,\,n+1} = F'\,\xi_{i,\,n} + G'\,\dot{\xi}_{i,\,n} + A'\,P_{i,\,n} + B'\,P_{i,\,n+1} \qquad (4)$$

$$\ddot{\xi}_{i,\,n+1} = \frac{P_{i,\,n+1}}{m_i} - 2\beta\,\dot{\xi}_{i,\,n+1} - \omega_o^2\,\xi_{i,\,n+1} \qquad (5)$$

Equation 3 is the solution for the modal displacement, Eq. 4 for the modal velocity, and Eq. 5 for the modal acceleration. A, B, G, F, A', B', G' and F' are the coefficients of integration for the ith mode. The derivation of these equations and the coefficients for underdamped, critically damped, overdamped and undamped rigid body modes can be found in Ref. 5. The simplicity of this solution is derived from the fact that the response for any one mode is uncoupled from all the rest. Again, with the one assumption that the force input varies linearly between solution time steps, the solution is mathematically exact. A distinct advantage of this solution approach is the ease with which the integration time step can be changed. A simple recalculation of the coefficients is all that is necessary. Although the following is based upon the above three equations, it could also proceed from any number of finite difference solution techniques.

In normal linear analyses, the generalized force input, P, is known. The right hand sides of the equations are defined, making the solution for the n+1 time step a trivial task. The introduction of a single nonlinearity will greatly complicate the task. The solution for the various modes now become coupled, each response depending upon the others. A finite difference technique is most often employed for such problems. However, the solution technique reported here will treat the local nonlinear forces the same as the known external forces. In general, these forces and the responses are interrelated (i.e., they are dependent upon each other) and they will both remain unknown until the solution for the n+1 time step is completely solved. In most cases the generalized modal mass is unity, i.e., $m_i = 1$ and this convention is adopted hereafter for convenience. Eqs. 3 through 5 are rewritten with the nonlinear forces remaining as force variables.

$$\xi_{i,n+1} = F\,\xi_{i,n} + G\,\dot{\xi}_{i,n} + A\,P_{i,n} + B\,P_{i,n+1} \\ + A\,[\phi_{ji}]^T\,\{f_j\}_n + B\,[\phi_{ji}]^T\,\{f_j\}_{n+1} \tag{6}$$

$$\dot{\xi}_{i,n+1} = F'\,\xi_{i,n} + G'\,\dot{\xi}_{i,n} + A'\,P_{i,n} + B'\,P_{i,n+1} \\ + A'\,[\phi_{ji}]^T\,\{f_j\}_n + B'\,[\phi_{ji}]^T\,\{f_j\}_{n+1} \tag{7}$$

$$\ddot{\xi}_{i,n+1} = P_{i,n+1} - 2\beta\,\dot{\xi}_{i,n+1} - \omega_o^2\,\xi_{i,n+1} + [\phi_{ji}]^T\,\{f_j\}_{n+1} \tag{8}$$

The variable f_j is the physical local nonlinear force(s) acting on the j physical DOFs, and ϕ_{ji} is the corresponding eigenvector jth physical DOF row partition(s) and the ith modal column partition. The matrix product, $[\phi_{ji}]^T\,\{f_j\}$, is simply the generalized nonlinear force. In general, f is dependent upon the response of all the modes, and is therefore an unknown variable in the solution. The above three equations are rewritten in short form by placing all the known right hand side variables together.

$$\xi_{i,n+1} = D_{n+1} + B\,[\phi_{ji}]^T\,\{f_j\}_{n+1} \tag{9}$$

$$\dot{\xi}_{i,n+1} = V_{n+1} + B'\,[\phi_{ji}]^T\,\{f_j\}_{n+1} \tag{10}$$

$$\ddot{\xi}_{i,n+1} = P_{i,n+1} - 2\beta\,\dot{\xi}_{i,n+1} - \omega_o^2\,\xi_{i,n+1} + [\phi_{ji}]^T\,\{f_j\}_{n+1} \tag{11}$$

III. Simple Structural Deadband

The first local nonlinearity to be discussed is that of structural deadbands. Deadbands, free-play, rattle space and/or slop can exist between many connecting structural components, either by design or through tolerance build-up. For example, the support structure between a payload and the Orbiter airframe must allow for longitudinal motion to minimize thermally induced loads. This is accomplished by allowing certain attach DOFs to slide. To prevent the possibility of binding, vertical deadbands are designed into the support structure. The worst-on-worst minimum deadband is 0.015 inch and the worst-on-worst maximum deadband is 0.0276 inch.

This small free-play is considered to be insignificant in limit loads analyses. However, the inclusion of the Orbiter-to-payload support hardware in a modal survey test fixture could severely taint the test results. At the low force levels typically used in modal survey testing, such deadbands can be very significant. Modal frequencies may appear to be a function of the test forcing level and accurate determination of modal damping may be impossible.

Figure 1 illustrates a simple deadband. To simulate this deadband, the finite element model would employ DOFs for both sides of the structure, most probably with the nodes located at the same physical locations, but with the DOFs remaining uncoupled through the eigensolution. One nodal point would represent side A of the structure and the other side B. The deadband, $\pm\delta$, is shown as being symmetric, although the following is valid for asymmetric situations.

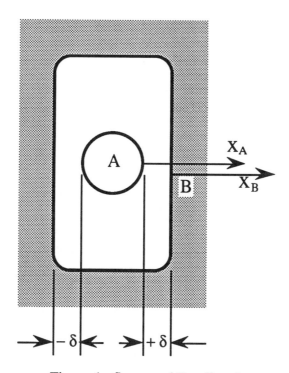

Figure 1. Structural Deadband

For small deformations, i.e., relative displacement $x_A - x_B < \pm\delta$, the system modal properties from the eigensolution apply and f is zero. But when $x_A - x_B > \pm\delta$, the solution is invalid and the system equations of motion must be modified by the inclusion of appropriate constraint equations. This then results in the necessary constraint forces, f. The displacements of nodal points A and B are found by transformation of all the modal displacements, Eq. 9, into the respective physical coordinates.

$$x_{A,\,n+1} = [\phi_{Ai}]\{\xi_i\}_{n+1} = [\phi_{Ai}]\{D_i\}_{n+1} + [\phi_{Ai}][B_{ii}][\phi_{ji}]^T\{f_j\}_{n+1} \tag{12}$$

$$x_{B,\,n+1} = [\phi_{Bi}]\{\xi_i\}_{n+1} = [\phi_{Bi}]\{D_i\}_{n+1} - [\phi_{Bi}][B_{ii}][\phi_{ji}]^T\{f_j\}_{n+1} \tag{13}$$

The sign convention on the force, f, is arbitrary. Note that B_{ii} is the diagonal matrix of the B coefficients of integration. The relative displacement is simply:

$$\Delta x = x_A - x_B \tag{14}$$

The constraint equation, for conditions of the deadband being "bottomed out", is given as:

$$\Delta x = \pm\delta \tag{15}$$

Assembling the above four equations into matrix form, dropping the n+1 subscript and placing all unknown variables on the left hand side results in the following equation for the n+1 time step:

$$\begin{bmatrix} I & 0 & 0 & -[\phi_{Ai}][B_{ii}][\phi_{ji}]^T \\ 0 & I & 0 & [\phi_{Bi}][B_{ii}][\phi_{ji}]^T \\ -I & I & I & 0 \\ 0 & 0 & I & 0 \end{bmatrix} \begin{Bmatrix} x_A \\ x_B \\ \Delta x \\ f_j \end{Bmatrix} = \begin{Bmatrix} [\phi_{Ai}]\{D_i\} \\ [\phi_{Bi}]\{D_i\} \\ 0 \\ \pm\delta \end{Bmatrix} \tag{16}$$

The sign on δ, plus or minus, is dependent upon the side of the deadband that the system is against. For the condition that $\Delta x < \pm\delta$, Eq. 16 is easily modified by removing the constraint.

$$\begin{bmatrix} I & 0 & 0 & -[\phi_{Ai}][B_{ii}][\phi_{ji}]^T \\ 0 & I & 0 & [\phi_{Bi}][B_{ii}][\phi_{ji}]^T \\ -I & I & I & 0 \\ 0 & 0 & 0 & I \end{bmatrix} \begin{Bmatrix} x_A \\ x_B \\ \Delta x \\ f_j \end{Bmatrix} = \begin{Bmatrix} [\phi_{Ai}]\{D_i\} \\ [\phi_{Bi}]\{D_i\} \\ 0 \\ 0 \end{Bmatrix} \tag{17}$$

Note that in this equation the local nonlinear force, f_j, is zero.

Equations 16 and 17 represent a small set of simultaneous equations, i.e., for the single DOF deadband of Fig. 1, there are only four variables, making the solution trivial. Once $f_{j,n+1}$ is known, the entire modal solution is solved by use of Eqs. 9, 10 and 11. The analysis then proceeds using the applicable equation. For $\Delta x < \pm\delta$, f_j is zero and Eq. 17 is active. The solution proceeds time step by time step until at some point Δx falls outside the bounds of $\pm\delta$. At this point, the physical system has bottomed out and the solution for this time step is invalid and must be discarded. Equation 16 becomes active. Again, the solution proceeds time step by time step. At some point f_j may experience a sign reversal. Since this is not possible (discounting the possibility of tacky surfaces) the solution for f_j is again discarded and Eq. 17 becomes active. In this manner, by transitioning between Eqs. 16 and 17, these local deadband DOFs are allowed to rattle around within the bounds defined by $+\delta$ and $-\delta$. These values need not be symmetric, e.g., $-\delta$ may be very small or even zero, and $+\delta$ could be large or even infinity. Thus, the method also lends itself to the solution of large modal problems involving local impacts and/or bounces.

Most practical problems will involve a number of deadbands, making the size of Eqs. 16 and 17 larger, i.e., four times the number of deadbands. At each solution of the active equation, each deadband's compatibility is checked against it's physical characteristics. The active equation is modified for each deadband as needed. Thus, for n number of deadbands, there are 2^n number of possible active equations. However, the system of equations is still quite small, making the necessary solutions an easy task. It may happen that two or more deadbands experience a state change at the same time step. Care should be taken in the analysis to only change one at a time. For example, at time step n+1, a solution is found and two deadbands show that a change in their respective state is warranted. The analyst's logic should then compare the two against the previous solution to determine which deadband appears to warrant the change the earliest. The active equation is then modified for that deadband only, and the solution for time step n+1 is resolved. Again, the solution is compared against the physical characteristics to see if another deadband warrants a state change. If so, the active equation is again modified and the solution is again resolved. This continues until the solution is physically compatible. Warning: once a state change has been made for a particular deadband, it should not be allowed to change back due to another deadband changing its state. This precaution will preclude the analyst from getting trapped in an endless loop. When the solution is physically compatible, the modal solution for time step n+1 is completed and the solution for time step n+2 is initiated.

The application of Eq. 16 can prove to be useful for the special condition of zero deadband, i.e., $\pm\delta = 0$. When sides A and B are totally uncoupled, i.e., two independent modal systems, Eq. 16 offers the means of substructuring, without the usual system eigensolution. For example, it is often required to perform parametric component transient analyses, each parametric condition resulting in a discrete set of system modes. However, the application of Eq. 16 avoids the costly execution of the system eigensolution. In the application of NSTS/payload parametric studies, where the payload structural and/or weight properties are varied, a further savings is achieved in that the forcing functions need to be generalized only once, i.e., the forcing functions act only upon the NSTS side of the interface. Another condition where Eq. 16, with $\pm\delta = 0$, is useful arises when the different components have discrete damping schedules. For example, NSTS/payload transient landing analyses typically use 1% critical modal damping, and this is most often deemed to be conservative. However, often the payload specific modal damping is considerably larger than one per cent. Transformation of component damping to the system modal level can result in analytic problems. Application of Eq. 16 avoids determination of the system modal damping parameters and allows each component its discrete damping schedule.

IV. Accuracy of the System Modal Properties

Solving transient problems by use of modes has the advantages of both uncoupling the equations of motion and allowing for problem reduction through modal truncation. As was stated in the introduction, an accurate solution involving local nonlinearities requires that the analyst account for the possible changes in the system's modal properties. In the above derivation there was no adjustment or change in the modal properties used to derive the solution. However, the application of the constraint equation to enforce the physical condition of the deadband being at one of its limits, $\pm\delta$, definitely does alter the physical system's modal characteristics. The same problem arises when including trunnion friction in the calculation of NSTS/payload transients, and also the booster to pad separation problem. The work detailed in Refs. 1, 2 and 4 was able to account for the resulting changes in system modal properties and yet use only one set of invariant modal characteristics in the solution. This was accomplished by employing the principles of modal synthesis.

One of the most widely used methods of modal synthesis is the improved component-mode representation given by S. Rubin (Ref. 6). This modal synthesis method accounts for the flexibility and inertia at subsystems' interfaces that was lost through modal truncation. With this, the subsystems can be coupled, and an accurate system eigensolution calculated. In the above discussion, the analyst could use Rubin's method of including the residual flexibility and inertia to correct for that which was truncated at the two deadband DOFs, x_A and x_B. This does require a few additional DOFs. With this, the analyst could then calculate a new set of system modes. These new modes will be identical to the original untruncated set, with the addition of a set of residual modes. With no further truncation, this new modal system will have all the necessary properties, on all the necessary DOFs, to calculate accurate system modes for any local state change. These residual modes are not real vibrational modes, but are eigenvalues and eigenvectors that correct for modal truncation effects at the desired physical DOFs. Again, accurate system modes, within the limits imposed by the original modal truncation, can then be calculated for any desired deadband state. Reference 6 suggests that the retained modal content be accurate up to 1.5 times the highest system frequency of interest. Thus, if the system response and/or the external forcing function, P, is of interest through 10 hertz (i.e., a standard linear analysis would be truncated above 10 hertz), the system modes should not be truncated below about 15 hertz. The necessary residual modes are then calculated and included in the analysis.

The solution approach presented in this chapter utilizes a set of structural invariant modal properties. However, with the use of residual modes, all the necessary structural information needed to calculate accurate system modal properties is contained within the solution's formulation.

The authors have used this approach to solve large NSTS/payload transients including trunnion friction effects (Ref. 1). As stated before,the residual modes do not represent real vibration modes. Experience has shown that they are best treated quasi-statically (i.e., massless). This is consistent with the fact that not including residual inertias in Rubin's method results in system eigensolutions almost identical to those accounting for the residual inertia. This method of modal synthesis is most often referred to as the Rubin-MacNeal method. Thus the residual inertia effects are typically of a lesser magnitude than the residual flexibility effects. For these reasons, residual inertia is used to calculate the residual modes, which are then treated quasi-statically in the transient solution. In this manner, the solution of dynamic systems with local nonlinearities, will be employing the Rubin-MacNeal method of modal synthesis. Treating modal DOFs quasi-statically is easily accomplished by using the following coefficients of integration: $F = G = A = 0$, and $B = 1/\omega^2$, with the modal velocity and acceleration equal to zero by definition.

V. V-Shaped Structural Deadband

To illustrate the utility of the method, a more complicated situation, a V-shaped deadband will be discussed. Figure 2 illustrates the physical properties of the V-shaped deadband.

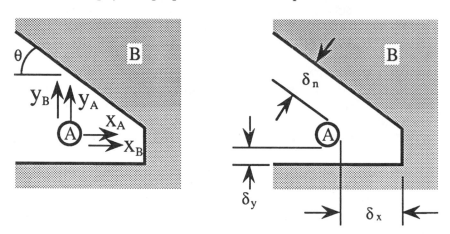

Figure 2. V-Shaped Deadband

To simplify the discussion, the δ_x limit will be considered to be large, thus the local system is limited by only two constraints, δ_y and δ_n. The three possible conditions are illustrated in Figure 3.

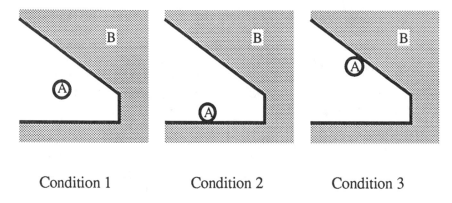

Condition 1   Condition 2   Condition 3

Figure 3. Three Possible Conditions

Conditions 1 and 2 are the same as was discussed earlier. Condition 3 however, presents a skewed surface constraint. For zero friction, the condition 3 bearing force between surfaces A and

B is normal to the contact surface of B and the following simple equations apply:

$$\Delta n = x_A \sin\theta + y_A \cos\theta - x_B \sin\theta - y_B \cos\theta \qquad (18)$$

$$\Delta n = \delta_n \qquad (19)$$

$$f_{nx} = f_n \sin\theta \qquad (20)$$

$$f_{ny} = f_n \cos\theta \qquad (21)$$

Some short hand will prove to be useful in the assembly of the necessary equations. In Eqs. 16 and 17 the force coupling terms to the discrete physical displacements is given by the matrix product $[\phi_{Ai}][B_{ii}][\phi_{ji}]^T$. This term is the f_j force's influence upon the discrete physical displacement A and as such is a flexibility term. We will introduce the following shorthand notation for convenience:

$$[\phi_{A_xi}][B_{ii}][\phi_{ji}]^T = F_{A_xj} \qquad (22)$$

where F_{A_xj} is the x DOF displacement of node A due to a unit load acting on DOF j. Furthermore, since the subscripts of F_{A_xj} are dictated by the term's position within the following matrix equations, the subscripts will be dropped. With this, the assembled equation for condition 1 can be written as follows:

$$
\begin{bmatrix}
I & 0 & 0 & 0 & 0 & 0 & -F & 0 & -F & -F \\
0 & I & 0 & 0 & 0 & 0 & -F & 0 & -F & -F \\
0 & 0 & I & 0 & 0 & 0 & F & 0 & F & F \\
0 & 0 & 0 & I & 0 & 0 & F & 0 & F & F \\
0 & -I & 0 & I & I & 0 & 0 & 0 & 0 & 0 \\
-\sin\theta & -\cos\theta & \sin\theta & \cos\theta & 0 & I & 0 & 0 & 0 & 0 \\
0 & 0 & 0 & 0 & 0 & 0 & I & 0 & 0 & 0 \\
0 & 0 & 0 & 0 & 0 & 0 & 0 & I & 0 & 0 \\
0 & 0 & 0 & 0 & 0 & 0 & 0 & -\sin\theta & I & 0 \\
0 & 0 & 0 & 0 & 0 & 0 & 0 & -\cos\theta & 0 & I
\end{bmatrix}
\begin{Bmatrix}
x_A \\ y_A \\ x_B \\ y_B \\ \Delta y \\ \Delta n \\ f_y \\ f_n \\ f_{nx} \\ f_{ny}
\end{Bmatrix}
$$

$$
= \begin{Bmatrix}
[\phi_{A_x}i]\{D_i\} \\
[\phi_{A_y}i]\{D_i\} \\
[\phi_{B_x}i]\{D_i\} \\
[\phi_{B_y}i]\{D_i\} \\
0 \\
0 \\
0 \\
0 \\
0 \\
0
\end{Bmatrix} \tag{23}
$$

Note that in condition 1 the forces f_y and f_n equate to zero. Equation 23 is easily modified to the constraints of condition 2 simply by repositioning one matrix component in the array and placing the displacement limit on the right hand side.

$$
\begin{bmatrix}
I & 0 & 0 & 0 & 0 & 0 & -F & 0 & -F & -F \\
0 & I & 0 & 0 & 0 & 0 & -F & 0 & -F & -F \\
0 & 0 & I & 0 & 0 & 0 & F & 0 & F & F \\
0 & 0 & 0 & I & 0 & 0 & F & 0 & F & F \\
0 & -I & 0 & I & I & 0 & 0 & 0 & 0 & 0 \\
-\sin\theta & -\cos\theta & \sin\theta & \cos\theta & 0 & I & 0 & 0 & 0 & 0 \\
0 & 0 & 0 & 0 & I & 0 & 0 & 0 & 0 & 0 \\
0 & 0 & 0 & 0 & 0 & 0 & 0 & I & 0 & 0 \\
0 & 0 & 0 & 0 & 0 & 0 & 0 & -\sin\theta & I & 0 \\
0 & 0 & 0 & 0 & 0 & 0 & 0 & -\cos\theta & 0 & I
\end{bmatrix}
\begin{Bmatrix}
x_A \\ y_A \\ x_B \\ y_B \\ \Delta y \\ \Delta n \\ f_y \\ f_n \\ f_{nx} \\ f_{ny}
\end{Bmatrix}
$$

$$
=
\begin{Bmatrix}
[\phi_{A_x}i]\{D_i\} \\
[\phi_{A_y}i]\{D_i\} \\
[\phi_{B_x}i]\{D_i\} \\
[\phi_{B_y}i]\{D_i\} \\
0 \\
0 \\
-\delta_y \\
0 \\
0 \\
0
\end{Bmatrix}
\tag{24}
$$

In the same manner, the equation for condition 3 is given by the following:

$$
\begin{bmatrix}
I & 0 & 0 & 0 & 0 & 0 & -F & 0 & -F & -F \\
0 & I & 0 & 0 & 0 & 0 & -F & 0 & -F & -F \\
0 & 0 & I & 0 & 0 & 0 & F & 0 & F & F \\
0 & 0 & 0 & I & 0 & 0 & F & 0 & F & F \\
0 & -I & 0 & I & I & 0 & 0 & 0 & 0 & 0 \\
-\sin\theta & -\cos\theta & \sin\theta & \cos\theta & 0 & I & 0 & 0 & 0 & 0 \\
0 & 0 & 0 & 0 & 0 & 0 & I & 0 & 0 & 0 \\
0 & 0 & 0 & 0 & 0 & I & 0 & 0 & 0 & 0 \\
0 & 0 & 0 & 0 & 0 & 0 & 0 & -\sin\theta & I & 0 \\
0 & 0 & 0 & 0 & 0 & 0 & 0 & -\cos\theta & 0 & I
\end{bmatrix}
\begin{Bmatrix}
x_A \\ y_A \\ x_B \\ y_B \\ \Delta y \\ \Delta n \\ f_y \\ f_n \\ f_{nx} \\ f_{ny}
\end{Bmatrix}
$$

$$
=
\begin{Bmatrix}
[\phi_{A_x}i]\{D_i\} \\
[\phi_{A_y}i]\{D_i\} \\
[\phi_{B_x}i]\{D_i\} \\
[\phi_{B_y}i]\{D_i\} \\
0 \\
0 \\
0 \\
\delta_n \\
0 \\
0
\end{Bmatrix}
\tag{25}
$$

With Eqs. 23, 24, and 25, the influence of the deadband depicted in Fig. 2 upon the system dynamics can be accurately modeled. Just as in the simple deadband analysis, each time step solution is checked for compatibility to the physical system before the system modal DOFs are calculated. Incompatible solutions are discarded and the equations are modified accordingly.

VI. Circular Structural Deadband

Figure 4 illustrates a circular structural deadband.

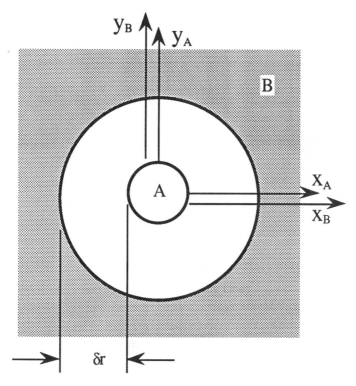

Figure 4. Circular Structural Deadband

As can been seen, the x and y DOFs at points A and B are uncoupled as long as the relative displacements do not exceed δ_r, i.e., condition 1 exists for the following:

$$\sqrt{(x_A - x_B)^2 + (y_A - y_B)^2} = \sqrt{\Delta x^2 + \Delta y^2} < \delta r \qquad (26)$$

Figure 5 illustrates condition 2 in which A bears against B. The force of contact, f_r, acts on A and B as an equal and opposite pair. The force on A is not shown for clarity.

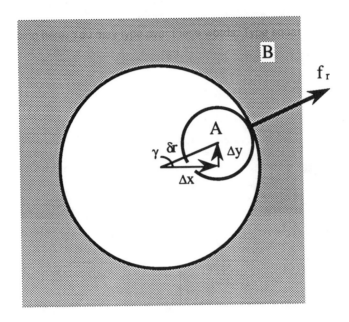

Figure 5. Circular Deadband "Bottomed Out"

For condition 2, the necessary constraint equation is given by the following:

$$\Delta x^2 + \Delta y^2 = \delta r^2 \tag{27}$$

The radial force of constraint is broken into its components through the following relations:

$$f_x = \cos\gamma\, f_r = \frac{\Delta x}{\delta r}\, f_r \tag{28}$$

$$f_y = \sin\gamma\, f_r = \frac{\Delta y}{\delta r}\, f_r \tag{29}$$

With the above, the equation that enforces displacement compatibility for condition 2 can be written.

$$
\begin{bmatrix}
I & 0 & 0 & 0 & 0 & 0 & 0 & 0 & -F & -F \\
0 & I & 0 & 0 & 0 & 0 & 0 & 0 & -F & -F \\
0 & 0 & I & 0 & 0 & 0 & 0 & 0 & F & F \\
0 & 0 & 0 & I & 0 & 0 & 0 & 0 & F & F \\
-I & 0 & I & 0 & I & 0 & 0 & 0 & 0 & 0 \\
0 & -I & 0 & I & 0 & I & 0 & 0 & 0 & 0 \\
0 & 0 & 0 & 0 & -\Delta x & -\Delta y & I & 0 & 0 & 0 \\
0 & 0 & 0 & 0 & 0 & 0 & I & 0 & 0 & 0 \\
0 & 0 & 0 & 0 & 0 & 0 & 0 & -\Delta x/\delta r & I & 0 \\
0 & 0 & 0 & 0 & 0 & 0 & 0 & -\Delta y/\delta r & 0 & I
\end{bmatrix}
\begin{Bmatrix}
x_A \\ y_A \\ x_B \\ y_B \\ \Delta x \\ \Delta y \\ \Delta r^2 \\ f_r \\ f_x \\ f_y
\end{Bmatrix}
$$

$$
= \begin{Bmatrix}
[\phi_{A_x}i]\{D_i\} \\
[\phi_{A_y}i]\{D_i\} \\
[\phi_{B_x}i]\{D_i\} \\
[\phi_{B_y}i]\{D_i\} \\
0 \\
0 \\
0 \\
\delta r^2 \\
0 \\
0
\end{Bmatrix}
\tag{30}
$$

At this point, it is convenient to employ the method of static condensation to Eq. 30 and eliminate (or reduce out) the x and y DOFs. This results in the following smaller set of equations:

$$
\begin{bmatrix}
I & 0 & 0 & 0 & -\mathcal{F} & -\mathcal{F} \\
0 & I & 0 & 0 & -\mathcal{F} & -\mathcal{F} \\
-\Delta x & -\Delta y & I & 0 & 0 & 0 \\
0 & 0 & I & 0 & 0 & 0 \\
0 & 0 & 0 & -\Delta x/\delta r & I & 0 \\
0 & 0 & 0 & -\Delta y/\delta r & 0 & I
\end{bmatrix}
\begin{Bmatrix}
\Delta x \\ \Delta y \\ \Delta r^2 \\ f_r \\ f_x \\ f_y
\end{Bmatrix} =
$$

$$
\begin{Bmatrix}
([\phi_{A_x}i]-[\phi_{B_x}i])\,\{D_i\} \\
([\phi_{A_y}i]-[\phi_{B_y}i])\,\{D_i\} \\
0 \\
\delta r^2 \\
0 \\
0
\end{Bmatrix}
\tag{31}
$$

The flexibility terms, $\mathcal{F}$, are now relative flexibilities, e.g., from Eq. 22, a relative flexibility would be the following:

$$\mathcal{F} = F_{A_x} + F_{B_x} = [\phi_{A_x i}][B_{ii}][\phi_{ji}]^T + [\phi_{B_x i}][B_{ii}][\phi_{ji}]^T \qquad (32)$$

Note the the nonlinear forces, f_x and f_y, act on nodes A and B as equal and opposite, thus making the relative flexibility additive.

Equation 31 is nonlinear. In the previous examples, the problem nonlinearity was the result of a state change in the system equations of motion. These problems are and were easily treated as piecewise linear. In Eq. 31, unknown variables appear in the left hand side square matrix. The analyst could choose a number of methods to solve such equations (e.g., successive iterations, predictor-corrector methods, Newton-Raphson method, etc.). For completeness, the Newton-Raphson method will be illustrated.

VII. Newton-Raphson Iterative Solution Method

To illustrate the Newton-Raphson iterative solution method, Eq. 31 is written in short form.

$$\mathbf{K} \mathbf{q} = \mathbf{P} \qquad (33)$$

Matrix brackets and braces have been dropped and bold face characters are used to denote matrices. A converged solution exists for the nth time step. The solution for the n+1 time step is estimated by using $\mathbf{K}_n$ and $\mathbf{P}_{n+1}$. $\mathbf{K}$ is then updated and the resulting load error is calculated.

$$\mathbf{f} = \mathbf{P} - \mathbf{K} \mathbf{q} \qquad (34)$$

The Newton-Raphson method is iterative and seeks to converge to the true solution by finding a $\mathbf{q}_{k+1}$ which results in $\mathbf{f} = 0$, where k is the iteration number. The following equations are applicable:

$$\mathbf{f}_k = \mathbf{P} - \mathbf{K}(\mathbf{q}_k) \mathbf{q}_k \qquad (35)$$

$$\mathbf{q}_{k+1} = \mathbf{q}_k + \Delta\mathbf{q} \qquad (36)$$

$$\mathbf{f}_{k+1} = 0 = \mathbf{P} - \mathbf{K}(\mathbf{q}_k + \Delta\mathbf{q})\{\mathbf{q}_k + \Delta\mathbf{q}\} \qquad (37)$$

Subtracting Eq. 37 from Eq. 35 results in

$$\mathbf{f}_k = -\mathbf{K}(\mathbf{q}_k) \mathbf{q}_k + \mathbf{K}(\mathbf{q}_k + \Delta\mathbf{q}) (\mathbf{q}_k + \Delta\mathbf{q})$$

$$= -\mathbf{Q}(\mathbf{q}_k) + \mathbf{Q}(\mathbf{q}_k + \Delta\mathbf{q}) \qquad (38)$$

The right hand side of Eq. 38 is a force vector and is a function of

q_k and Δq. $Q(q_k + \Delta q)$ is approximated by a Taylor series expansion, truncated after two terms:

$$Q(q_k + \Delta q) = Q(q_k) + \left[\frac{\partial Q(q)}{\partial q}\right] \Delta q + \ldots\ldots$$

$$= Q(q_k) + K^* \Delta q \qquad (39)$$

where:

$$K^*_{ij} = \frac{\partial Q_i}{\partial q_j} \qquad (40)$$

Substituting Eq. 39 into 38 and solving for Δq yields the following:

$$\Delta q = K^{*-1} f_k \qquad (41)$$

Thus, the iterative Newton-Raphson method follows the path of Eqs. 35, 40, 41 and then 36. The solution is converged when Δq is zero. Note that K^* can be referred to as the tangent stiffness matrix. With this, the tangent stiffness matrix for Eq. 31 becomes:

$$K^* = \begin{bmatrix} I & 0 & 0 & 0 & -\mathcal{F} & -\mathcal{F} \\ 0 & I & 0 & 0 & -\mathcal{F} & -\mathcal{F} \\ -2\Delta x & -2\Delta y & I & 0 & 0 & 0 \\ 0 & 0 & I & 0 & 0 & 0 \\ -fr/\delta r & 0 & 0 & -\Delta x/\delta r & I & 0 \\ 0 & -fr/\delta r & 0 & -\Delta y/\delta r & 0 & I \end{bmatrix} \qquad (42)$$

The formation of this matrix is easy and for problems with a few number of circular deadbands, the necessary computational effort should prove to be manageable. The iterative solution at each time step is necessary for condition 2. The solution reverts to the standard noniterative linear solution for condition 1. As before, when solving a time step for condition 2, after a converged solution to Eq. 31 is reached and it is determined that this solution is compatible with the physical system (i.e., deadband should remain "bottomed out"), the modal solution for the entire system is simply generated by use of Eqs. 9, 10 and 11. The effort then proceeds to the next time step.

VIII. Separation from a Spherical Fitting

Structural components are often coupled together by means of spherical fittings, a good example being launch vehicles (or boosters) at the launch pad interfaces. In Ref. 2, the authors presented a methodology for treatment of booster/launch pad separation. The method presented employs constraint equations to enforce displacement compatibility between the two components. As the booster thrust builds, the vertical constraint force decays to zero, this being the point where the two systems start to separate (assuming the tie down bolts have been severed). In the method of Ref. 2, the constraint equations, both vertical and lateral, are removed at this point. Since the lateral forces may be rather large, the analyst may desire to decay these forces over some time increment or even over some booster/launch pad relative displacement increment. However, with the above method for treating circular deadbands, the extension to the treatment of separation from a spherical fitting is straightforward. Figure 6 illustrates the geometry in question.

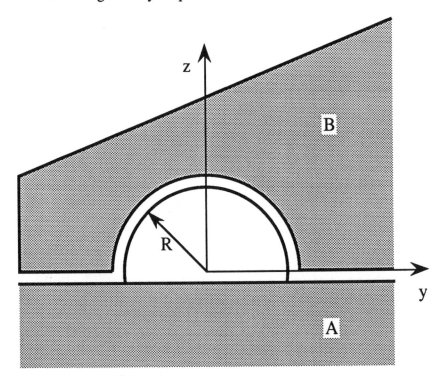

Figure 6. Spherical Fitting

In Fig. 6, the x DOF is out of and perpendicular to the page. As B rises away from A, the lateral dead band will increase. The geometry is illustrated in Fig. 7.

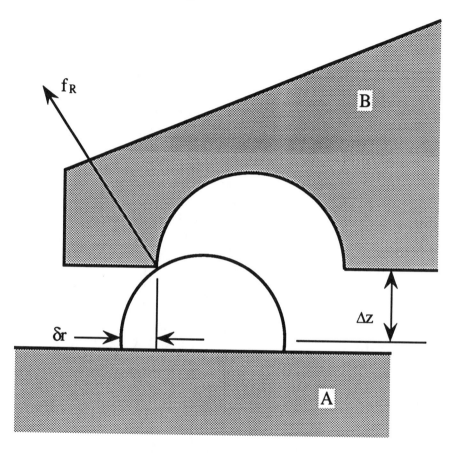

Figure 7. Spherical Fitting Separation Geometry

For the condition of Fig. 7, i.e., contact remaining between A and B, the relative displacement between nodes A and B is along the surface of a torus with radius of R, and a revolute radius of R, centered on the center of the spherical fitting. The following equation holds:

$$\Delta x^2 + \Delta y^2 + \Delta z^2 = 2R\sqrt{\Delta x^2 + \Delta y^2} \tag{43}$$

The force components are given by the following relations:

$$f_x = \frac{-\Delta x \sqrt{R^2 - \Delta z^2}}{R \sqrt{\Delta x^2 + \Delta y^2}} f_R \tag{44}$$

$$f_y = \frac{-\Delta y \sqrt{R^2 - \Delta z^2}}{R \sqrt{\Delta x^2 + \Delta y^2}} f_R \tag{45}$$

$$f_z = \frac{\Delta z}{R} f_R \tag{46}$$

Assembly of these equations leads to the following matrix expression:

$$
\begin{bmatrix}
I & 0 & 0 & 0 & 0 & -\mathcal{F} & -\mathcal{F} & -\mathcal{F} \\
0 & I & 0 & 0 & 0 & -\mathcal{F} & -\mathcal{F} & -\mathcal{F} \\
0 & 0 & I & 0 & 0 & -\mathcal{F} & -\mathcal{F} & -\mathcal{F} \\
-\Delta x & -\Delta y & -\Delta z & I & 0 & 0 & 0 & 0 \\
0 & 0 & 0 & I & 0 & 0 & 0 & 0 \\
0 & 0 & 0 & 0 & \Delta x Q & I & 0 & 0 \\
0 & 0 & 0 & 0 & \Delta y Q & 0 & I & 0 \\
0 & 0 & 0 & 0 & -\Delta z/R & 0 & 0 & I
\end{bmatrix}
\begin{Bmatrix}
\Delta x \\ \Delta y \\ \Delta z \\ \mathcal{R} \\ f_R \\ f_x \\ f_y \\ f_z
\end{Bmatrix} =
$$

$$
\begin{Bmatrix}
([\phi_{A_x}]-[\phi_{B_x}]) \{D_i\} \\
([\phi_{A_y}]-[\phi_{B_y}]) \{D_i\} \\
([\phi_{A_z}]-[\phi_{B_z}]) \{D_i\} \\
0 \\ 0 \\ 0 \\ 0 \\ 0
\end{Bmatrix} +
\begin{Bmatrix}
0 \\ 0 \\ 0 \\ 0 \\ 2R\sqrt{\Delta x^2 + \Delta y^2} \\ 0 \\ 0 \\ 0
\end{Bmatrix} \tag{47}
$$

where:

$$Q = \frac{\sqrt{R^2 - \Delta z^2}}{R \sqrt{\Delta x^2 + \Delta y^2}} \tag{48}$$

Note that the variable $\mathcal{R}$ has been introduced and forced to equal the right hand side of Eq. 43. Equation 47 differs from Eq. 31 in character in that there is now a nonlinear vector on the right hand side; however, Eq. 47 can still be solved by the iterative Newton-Raphson method in the same manner as Eq. 31. For the case of a spherical fitting, the tangent stiffness matrix becomes:

$$
K^* = \begin{bmatrix}
I & 0 & 0 & 0 & 0 & -\mathcal{F} & -\mathcal{F} & -\mathcal{F} \\
0 & I & 0 & 0 & 0 & -\mathcal{F} & -\mathcal{F} & -\mathcal{F} \\
0 & 0 & I & 0 & 0 & -\mathcal{F} & -\mathcal{F} & -\mathcal{F} \\
-2\Delta x & -2\Delta y & -2\Delta z & I & 0 & 0 & 0 & 0 \\
\Delta x \mathcal{P} & \Delta y \mathcal{P} & 0 & I & 0 & 0 & 0 & 0 \\
\mathcal{A} & \mathcal{B} & \mathcal{C} & 0 & \Delta x \mathcal{Q} & I & 0 & 0 \\
\mathcal{B} & \mathcal{D} & \mathcal{E} & 0 & \Delta y \mathcal{Q} & 0 & I & 0 \\
0 & 0 & -f_R/R & 0 & -\Delta z/R & 0 & 0 & I
\end{bmatrix} \tag{49}
$$

where:

$$
\mathcal{P} = \frac{-2\,R}{\sqrt{\Delta x^2 + \Delta y^2}} \tag{50}
$$

$$
\mathcal{A} = \mathcal{Q} f_R - \frac{\Delta x^2\, \mathcal{Q}\, f_R}{(\Delta x^2 + \Delta y^2)} \tag{51}
$$

$$
\mathcal{B} = \frac{-\Delta x\, \Delta y\, \mathcal{Q}\, f_R}{(\Delta x^2 + \Delta y^2)} \tag{52}
$$

$$
\mathcal{C} = \frac{-\Delta x\, \Delta z\, \mathcal{Q}\, f_R}{(R^2 - \Delta z^2)} \tag{53}
$$

$$
\mathcal{D} = \mathcal{Q} f_R - \frac{\Delta y^2\, \mathcal{Q}\, f_R}{(\Delta x^2 + \Delta y^2)} \tag{54}
$$

and:

$$
\mathcal{E} = \frac{-\Delta y\, \Delta z\, \mathcal{Q}\, f_R}{(R^2 - \Delta z^2)}\;. \tag{55}
$$

With Eqs. 47 and 49, the solution for the case of separation from a spherical fitting can be solved. Separation occurs when f_R decays to zero, or when Δz equals R. The constraint of Eq. 43 is "turned off" simply by moving the I in the f_R row over one column. For the case of NSTS liftoff transient, eight spherical fittings are disengaged, thus the nonlinear iterative solution would be necessary until all eight f_R's have decayed. In this case, the equations would encompass a total of 64 DOFs (8 x 8). This launch vehicle to launch pad separation occurs very rapidly, thus the iterative solution is necessary only for a very short duration. As mentioned before, a smaller time step is easily changed by simply recalculating the coefficients of integration. The above solution does not include friction effects.

IX. Component Interface Forces

In the above examples, the resulting constraint force comprised the local nonlinearity. Often, nonlinear forces are determined by the magnitudes of component interface forces, e.g., payload trunnion friction effects at the Orbiter interface. In order to treat such cases in a transient solution, the interface forces can be carried as part of the solution, i.e., they become solution DOFs at every time step. In linear transient solutions, these forces are system internal loads and are recovered only after the entire transient has been solved. The component interface forces of interest can be determined by looking at the component's governing equilibrium equation:

$$\begin{bmatrix} m_{bb} & m_{bi} \\ m_{ib} & m_{ii} \end{bmatrix} \begin{Bmatrix} \ddot{x}_b \\ \ddot{x}_i \end{Bmatrix} + \begin{bmatrix} k_{bb} & k_{bi} \\ k_{ib} & k_{ii} \end{bmatrix} \begin{Bmatrix} x_b \\ x_i \end{Bmatrix} = \begin{Bmatrix} f_b \\ f_i \end{Bmatrix} \quad (56)$$

where k and m are the component respective stiffness and mass matrices used in the synthesis of the system modal properties. The subscripts b and i refer to boundary and internal DOFs respectively. The structural component's coordinates, x, can be discrete physical, generalized, or even mixed, but the boundary DOFs must be physical if f_b is to represent the physical interface forces. Note that the component's damping properties are treated as being negligible. The f_i term is the internal force vector acting on the component which could be comprised of internal nonlinear forces and/or externally applied forces. Eq. 56 is expressed in system modal coordinates by employing the component's eigenvector partition:

$$\begin{bmatrix} m_{bb} & m_{bi} \\ m_{ib} & m_{ii} \end{bmatrix} \begin{bmatrix} \phi_b \\ \phi_i \end{bmatrix} \begin{Bmatrix} \ddot{\xi} \end{Bmatrix} + \begin{bmatrix} k_{bb} & k_{bi} \\ k_{ib} & k_{ii} \end{bmatrix} \begin{bmatrix} \phi_b \\ \phi_i \end{bmatrix} \begin{Bmatrix} \xi \end{Bmatrix} = \begin{Bmatrix} f_b \\ f_i \end{Bmatrix}$$

$$(57)$$

Once the modal response is known, the forces can be directly solved. However, experience has shown that the internal forces, f_i, will not be the same as used to determine the system modal response. For example, in NSTS/payload linear transient dynamics, the internal forces acting on the payload are zero; but, the use of Eq. 57 will yield nonzero f_i. This inaccuracy is a modal truncation effect. Basically, since the system modes where truncated, not all of the information imbedded in the mass and stiffness matrices was included in the transient modal solution. Put another way, the modal truncation was an imposed constraint upon the system, and the nonzero f_i reflects the forces necessary to enforce these constraints.

Experience has shown that the physical accelerations recovered from a truncated modal solution (vs. number of modes truncated) converge faster than the same physical displacements. Thus it is dangerous to accept the above physical displacements, x_b and x_i. Use of Eq. 57 is most often referred to as the modal displacement method and its use should be avoided. A better approach is to use the internal forces, f_i, as an input and to recalculate the internal displacements, x_i. With this, the boundary forces become a function of the modal accelerations, the internal component forces, and the boundary displacements:

$$\{f_b\} = [m_{bb} - k_{bi} k_{ii}^{-1} m_{ib}] [\phi_b] \{\ddot{\xi}\}$$

$$+ [m_{bi} - k_{bi} k_{ii}^{-1} m_{ii}] [\phi_i] \{\ddot{\xi}\}$$

$$+ [k_{bb} - k_{bi} k_{ii}^{-1} k_{ib}] [\phi_b] \{\xi\}$$

$$+ [k_{bi} k_{ii}^{-1}] \{f_i\} \tag{58}$$

Note that in this equation, the boundary displacements are multiplied by the component's stiffness matrix reduced down to the boundary DOFs. Thus for a simply supported component, the modal displacements contribute zero to the boundary or interface forces. It is standard practice to use Eq. 58 to recover NSTS/payload interface forces, the displacement modal truncation effects being considered small for 1 or 2 DOFs indeterminant systems.

For indeterminant interfaces, the use of Eq. 58 still suffers from modal truncation inaccuracies. A better approach is to resolve for the physical boundary displacements as system internal displacements, i.e., repeat the same logic as between Eqs. 57 and 58, except at the system level and with discrete physical DOFs. In this manner, only the physical accelerations (based on truncated modal accelerations) are accepted as accurate. This method is most often referred to as the modal acceleration method (the method of Eq. 58 has been referred to as the hybrid method). The modal acceleration method recovers all of the system truncated flexibility at the boundary DOFs. This recovered flexibility is added back into the analysis as a quasi-static effect.

However, for large problems, its use is quite cumbersome and expensive. The same flexibility correction can be incorporated into the analysis by the use of residual modes. For a boundary consisting of n number of DOFs, n number of residual modes would be necessary. Earlier it was mentioned that residual modes could be used to account for changes in system modal properties due to changing constraint conditions. Their use in lieu of the modal acceleration method would be exactly the same.

The same truncated modal displacement inaccuracies exist in the calculation of component relative displacements. Historically, clearance loss calculations have been based on the truncated modal displacements. It must be remembered that truncating a mode is a constraint on the system, i.e., the system is no longer allowed to deform in that mode shape and is therefore analytically stiffer. The physical discrete stiffness matrix is typically test verified via static influence coefficient testing. Therefore, the analytic flexibility should be quite accurate. However, experience has shown that modal truncation typically results in flexibility losses of upward to 50 per cent, thus destroying that accuracy acquired through detailed finite element modeling and expensive hardware verification testing. The error induced by use of the modal displacement method in the calculation of clearance losses is not only unknown, but is also almost always unconservative. It is therefore recommended to use residual modes to correct for truncated flexibility at all DOFs used in the calculations of clearance loss.

X. Kinetic and Static Friction

The treatment of kinetic (sliding) and static (stuck or stiction) friction is handled in the same manner as the simple deadband presented in Section III. In the stuck condition, the frictional force constitutes an internal load, i.e., the load acts the same as if the sliding interface was pinned together. Analytically, these forces are the result of the application of the appropriate constraint equation, just as in the "bottom out" condition of the simple deadband of Section III, Eq. 13. Sliding friction results when the static friction exceeds the maximum friction force available, i.e., μN. At this point, the friction joint is allowed to slide, the same as in the no contact condition of the simple deadband except that a resistive friction force is applied. Eq. 59 has the same form as the "bottom out" deadband, Eq. 16, except that δ is now the relative displacement at which the joint became locked:

$$
\begin{bmatrix}
I & 0 & 0 & -[\phi_{Ai}][B_{ii}][\phi_{ji}]^T \\
0 & I & 0 & [\phi_{Bi}][B_{ii}][\phi_{ji}]^T \\
-I & I & I & 0 \\
0 & 0 & I & 0
\end{bmatrix}
\begin{Bmatrix}
x_A \\
x_B \\
\Delta x \\
f_s
\end{Bmatrix}
=
\begin{Bmatrix}
[\phi_{Ai}]\{D_i\} \\
[\phi_{Bi}]\{D_i\} \\
0 \\
\delta
\end{Bmatrix}
\quad (59)
$$

The analysis would use this equation to solve for the static friction force, f_s, then Eqs. 9, 10, and 11 to solve for the modal response.

The boundary forces can then be determined from Eq. 58. If the static friction force exceeds μN, the solution to that time step is discarded, Eq. 59 is modified to remove the constraint, and a resistive kinetic friction force is applied as follows:

$$
\begin{bmatrix}
I & 0 & 0 & -[\phi_{Ai}][B_{ii}][\phi_{ji}]^T \\
0 & I & 0 & [\phi_{Bi}][B_{ii}][\phi_{ji}]^T \\
-I & I & I & 0 \\
0 & 0 & 0 & I
\end{bmatrix}
\begin{Bmatrix}
x_A \\
x_B \\
\Delta x \\
f_k
\end{Bmatrix}
=
\begin{Bmatrix}
[\phi_{Ai}]\{D_i\} \\
[\phi_{Bi}]\{D_i\} \\
0 \\
f_k
\end{Bmatrix}
\tag{60}
$$

For most applications, it is probably accurate enough to base f_k upon the previous time step, i.e., let the kinetic friction force lag one time step. The integration time steps are usually small and the exact coefficients of sliding friction, μ, are not precisely known. In fact, the μ most often used is based upon an envelope of test data. The tendency for f_k to be somewhat larger part of the integration time, and then somewhat smaller than a converged solution, should average out any gross errors induced by this approximation. The solution now continues until the relative motion reverses direction. The joint now change states, and becomes stuck again and Eq. 59 is employed. Care must be taken to insure that the kinetic friction force applied to the system always resists the relative motion, Δx.

Some applications may require that the kinetic friction force not lag one time step. This can be accomplished by including the interface forces as a part of the solution. Eqs. 9, 10, and 11 are substituted in Eq. 58, resulting in the following:

$$
\{f_b\} = [\mathcal{K}] \{f_k\} + \{\mathcal{V}\}
\tag{61}
$$

where $\mathcal{K}$ is the boundary force coupling from the friction forces and $\mathcal{V}$ is derived from the forcing function and the previous solution. Incorporation of Eq. 61 into Eq. 60 yields the following:

$$
\begin{bmatrix}
I & 0 & 0 & -[\phi_{Ai}][B_{ii}][\phi_{ji}]^T & 0 \\
0 & I & 0 & [\phi_{Bi}][B_{ii}][\phi_{ji}]^T & 0 \\
-I & I & I & 0 & 0 \\
0 & 0 & 0 & I & 0 \\
0 & 0 & 0 & -\mathcal{K} & I
\end{bmatrix}
\begin{Bmatrix}
x_A \\
x_B \\
\Delta x \\
f_k \\
f_b
\end{Bmatrix}
=
\begin{Bmatrix}
[\phi_{Ai}]\{D_i\} \\
[\phi_{Bi}]\{D_i\} \\
0 \\
f_k \\
\mathcal{V}
\end{Bmatrix}
\tag{62}
$$

This equation can be solved iteratively to reach a converged kinetic friction force at each time step.

The difference between a linear analysis and a friction analysis is that friction provides a mechanism by which energy can be transferred between modes. Although friction does result in greater over all system damping, this redistribution of energy can result in higher component responses. Because of the greater damping, many responses (maybe most) will be less severe; however, since, in most cases, friction cannot be depended upon, friction analyses cannot be used for any load relief.

Again, the x_A and x_B DOFs contain their total flexibility by virtue of the dynamic modes retained in the system and the residual modes used to recover the truncated flexibility. In this manner, the transitions between Eqs. 59 (static friction) and 60 (sliding friction) continually observe the principles of modal synthesis laid down in Ref. 6.

XI. Numerical Stability

The above discussion presented a powerful analytic method to the solution of a set of complicated problems. The method is based upon the standard linear modal solution technique and thus avoids many of the complications associated with finite difference approaches. However, some care must be taken to insure a numerically stable solution. This is accomplished by understanding the mechanisms of the solution technique.

As mentioned earlier, the solution technique assumes that the external forces vary linearly between the solution time steps. The local nonlinearities were induced into the system as external forces which were generated by application of various constraint equations. It must be realized that no assumptions have been made on the nature of the constraints between the solution time steps, i.e., the nonlinear forces are the result of enforcement of the constraint equations at the solution times. Mathematically, this implies that the motion of the constraint DOFs are not restrained to following the exact same trajectory between the solution time steps and therefore, the constraint forces can add or substract energy (i.e., it is possible for these forces to perform net work). Obviously, this "constraint force work" must be negligible in order for the solution to be accurate. This is accomplished by having a sufficient number of solution time steps per the shortest dynamic modal period included in the analysis.

At the same time, system modal accuracy must also be insured. This is accomplished by use of a sufficient number of component modes, along with the necessary residual modes, so as to observe the principles of modal synthesis (Ref. 6). High frequency modes (residual modes are very high) can be and should be treated quasi-statically.

With the above, the question of numerical stability and system modal accuracy boils down to the determination of the highest frequency component mode to be treated dynamically and selection of a time step small enough for its accurate treatment. In linear modal analyses, the time step is usually selected to be 1/16 to 1/20 the period of the highest frequency mode of interest. Since contact/recontact forces (and other nonlinear forces) can vary quite rapidly, it should be assumed that the highest frequency dynamic mode will be of interest in determination of the resulting local nonlinear forces. With a time step of 1/16 to 1/20 this shortest modal period, the differences between the constraint DOFs' trajectories should be so small that the "constraint force work" will be negligible. Also, convergence of the solution is easily tested by use of a smaller time step. In analyses in which the time interval of rapid change in local nonlinear forces is known, (e.g., launch vehicle/launch pad separation sequence), the analyst may elect to transition to a smaller integration time step during the subject interval. Again, this is easily accomplished by recalculation of the coefficients of integration.

XII.   Nomenclature

A, A', B, B'	=	coefficients of integration
b	=	modal damping
D	=	all known variables to the modal displacement solution
F, F'	=	coefficients of integration
f	=	physical force induced by the local nonlinearity
G, G'	=	coefficients of integration
h	=	time step size
I	=	identity matrix
k	=	generalized modal stiffness
m	=	generalized modal mass
N	=	normal force
P	=	generalized force
V	=	all known variables to the modal velocity solution
x, y, z	=	discrete physical degrees of freedom
β	=	modal damping coefficient
Δ	=	displacement difference
δ	=	displacement limit
m	=	coefficient of friction
ξ	=	modal displacement degree of

		freedom
ϕ	=	eigenvectors
ω	=	eigenvalues
0	=	null matrix

Superscript

T	=	matrix transpose

Subscript

i	=	modal degree of freedom
j	=	physical degree of freedom number
k	=	iteration number
n	=	time step number
o	=	eigenvalue/generalized mass

XIII.   References

1. Henkel, E. E., Misel, J. E., and Frederick, D. H., *A Methodology to Include Static and Kinetic Friction Effects in Space Shuttle Payload Transient Loads Analysis*, Proceedings of the AIAA Shuttle Environment and Operations Meeting, AIAA, Washington, D. C., October 31 - November 2, 1983, pp 171 - 177, No. 83-2654.

2. Henkel, E. E., and Mar, R., *Improved Method for Calculating Booster to Launch Pad Interface Transient Forces,* AIAA Journal of Spacecraft and Rockets, November - December 1988, Vol. 25, pp 433 - 438.

3. Bowden, A. M., and Alexander, R. M., "A General Approach to Modal Analysis for Time-Varying Systems", AIAA Paper 88-2356, April 1988.

4. Henkel, E. E., Hewlett, R. A., and Mar, R., *Transient Solution of Time-Variant Structural Systems Using Invariant Modal Properties*, AIAA Journal of Guidance, Control, and Dynamics, July-August 1991, Vol. 14, No. 4, pp. 761-769.

5. The NASTRAN Theoretical Manual (Level 16.0), NASA SP-221(03), general release March 1, 1979.

6. Rubin, S., *Improved Component-Mode Representation for Structural Dynamic Analysis*, <u>AIAA Journal</u>, Vol 13, Number 8, August 1975, pp 995 - 1006.

7. Bathe, Klaus-Jurgen, and Sheryl Gracewski, *On Nonlinear Dynamic Analysis Using Substructuring and Mode Superposition*, <u>Computers and Structures</u>, Vol 13 (1981),pp 699-707.

BALANCED SYSTEMS AND STRUCTURES:
REDUCTION, ASSIGNMENT, AND PERTURBATIONS

Wodek Gawronski

Jet Propulsion Laboratory
California Institute of Technology
Pasadena, California 91109

1. Introduction

New results of properties of balanced linear systems and structures are presented. Balanced representation is defined for systems with poles at an imaginary axis or at the origin. The grammians do not exist in this case, but introduced antigrammians exist, making balanced reduction possible.

System grammians of specified properties are obtained through assigning the sensor and actuator configuration. This configuration is determined using input-output assignment procedures introduced for general systems and specified to structures. A system is said to be uniformly balanced if all its Hankel singular values are equal. It can always be assigned through shaping the input or output configuration, and its properties are analyzed in this paper.

Structures with small Hankel singular values (which reflect low level dynamics of the system) cause difficulties in identification in the presence of noise. A method of varying the system input or output configuration allows the "generic" low-level dynamics to be distinguished from the low dynamics due to ill-positioned sensor or actuators.

2. Balanced Representation and Reduction

In this paper a linear time invariant system

$$\dot{x}=Ax+Bu, \qquad y=Cx+Du, \qquad x(0)=x_0 \tag{1}$$

is considered, where $x \in R^n$ is the system state, $u \in R^p$ is the input, $y \in R^q$ is the output, and (A,B,C,D) is the system state space representation. A structure is a system given by representation (1) in modal form

$$A=diag(A_i), \qquad A_i = \begin{bmatrix} -2\zeta_i\,\omega_i & -\omega_i \\ \omega_i & 0 \end{bmatrix}, \qquad i=1,\ldots,n/2, \tag{2}$$

where $\zeta_i \ll 1$ is a modal damping of the i-th mode, and ω_i is the i-th natural frequency, and $\omega_i \neq \omega_j$, $i \neq j$, for $i,j=1,\ldots,n/2$.

For stable systems the balanced reduction is well developed, and computationally efficient algorithms are available [1, 2, 3, 4, 5]. For unstable systems many approaches and solutions already exist [5, 6, 7, 8]. For systems with poles at zero or on an imaginary axis (e.g. systems with integrators, or structures with rigid-body modes) the balanced reduction technique has not yet been developed. For this important class of systems balanced reduction cannot be applied for the obvious reason: the grammians do not exist. Recall that the grammians are the base of the system reduction of stable and unstable systems (grammians of unstable systems exist, although are not positive definite). In this paper antigrammians are introduced as a tool for system balancing. It will be shown that the obtained balanced systems with poles at zero can be reduced in a manner similar to stable systems.

2.1. Stable Systems and Structures

A stable system representation (1) is considered. Its controllability and observability grammians W_c and W_o are solutions of the Lyapunov equations

$$AW_c + W_c A^T + BB^T = 0, \qquad A^T W_o + W_o A + C^T C = 0 \tag{3}$$

The system representation is balanced if its controllability and observability grammians are diagonal and equal [3]. Hence, for the balanced representation $(A_b, B_b, C_b) = (T^{-1}AT, \; T^{-1}B, \; CT)$ the following is true

$$W_c = W_o = \Gamma^2, \quad \Gamma = diag(\gamma_1, \ldots, \gamma_n), \quad \gamma_i \geq 0, \quad i = 1, \ldots, n \tag{4}$$

where T is a linear transformation, and γ_i is the ith Hankel singular value of the system. The transformation T is determined as follows [5]

$$T = PU\Gamma^{-1}, \qquad T^{-1} = \Gamma^{-1}V^T Q \tag{5}$$

The matrices Γ, V, U are obtained from the singular value decomposition of the matrix H

$$H = V\Gamma^2 U^T \tag{6}$$

where

$$H = QP, \tag{7a}$$

and P, Q are obtained from the decomposition of the grammians

$$W_c = PP^T, \qquad W_o = Q^T Q, \tag{7b}$$

e.g. Cholesky, or singular value decomposition.

The system reduction is based on the properties of grammians. Let the balanced grammian Γ be ordered in the decreasing order:

$$\Gamma = diag(\gamma_1, \; \gamma_2, \; \ldots, \; \gamma_{n-1}, \; \gamma_n) \tag{8}$$

where $\gamma_i \geq 0$, $\gamma_i \geq \gamma_{i+1}$, $i=1,\dots,n$. The system is reduced by truncating the last $n-k$ states of the balanced representation, thus leaving its first k states. Let the matrices A_b, B_b, C_b be partitioned conformably

$$A_b = \begin{bmatrix} A_{11} & A_{12} \\ A_{21} & A_{22} \end{bmatrix}, \qquad B_b = \begin{bmatrix} B_1 \\ B_2 \end{bmatrix}, \qquad C_b = [C_1 \ C_2] \qquad (9)$$

then the reduced system representation (A_r, B_r, C_r, D_r) be $A_r = A_{11}$, $B_r = B_1$, $C_r = C_1$, and D unchanged. The system can also be reduced by the low-frequency approximation, see [9], [10]

$$A_r = A_{11} - A_{12}A_{22}^{-1}A_{21}, \qquad B_r = B_1 - A_{12}A_{22}^{-1}B_2,$$

$$C_r = C_1 - C_2A_{22}^{-1}A_{21}, \qquad D_r = D - C_2A_{22}^{-1}B_2 \qquad (10)$$

For structures with small damping and distinct poles the modal representation is almost balanced, c.f. [5, 11]. Each mode has almost the same controllability and observability property, hence each mode is considered for reduction separately. For a structure with m modes, matrix B has $2m$ rows, and C has $2m$ columns. Denote C_q the first m columns of C, C_r the last m columns of C, then b_i is the ith row of B and c_{qi} is the ith column of C_q, c_{ri} is the ith column of C_r. The Hankel singular value for the ith mode is given by [5]

$$\gamma_{2i} = \gamma_{2i-1} = \frac{\sqrt{(b_i b_i^T)(c_{qi}^T c_{qi} + \omega_i^2 c_{ri}^T c_{ri})}}{4\zeta_i \omega_i^2} \qquad (11)$$

$i=1,\dots,n/2$, and serves as ranking of each mode in the reduction procedure.

2.2. Systems with Integrators and Free Structures

The above model order reduction procedure is based on the system's joint controllability and observability properties, through balancing its grammians. For linear unstable systems various techniques based on properties of the controllability and observability grammians have been developed [5,6,7,8]. Consider a system with integrators, whose output is an integral of an input. It has a pole (or poles) at zero, consequently, its controllability and observability grammians do not exist, and model reduction based on its grammian properties cannot be executed. It remains unquestionable, however, that systems with integrators are controllable and observable, hence these properties can still be used in model reduction. A system with integrators is a sub-class of linear systems defined as follows.

Definition 1. An observable and controllable system with $n-m$ poles stable, the remaining m poles at zero, and A nondefective (geometric multiplicity of poles at zero is m) is called a system with integrators.

The system with m integrator is often called a type m system, see [12, 13]. The reduction problem for systems with integrators is solved by introducing antigrammians [14].

Definition 2. For a controllable and observable triple (A,B,C), the matrices V_c, V_o satisfying the following Riccati equations:

$$V_c A + A^T V_c + V_c B B^T V_c = 0, \qquad V_o A^T + A V_o + V_o C^T C V_o = 0 \qquad (12)$$

are the controllability and observability antigrammians.

For stable, controllable, and observable systems $V_c = W_c^{-1}$, $V_o = W_o^{-1}$, where W_c, W_o are controllability and observability grammians which satisfy the Lyapunov equations (3). Note that the grammians for a system with integrators do not exist, but antigrammians do; for an unobservable or uncontrollable system antigrammians do not exist, but grammians do. The existence of antigrammians is exploited for the balancing and model reduction of systems with integrators.

For a stable, controllable, observable, and balanced system

the grammians as well as the antigrammians are equal and diagonal

$$W_c = W_o = \Gamma, \qquad V_c = V_o = \Pi, \quad \text{and} \quad \Pi = \Gamma^{-1} \qquad (13)$$

where $\Gamma = diag(\gamma_i)$, $\Pi = diag(\pi_i)$, $i = 1, 2, \ldots, n$, and satisfy the following equations

$$\Pi A_b + A_b^T \Pi + \Pi B_b B_b^T \Pi = 0, \qquad \Pi A_b^T + A_b \Pi + \Pi C_b^T C_b \Pi = 0 \qquad (14a)$$

$$A_b \Gamma + \Gamma A_b^T + B_b B_b^T = 0, \qquad \Gamma A_b + A_b^T \Gamma + C_b^T C_b = 0, \qquad (14b)$$

The representation (A_b, B_b, C_b, D) is balanced.

Theorem 1. For a balanced system with integrators $\Pi = diag(0_m\ \Pi_o)$, where 0_m is a $m \times m$ zero matrix, and A_b is block-diagonal, $A_b = diag(0_m\ A_{bo})$.

Proof. Consider A, B, C in the form

$$A = diag(0_m\ A_o), \qquad B^T = [B_r^T\ B_o^T], \qquad C = [C_r\ C_o]. \qquad (15)$$

Matrix A in the form (15) always exists due to m poles at zero, and B, C exist due to nondefectiveness of A. From Eq.(12) it follows that

$$V_{crr} B_r = 0, \quad C_r V_{orr} = 0, \quad V_{cro} = 0, \quad V_{oro} = 0 \qquad (16)$$

where V_c, V_o are divided conformably to A:

$$V_c = \begin{bmatrix} V_{crr} & V_{cro} \\ V_{cro}^T & V_{coo} \end{bmatrix}, \qquad V_o = \begin{bmatrix} V_{orr} & V_{oro} \\ V_{oro}^T & V_{ooo} \end{bmatrix}$$

For a controllable and observable system (by Definition 1) the matrices B_r and C_r are of full rank, thus from Eq.(16) one obtains $V_{crr} = 0$, $V_{orr} = 0$, and $V_c = diag(0_m\ V_{coo})$, $V_o = diag(0_m\ V_{ooo})$, which in balanced coordinates gives $\Pi = diag(0_m\ \Pi_o)$. $\square$

Theorem 2. A balanced representation of a system with integrators (A_b, B_b, C_b, D) is obtained by the transformation T_b

$$A_b = T_b^{-1} A T_b, \quad B_b = T_b^{-1} B, \quad C_b = C T_b \tag{17}$$

where

$$T_b = T_1 T_2. \tag{18}$$

The transformation T_1 turns A into block-diagonal form $A_1 = diag(0_m, A_o)$, e.g. into real modal form

$$A_1 = T_1^{-1} A T_1, \tag{19}$$

and the transformation T_2 is in the form $T_2 = diag(I_m, T_{bo})$ where I_m is an identity matrix of order m, and T_{bo} balances A_o

$$A_{bo} = T_{bo}^{-1} A_o T_{bo} \tag{20}$$

Proof. Immediate, by introducing Eq.(18) to (14a).□

The antigrammian of the balanced system with integrators is ordered increasingly:

$$\Pi = diag(\pi_1, \pi_2, ..., \pi_{n-1}, \pi_n) \tag{21}$$

where $\pi_i \geq 0$, $\pi_{i+1} \geq \pi_i$, $i = 1, ..., n$, with the first m singular values at zero, $\pi_i = 0$, $i = 1, ..., m$. The system is reduced by truncating the last $n-k$ states of the balanced representation and leaving its first k states. The system can also be reduced by the low-frequency approximation, as in Eq.(10).

For free structures with poles on an imaginary axis the Hankel singular values for the rigid body modes tend to infinity, see Eq.(11), while the Hankel singular values for stable modes are finite. In this case the procedure described in Section 2.1 can be applied with rigid body modes always included in the reduced model.

Example 1. A system with integrators and the state-space representation

$$A = \begin{bmatrix} -5 & 0 & 0 & 0 & 0 & 0 \\ 1 & 0 & 0 & 0 & 0 & 0 \\ 0.5 & 5 & -20 & -2 & 0 & 0 \\ 0 & 0 & 1 & 0 & 0 & 0 \\ 0 & 0 & 40 & 40 & -0.2 & -1 \\ 0 & 0 & 0 & 0 & 1 & 0 \end{bmatrix}$$

$B^T = [1 \; 0 \; 0 \; 0 \; 0 \; 0]$, $C = [0 \; 0 \; 0 \; 0 \; 0 \; 10]$, has one pole at zero. Although the system is controllable and observable, its controllability and observability grammians do not exist. The balanced representation is obtained from Eq.(17)

$$A_b = \begin{bmatrix} 0 & 0 & 0 & 0 & 0 & 0 \\ 0 & -0.1063 & -0.0549 & -0.0527 & 0.0091 & 0.0011 \\ 0 & -0.0549 & -0.1055 & -0.9949 & 0.0324 & 0.0040 \\ 0 & 0.0527 & 0.9949 & -0.8882 & 0.0328 & 0.0040 \\ 0 & 0.0091 & 0.0324 & -0.0328 & -5.1058 & -1.2717 \\ 0 & 0.0011 & 0.0040 & -0.0040 & -1.2517 & -1.9794 \end{bmatrix}$$

$B_b^T = [20 \quad 13.9642 \quad 3.8892 \quad -3.2384 \quad -0.5972 \quad -0.0735]$

$C_b = [9.9964 \; -13.9642 \; -3.8892 \; -3.2384 \quad 0.5972 \quad 0.0735]$

and the balanced antigrammian

$\Pi = diag(0, \; 0.0011, \; 0.0139, \; 0.0169, \; 28.6333, \; 7330.82)$

satisfies Eq.(14a). Its last two (largest) singular values indicate that the system can be reduced to a system of order 4. Indeed, as shown in Figs.1 and 2, the output of the reduced model, obtained either by truncation or low-frequency approximation procedure, is close to the output of the full-order model, as is illustrated by the impulse responses

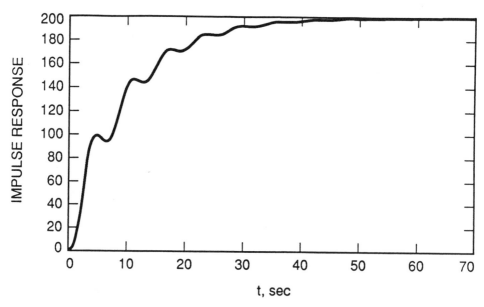

Fig.1. Impulse responses of the full-order and reduced system with integrators.

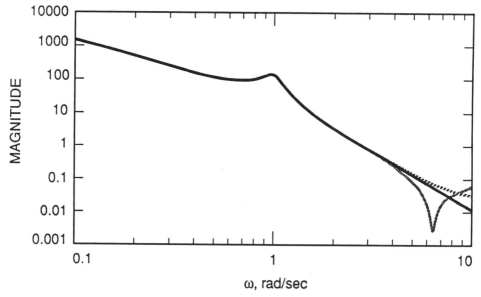

Fig.2. Magnitude of transfer function of the full-order and reduced system with integrators.

and the transfer functions of the full (solid line) and the reduced order models (dashed line = truncation, dotted line = low frequency approximation, curves in Fig.1 are overlapped).

Example 2. Consider the NASA Deep Space Network DSS-13 antenna (presented in Fig.3) and its rate loop model (Fig.4; for details of the model see [15]). The rate loop model consists of the antenna structural model and models of elevation and azimuth drives. Two poles of the structural model are on the imaginary axis, each drive has a pole at zero, and the rate loop model has three poles at zero. The state-space representation of the rate loop model has $n = 90$ states. The model is reduced using antigrammian balancing in two steps. First, each subsystem (the structure, elevation drive, and azimuth drives) is reduced, next the rate loop system itself is reduced. The Hankel singular values are presented in Fig.5 for a 21-mode structural model (the plot presents flexible modes only, since the Hankel singular values for the rigid body modes are infinitely large). Ten modes were selected in the reduced structural model: two rigid body, and eight flexible modes (shadowed in Fig.5). Singular values of the balanced antigrammians of the elevation and azimuth drives are shown in Fig.6. Five states, one with the pole at zero, and four with the smallest singular values, have been selected for each reduced drive model. The rate loop model, combined from the reduced subsystems, now has 35 states, including three states with poles at zero. This model is further reduced using balanced antigrammians: their singular values are shown in Fig.7. By deleting the states with the largest singular values one obtains the 27-state rate loop model. As a result, the model has shrunk from a 90-state to a 27-state model. The reduced model preserves the full-model properties, as is illustrated in Fig. 8 (frequency response) and Fig.9 (step response).

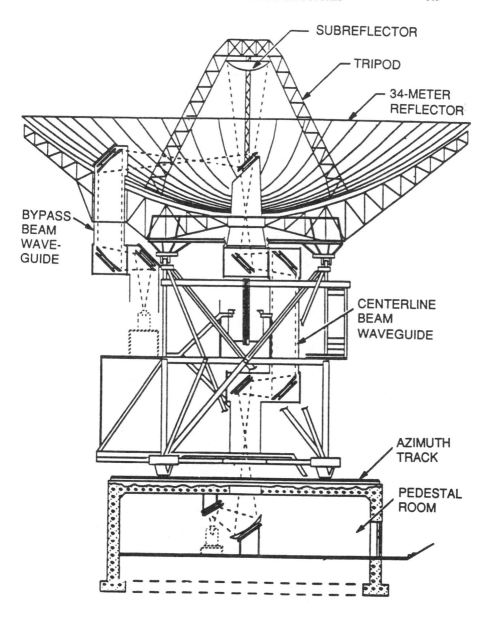

Fig.3. The DSS-13 antenna.

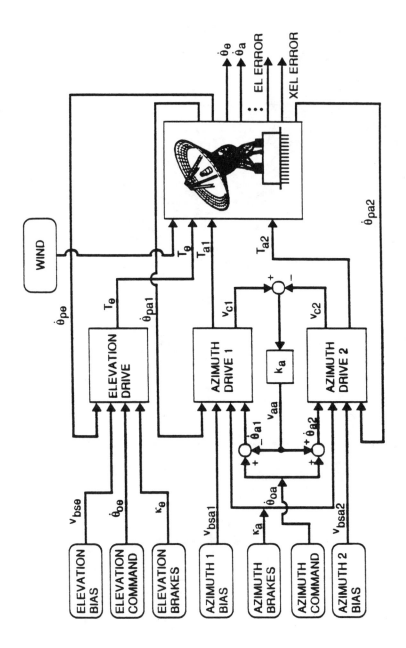

Fig. 4. Rate loop control system of the DSS-13 antenna.

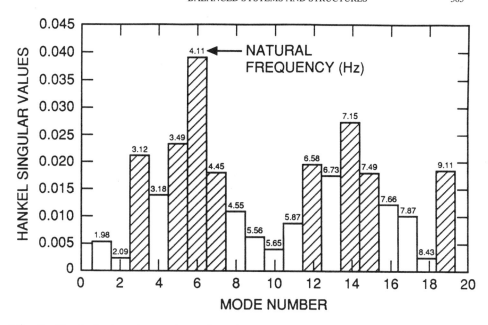

Fig.5. Hankel singular values for the antenna structure.

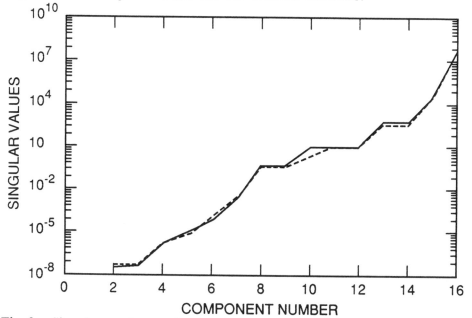

Fig.6. Singular values of the balanced antigrammians for the elevation and azimuth drives.

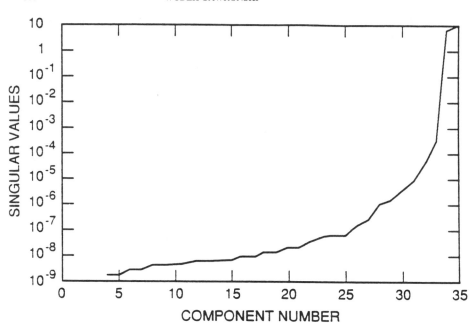

Fig.7. Singular values of the balanced antigrammians for the rate loop model.

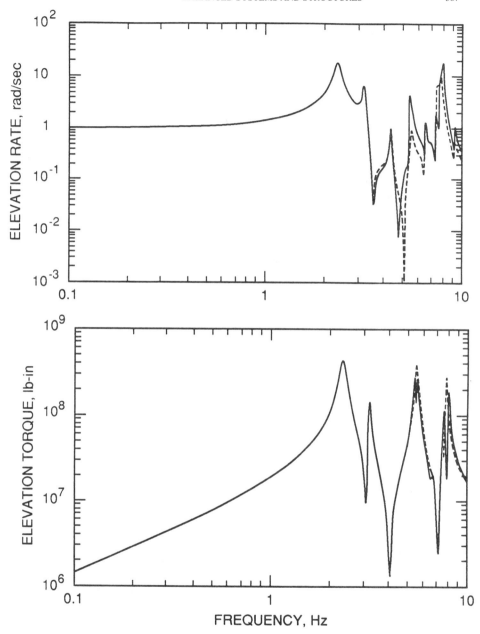

Fig.8. Magnitudes of the transfer functions for the full and reduced rate loop model: a) elevation rate, rad/sec, b) elevation torque, lbs-in.

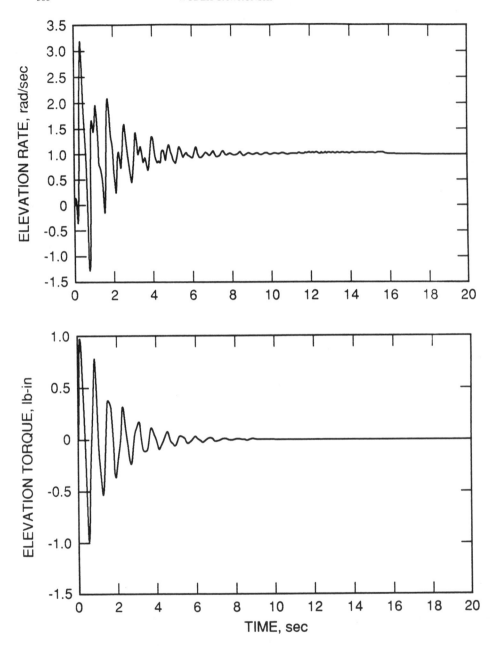

Fig.9. Step responses of the full and the reduced rate loop model: a) elevation rate, rad/sec, b) elevation torque, lbs-in.

3. Balanced Input-Output Assignment

In this section the controllability and observability properties of a linear system, in terms of sensor and actuator locations, are investigated. Typically, for a given state-space representation these properties are reviewed, rather than actively shaped. In many cases, however, one would like to prescribe controllability and observability properties. This can be done in two ways: by modifying the system structural properties, such as by feedback assignment [17], or by determining proper sensor and/or actuator configurations [18]. The latter method, called an input-output assignment problem, is addressed here. This approach has been found useful in control design (to avoid spillover), or identification (placing actuators and sensors to recover the full order system).

3.1. System Assignment

Consider a stable system (1), and the following problem. For a given positive definite matrix W, find a state-space representation such that its grammians are equal to W. It can be solved by finding $\bar{B}$ and/or $\bar{C}$ such that grammians of $(A, \bar{B}, \bar{C})$ are equal to W. It is divided into three separate problems.

Problem 1. Find $\hat{C}$ and the transformation T such that $\bar{W}_c = \bar{W}_o = W$ for the representation $(\bar{A}, \bar{B}, \hat{C})$, where $\bar{A} = T^{-1}AT$, and $\bar{B} = T^{-1}B$.

Problem 2. Find $\hat{B}$ and the transformation T such that $\bar{W}_c = \bar{W}_o = W$ for the representation $(\bar{A}, \hat{B}, \bar{C})$, where $\bar{A} = T^{-1}AT$, and $\bar{C} = CT$.

Problem 3. Find $\hat{B}$ and $\hat{C}$ and the transformation T such that $\bar{W}_c = \bar{W}_o = W$ for the representation $(\bar{A}, \hat{B}, \hat{C})$, where $\bar{A} = T^{-1}AT$.

Not every positive definite grammian can be obtained through sensor or actuator placement; the solution of the assignment problem does not exist for every positive semi-definite matrix W. A matrix W, for which the solution exist is an assignable one.

Definition 3. A symmetric positive definite matrix W is c-assignable with respect to (A, C) if there exist a nonsingular T and input matrix B, such that the controllability grammian of (T^1AT, T^1B) is equal to W. A symmetric positive-definite matrix W is o-assignable with respect to (A, B) if there exist a nonsingular T and output matrix C, such that the observability grammian of (T^1AT, CT) is equal to W. A symmetric positive-definite matrix W is assignable with respect to A if there exist a nonsingular T and matrices B, C such that the controllability and observability grammians of (T^1AT, T^1B, CT) are equal to W.

Theorem 3. A symmetric positive semi-definite matrix W is c-assignable with respect to A, if R_c is positive semi-definite. It is o-assignable with respect to A if R_o is positive semi-definite, where

$$R_c = -AW-WA^T, \qquad R_o = -A^TW-WA. \tag{22}$$

If it is c- and o-assignable, it is assignable.

Proof. Straightforward.□

Assuming c- and/or o-assignable W, the following algorithms solve problems *1, 2, 3*, respectively.

Algorithm 1. For given A, B, and matrix W o-assignable and positive-definite:
1. Determine W_c from the Lyapunov equation (3).
2. Find P_c and P from the decomposition of W_c and W: $W_c = P_c P_c^T$, $W = PP^T$.
3. Find $\bar{A}$, $\bar{B}$: $\bar{A} = T^1AT$, $\bar{B} = T^1B$, $T = P_c P^{-1}$.
4. Determine $\hat{C}$ from the decomposition of $R_o = -\bar{A}^TW-W\bar{A}$: $R_o = \hat{C}^T\hat{C}$ (R_o is positive semi-definite for W o-assignable).

In order to check the algorithm, note from step 4 that $\bar{A}^TW+W\bar{A}+\hat{C}^T\hat{C}=0$ (i.e. W is the assigned observability grammian). For $\bar{A}$, $\bar{B}$ from step 3 one obtains $\bar{A}W+W\bar{A}^T+BB^T=0$, hence W is the assigned controllability grammian.

A similar algorithm is obtained for determination of actuator locations.

Algorithm 2. For given A, C, and matrix W c-assignable and positive-definite:
1. Determine W_0 from the Lyapunov equation (3).
2. Find P_0 and P such that $W_0 = P_0^T P_0$, $W = P^T P$.
3. Find $\bar{A}$, $\bar{C}$: $\bar{A} = T^1 A T$, $\bar{C} = CT$, $T = P_0^{-1} P$.
4. Determine $\hat{B}$ from the decomposition of $R_c = -\bar{A}W - W\bar{A}^T$: $R_c = \hat{B}\hat{B}^T$ (R_c is positive semi-definite for W c-assignable).

Problem 3 is solved by the following algorithm:

Algorithm 3. For W assignable and positive definite the matrices B and C obtained from the decomposition of R_c and R_0: $R_c = \hat{B}\hat{B}^T$, $R_0 = \hat{C}^T\hat{C}$ are the solution of Problem 3.

3.2. Structure Assignment

The balanced grammians for structures have the following form

$$W_c = W_0 = diag(\gamma_1, \ \gamma_1 + \varepsilon_1, \ \gamma_2, \ \gamma_2 + \varepsilon_2, ..., \ \gamma_n, \ \gamma_n + \varepsilon_n) \qquad (23)$$

where $|\varepsilon_i| \ll \gamma_i$, $i = 1,...,n$ (i.e., each pair of states is described by two almost equal Hankel singular values). The following theorem gives a sufficient condition of assignability of structures for A_i as in (2).

Theorem 4. Let $W_i = \gamma_i I_2$, $\gamma_i \geq 0$, $A = diag(A_i)$, where A_i is as in (2), then $W = diag(W_i)$, $i = 1,...,n/2$ is assignable.

Proof. In this case $R_c = R_0 = diag(R_i)$, $R_i = 4\zeta_i\omega_i\gamma_i \ diag(1,0)$, hence R_c and R_0 are positive-definite, and W is assignable.□

The sufficient condition is seldom met, but is useful for approximate assignment, defined as follows.

Definition 4. Let $\lambda^- = [\lambda_1^-, \ldots, \lambda_p^-]$ be a vector of all negative eigenvalues of a symmetric matrix R, and $\lambda^+ = [\lambda_1^+, \ldots, \lambda_q^+]$ be a vector of all non-negative eigenvalues of R. The matrix R is almost positive semi-definite if $\|\lambda^-\| \ll \|\lambda^+\|$, where $\|.\|$ is the Euclidean norm.

Definition 5. A symmetric positive semi-definite matrix W is approximately c-assignable with respect to A if R_c is almost positive semi-definite; it is approximately o-assignable with respect to A if R_o is almost positive semi-definite. If it is approximately c- and o-assignable, it is approximately assignable.

The error ε of the approximate assignment

$$\varepsilon = \|\hat{W} - W\| / \|W\|, \qquad \hat{W} = (W_c W_o)^{1/2} \tag{24}$$

quantifies the accuracy of the assignment.

It is well known, see [5, 11], that modal representation of a structure (with small damping and separate poles, assumed as above) is almost balanced, with the grammians diagonally dominant (off diagonal terms much smaller than diagonal terms), and with the diagonal terms as in (23). Thus, it follows from Theorem 4 that the grammian

$$W = \operatorname{diag}(\gamma_i I_2), \quad i = 1, \ldots, n/2, \tag{25}$$

is approximately assigned.

Consider now an assignment of each mode separately, i.e. assign a grammian

$$W_{oi} = \operatorname{diag}(0, 0, \ldots, \gamma_i I_2, \ldots, 0) \tag{26}$$

Each mode can be approximately assigned, and the combined assignment is a sum of assignments of each individual mode. Denote $B = [B_1, B_2, \ldots, B_{n/2}]$, $C = [C_1, C_2, \ldots, C_{n/2}]$, where B_i is a two-row block of B, C_i is a two–column block of C,

$$B_{oi}^T = [0, 0, \ldots, B_i^T, \ldots, 0], \quad C_{oi} = [0, 0, \ldots, C_i, \ldots, 0]. \tag{27}$$

and W, W_{oi} as in (25), and (26) respectively, then

Theorem 5. If (A,B,C) is approximately assigned with W, then (A,B_{oi},C_{oi}) is approximately assigned with W_{oi}. Conversely, if (A,B_{oi},C_{oi}), $i=1,...,n/2$ are approximately assigned with W_{oi}, then (A,B,C) is approximately assigned with W.

Proof. From the definition (2) of a structure and the Lyapunov equations (3) one obtains

$$B_i B_i^* \approx C_i^* C_i \approx -\gamma_i(A_i^* + A_i), \quad B_i B_j^* \approx C_i^* C_j \approx 0, \ i \neq j, \tag{28}$$

hence

$$B_{oi} B_{oi}^* \approx C_{oi}^* C_{oi} \approx -W_{oi}(A^* + A) \approx -(A^* + A)W_{oi} \tag{29}$$

i.e. W_{oi} is approximately assigned. To prove the second part of the theorem, note that $B = \sum_{i=1}^{n/2} B_{oi}$, $C = \sum_{i=1}^{n/2} C_{oi}$. It follows from (28) that the components of B are almost orthogonal, $B_{oi} B_{oj}^* \approx 0$, $i \neq j$, and that the components of C are almost orthogonal, $C_{oi}^* C_{oj} \approx 0$, $i \neq j$, hence $W = \sum_{i=1}^{n/2} W_{oi}$. The latter fact shows that W is approximately assigned if W_{oi}, $i=1,..,n/2$, are approximately assigned.□

The above theorem allows one to assign each mode separately, and obtain results for a combination of modes. Conversely, by determining the assignment for all modes, one can approximately assign a single mode by specifying B, C as in (27), and as illustrated in the example below.

The approximate assignment can be determined using modal parameters ζ_i, ω_i. For the Hankel singular values γ_i, from (28)

$$B_i B_i^* \approx C_i^* C_i \approx -\gamma_i(A_i^* + A_i) = diag(4\zeta_i \omega_i \gamma_i \ 0)$$

therefore

$$B_i^T = C_i = [b_{oi}\ 0], \qquad b_{oi} = 2(\zeta_i \omega_i \gamma_i)^{1/2} \tag{30}$$

is the approximate assignment obtained for collocated inputs and outputs.

The following iterative procedure reduces the approximation error.

Algorithm 4. Given approximation accuracy ε_o, positive semi-definite W, and system matrices A, B: assume C. Denote this triple (A_1, B_1, C_1) and its balanced grammian W_1. Determine P: $W = PP^T$. For $i = 1, 2, \ldots$

1. Determine P_{ci} from the decomposition of W_i: $W_i = P_{ci} P_{ci}^T$.
2. Determine A_{i+1}, B_{i+1}: $A_{i+1} = T_i^{-1} A_i T_i$, $B_{i+1} = T_i^{-1} B_i$, $T_i = P_{ci} P^{-1}$.
3. Determine R_i: $R_i = -A_{i+1}^T W - W A_{i+1}$.
4. Decompose R_i: $R_i = U_i \Lambda_i U_i^T$, where $\Lambda_i = diag(\lambda_k)$, $k = 1, \ldots, n$, and λ_k is the k-th eigenvalue of R_i.
5. Find the positive semi-definite replacement R_{pi} of R_i by setting all negative eigenvalues of R_i to zero: $R_{pi} = U_i \Lambda_{pi} U_i$, where $\Lambda_{pi} = 0.5(\Lambda_i + |\Lambda_i|)$.
6. Decompose R_{pi}: $R_{pi} = Q_i^T Q_i$, $C_{i+1} = Q_i$.
7. Balance $(A_{i+1}, B_{i+1}, C_{i+1})$, obtaining W_{i+1}.
8. Check convergence, determining ε_{i+1}: $\varepsilon_{i+1} = \|W - W_{i+1}\| / \|W\|$. If $\varepsilon_{i+1} < \varepsilon_o$ stop, otherwise set $i = i + 1$ and go to 1.

Example 3. For a system with

$$A = \begin{bmatrix} 0 & 0 & 0 & 1 & 0 & 0 \\ 0 & 0 & 0 & 0 & 1 & 0 \\ 0 & 0 & 0 & 0 & 0 & 1 \\ -150 & 50 & 0 & -0.15 & 0.05 & 0 \\ 50 & -60 & 10 & 0.05 & -0.06 & 0.01 \\ 0 & 10 & -110 & 0 & 0.01 & -0.11 \end{bmatrix},$$

$$B^T = \begin{bmatrix} 0 & 0 & 0 & 0.1 & 0.2 & -0.5 \\ 0 & 0 & 0 & -0.2 & -0.3 & 0.1 \end{bmatrix}$$

the grammian $W = diag(10, \ 10, \ 7, \ 7, \ 1, \ 1)$ is assigned by determining the matrix C of sensor locations. Using Algorithm 1, one obtains the triple $(\bar{A}, \bar{B}, \hat{C})$

$$\bar{A} = \begin{bmatrix} -0.0363 & 6.0463 & -0.0499 & -0.0019 & 0.0063 & 0.0004 \\ -6.0526 & -0.0003 & 0.0030 & 0.0003 & -0.0006 & 0.0000 \\ 0.1042 & 0.0042 & -0.1105 & -10.5229 & 0.0351 & 0.0024 \\ -0.0034 & -0.0006 & 10.5354 & -0.0004 & 0.0020 & 0.0001 \\ 0.0318 & 0.0027 & 0.0109 & -0.0001 & -0.1718 & -13.1238 \\ -0.0022 & -0.0002 & -0.0016 & 0.0000 & 13.1468 & -0.0008 \end{bmatrix}$$

$$\bar{B} = \begin{bmatrix} 0.3620 & 0.0349 & -1.1979 & 0.0640 & 0.2423 & -0.0170 \\ -0.7715 & -0.0649 & 0.3354 & -0.0315 & 0.5337 & -0.0354 \end{bmatrix}$$

$$\hat{C} = \begin{bmatrix} 0.2896 & 0.0561 & -1.2308 & 0.0709 & 0.2188 & 0.0078 \\ -0.7874 & -0.0559 & -0.1387 & -0.0361 & 0.2881 & -0.0040 \\ 0.1495 & 0.0149 & 0.1152 & -0.0405 & 0.4612 & -0.0502 \\ -0.0039 & 0.0370 & 0.0062 & 0.0655 & 0.0082 & 0.0365 \\ 0.0008 & -0.0338 & 0.0007 & 0.0377 & 0.0003 & -0.0335 \\ 0.0030 & -0.0289 & 0.0001 & -0.0001 & 0.0031 & 0.0291 \end{bmatrix}$$

For this representation the grammians are $W_c = W_o = diag(10.0610,$ $10.0609, \quad 7.0391, \quad 7.0390, \quad 1.0131, \quad 1.0130)$; i.e., it is approximately assigned, with the approximation error $\varepsilon = 0.0030$. This assignment is improved iteratively with Algorithm 4, obtaining the assignment error as in Fig.10. It shows that with 21 iterations the error improved from 0.003 to 0.000699.

Next, the assignment for $W_1 = diag(1,1,0,0,0,0)$ and $W_2 = diag(1,1,0,0,1,1)$ is examined. In the first case only the first mode is excited and observed, while in the second case the first and the third modes are excited and observed. It follows from Theorem 5 that the matrix $\bar{A}$ as above assigns W_1 and W_2, with the matrices $\bar{B}_1$, $\hat{C}_1$ and $\bar{B}_2$, $\hat{C}_2$ obtained from $\bar{B}$, $\hat{C}$ (given above) by setting to zero or scaling appropriate rows of $\bar{B}$ and columns of $\hat{C}$. In the first case (3,4,5,6) columns of $\hat{C}$

and rows of $\bar{B}$ are set to zero and (1,2) columns of $\hat{C}$ and rows
of $\bar{B}$ are scaled by 10. In the second case (3,4) columns of $\hat{C}$
and rows of $\bar{B}$ are set to zero and (1,2) columns of $\hat{C}$ and rows
of $\bar{B}$ are scaled by 10. The plots of power spectrum for the
first case are shown in Fig.11, and for the second case in
Fig.12. The figures show clearly that in the first case, only
the first mode is excited and observed, and in the second case,
the first and the third modes are excited and observed. The
assignment error is $\varepsilon=0.0197$ in the first case, and $\varepsilon=0.0419$ in
the second case.

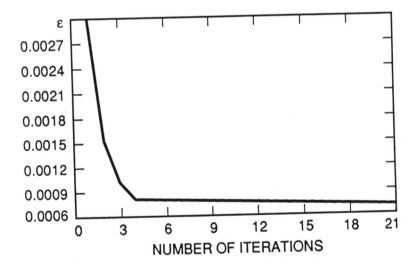

Fig.10. Assignment error of the iterative procedure.

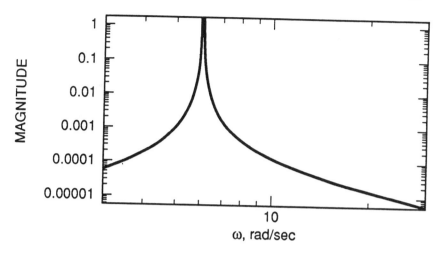

Fig.11. Magnitude of the transfer function for a single
controllable and observable mode.

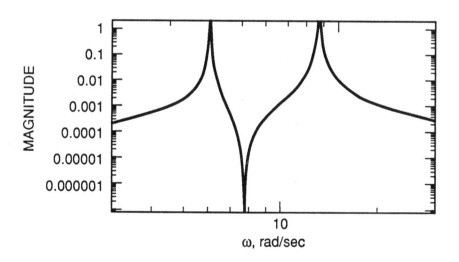

Fig.12. Magnitude of the transfer function for a pair of
controllable and observable modes.

Example 4. A flexible truss from [5] shown in Fig.13 is assigned with $W=I$ and $W=diag(1,1,0,0,...,0)$. In the first case all its modes are equally controllable and observable, in the second case only its first mode is controllable and observable. The results of the assignment are shown in Fig.14a,b. For modes equally controllable and observable the related resonant peaks are equal to two. For instance, the resonant peak for the only controllable and observable mode is equal to two, while all other resonances have been eliminated.

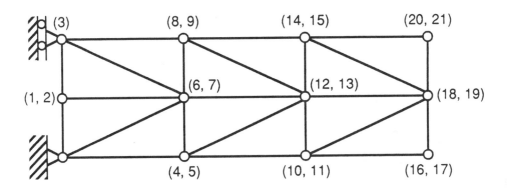

Fig.13. Flexible truss.

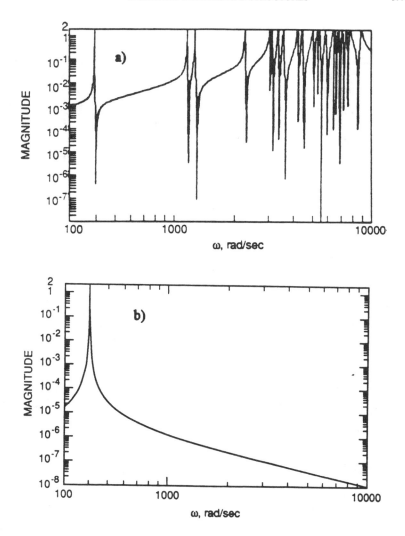

Fig.14. Magnitude of the truss transfer function: a) all modes equally controllable and observable, b) only the first mode controllable and observable.

4. Uniformly Balanced Systems and Structures

A uniformly balanced system has its states equally controllable and observable, i.e., all its Hankel singular values are equal, with grammians $W_c = W_o = \gamma I_n$, $\gamma > 0$. For stable A the uniformly balanced system is always assignable. Indeed, in this case $BB^T = C^T C = -\gamma(A + A^T)$ and $-A - A^T$ is positive-definite, hence $W = \gamma I_n$ is assignable. The uniformly balanced system is determined as follows.

Algorithm 5. For given A, B, and the controllability grammian W_c, the grammian $W = \gamma I_n$ is assigned with $C = B^T W_c^{-1} \gamma$.

Algorithm 6. For given A, C, and the observability grammian W_o, the grammian $W = \gamma I_n$ is assigned with $B = W_o^{-1} C^T \gamma$.

Proof. By inspection of the Lyapunov equations (3).□
Theorem 6. The transfer function $G(j\omega) = C(j\omega I - A)^{-1} B$ of a uniformly balanced system (A, B, C) has the following properties:
(a) G is stable,
(b) $\|G\|_{ui} \leq 2\gamma$, where $\|.\|_{ui}$ is a unitarily invariant norm,
(c) If the system is square (the number of inputs is equal to the number of outputs $= m$) then G is positive real, and $G_o = \gamma^{-1} G - I_m$ is all pass.

Proof.
(a) G satisfies eqs. (A1) and (A2) of Appendix, thus is stable.
(b) From the left-hand side of (A1) it follows that $\|GU + U^* G^*\|_{ui} \leq 2\|GU\|_{ui} = 2\|G\|_{ui}$ ($\|.\|_{ui}$ is a unitarily invariant norm e.g., $\|G\|_\infty = \sup_\omega \bar{\sigma}(G(j\omega))$ $\bar{\sigma}(G)$ - largest singular value of G, or $\|G\|_2^2 = (2\pi)^{-1} \int_{-\infty}^{\infty} tr(GG^*) d\omega$). From the right-hand side of (A1) $\|GG^*\|_{ui}/\gamma = \|G\|_2^2/\gamma$, thus $\|G\|_{ui}^2/\gamma \leq 2\|G\|_{ui}$, or $\|G\|_{ui} \leq 2\gamma$. (c) The positive-realness of the square uniformly balanced system follows from (A1) and the definition of positive-realness of a transfer function, see [19]. Introducing $G = \gamma(G_o + I_m)$ to (A1) or (A2) one obtains $G_o^* G_o = G_o G_o^* = I_m$, hence G_o is all-pass.□

The plot of the transfer function for a single-input single-output uniformly balanced system is a circle of radius γ, and center $(\gamma,0)$.

Example 5. The uniformly balanced representation with $\gamma=1$ is determined for A as in Example 3, and $B^T=[0\ 0\ 0\ 1\ 0\ 0]$. From Algorithm 5 we arrive at the triple $(\bar{A},\bar{B},\bar{C})$

$$\bar{A}=\begin{bmatrix} 0.0000 & 6.0485 & 0.1659 & 0.0085 & 0.0003 & 0.0484 \\ -6.0510 & -0.0377 & -0.0913 & -0.0296 & 0.0596 & -0.0589 \\ -0.1716 & -0.0755 & -0.1848 & -13.1416 & 0.2270 & -0.1477 \\ -0.0090 & 0.0163 & 13.1121 & -0.0012 & 0.0305 & -0.0819 \\ -0.0005 & -0.0650 & -0.2389 & -0.0314 & -0.0002 & 10.5323 \\ -0.0525 & -0.0615 & -0.1189 & 0.0606 & -10.5409 & -0.0962 \end{bmatrix}$$

$$\bar{B}^T=\bar{C}=[0.0094\ \ 0.2744\ \ 0.6079\ 0.0485\ \ 0.0195\ \ 0.4387]$$

which is uniformly balanced. The plot of the transfer function of $(\bar{A},\bar{B},\bar{C})$ in Fig.15 shows that that the property (b) in Theorem 6 is satisfied. The uniformly balanced truss structure has been presented in Example 4. The DSS-13 antenna model is uniformly balanced in the following example.

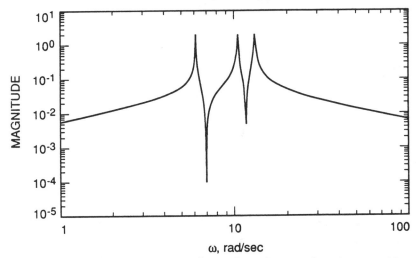

Fig.15. Magnitude of transfer function of the uniformly balanced system with $\gamma=1$.

Example 6. The previously presented model of the DSS-13 antenna is uniformly balanced. The elevation rate command is selected as a single input to the antenna. The Hankel singular values of all states are set equal to one, except for states with poles at zero - their Hankel singular values are infinitely large. These states are removed from the model for clarity of presentation. The results are presented in Fig.16, where the magnitude of the transfer function of the uniformly balanced antenna is presented; the magnitude, according to Theorem 7, is less then two.

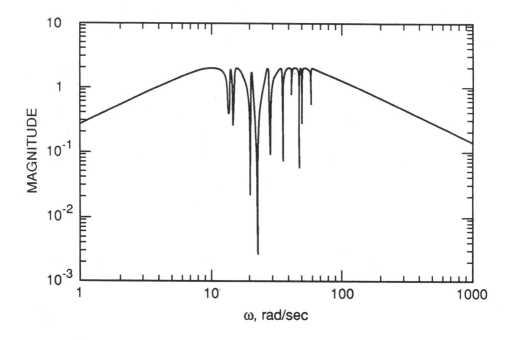

Fig.16. Magnitude of the transfer function of the uniformly balanced antenna control system.

5. Balanced Identification of Low-Level Structural Dynamics

In structural dynamics some modes are excited and sensed intensively, while others are either barely visible or not present in the output data. There are two reasons for the presence of weak modes in the data. First, there could be generic low-level dynamics independent of sensor-actuator configuration. Second, there could be low-level dynamics caused by a particular sensor and/or actuator configuration which hides or weakens the visibility of some modes. The need for distinguishing such configuration-dependent low-level dynamics from generic low-level dynamics is particularly important in identification, model reduction, and control of structures. While generic low-level dynamics can be neglected in determining system model order, neglecting low-level dynamics caused by either sensor or actuator configuration causes significant modeling errors or spillover effects. Based on [20], the case of configuration-dependent low-level dynamics is considered in this Section.

5.1. Hankel Singular Values for Structures

Consider a system representation (A,B,C), with matrix A having separated complex poles with small real parts. It exhibits the characteristics of a structure. Hankel singular values of structures given by (11) are expressed as

$$\gamma_i = a_i \| c_{bi} b_{bi} \| \tag{31}$$

where $\|.\|$ is spectral norm of a matrix, $a_i = 1/(4\zeta_i \omega_i^2)$, and c_{bi} is the i-th column of C_b, b_{bi} is the i-th row of B_b

$$b_{bi} = s_i B, \quad c_{bi} = C t_i \tag{32}$$

and s_i is i-th row of T^1, t_i is the i-th column of T. For a balanced system $b_{bi} b_{bi}^T = c_{bi}^T c_{bi}$, see [4, 5]. Left multiplication of the above by b_{bi}^T and right multiplication by b_{bi} gives $b_{bi}^T b_{bi} b_{bi}^T b_{bi} = b_{bi}^T c_{bi}^T c_{bi} b_{bi}$, or

$$\| c_{bi} b_{bi} \| = \| b_{bi} \|^2 = \| c_{bi} \|^2 \tag{33}$$

Combining (31), (32) and (33), the Hankel singular value is expressed in terms of matrices B and C

$$\gamma_i = a_i \|Ct_i\|^2 = a_i \|s_i B\|^2. \tag{34}$$

5.2. Variations of Sensor/Actuator Configuration

Matrices B and C of the triple (A,B,C) describe the location and amplification level of each actuator and sensor. The impact of C and B variations on Hankel singular values is analyzed in this section. Let variations ΔB and ΔC of B and C be from a ball with the radius $\rho \geq 0$

$$\mathscr{B}_b(\rho) = \{\Delta B : \|\Delta B\| \leq \rho \|B\|\}, \text{ or } \mathscr{B}_c(\rho) = \{\Delta C : \|\Delta C\| \leq \rho \|C\|\}. \tag{35b}$$

Amplifications of the i-th Hankel singular value is measured by the factors $\delta_{\gamma i}$ and $\Delta_{\gamma i}$.

Definition 6. For a simple Hankel singular value γ_i

$$\delta_{\gamma i} = |\gamma_i(C + \Delta C) - \gamma_i(C)| / \gamma_i(C), \quad \delta_{\gamma i} = |\gamma_i(B + \Delta B) - \gamma_i(B)| / \gamma_i(B) \tag{36a}$$

$$\Delta_{\gamma i} = \max_{\|\Delta C\| \in \mathscr{B}} (\delta_{\gamma i}(\Delta C)), \quad \Delta_{\gamma i} = \max_{\|\Delta B\| \in \mathscr{B}} (\delta_{\gamma i}(\Delta B)) \tag{36d}$$

are amplification factors of the i-th Hankel singular value due to variations ΔC or ΔB.

The factors $\delta_{\gamma i}$ and $\Delta_{\gamma i}$ indicate the actual and the largest variation of γ_i due to variations of ΔB or ΔC from the ball $\mathscr{B}_b$ or $\mathscr{B}_c$, respectively. The cross-complement (see the Appendix) is used for determination of the amplification factor. As shown in the Appendix, Eqs.(A5), (A6), the cross-complement δC_i is a solution of the least-squares problem, hence

$$\delta C_i = -C \, t_i \, t_i^T / \|t_i\|^2 \tag{37}$$

Denote $\gamma_i(C)$, or $\gamma_i(B)$ the i-th Hankel singular value for given C or B matrix, and δC_i, δB_i the cross-complements of C and B

with respect to t_i and s_i respectively, and let $\Delta C \in \mathcal{B}_c$, $\Delta B \in \mathcal{B}_b$.

Theorem 7. For a structure the maximal variation of the i-th Hankel singular value is due to variations $\Delta C = -\alpha\ \delta C_i$, or $\Delta B = -\alpha\ \delta B_i$ where α is a positive scalar, $\alpha \geq 0$. In this case one obtains

$$\gamma_i(C-\alpha\ \delta C_i)/\gamma_i(C) = (\alpha+1)^2, \tag{38a}$$

$$\gamma_i(B-\alpha\ \delta B_i)/\gamma_i(B) = (\alpha+1)^2 \tag{38b}$$

$$\Delta_{\gamma i} = \alpha_o(\alpha_o+2), \tag{39}$$

where $\alpha_o = \rho\|C\|/\|\delta C_i\|$, or $\alpha_o = \rho\|B\|/\|\delta B_i\|$.

Proof. From (34)

$$\gamma_i(C-\alpha\ \delta C_i) = a_i\|(C-\alpha\ \delta C_i)t_i\|^2 = a_i\|(C+\delta C_i)t_i + (-\alpha-1)\delta C_i t_i\|^2$$

and from Definition A1 in Appendix $(C+\delta C_i)t_i = 0$, thus

$$\gamma_i(C-\alpha\ \delta C_i) = a_i\|(\alpha+1)Ct_i\|^2 = (\alpha+1)^2\gamma_i(C)$$

consequently (38) is proved. According to Lemma A2 in Appendix

$$\gamma_i(-\alpha\ \delta C_i) = a_i\|-\alpha\ \delta C_i t_i\|^2 = a_i\|-\alpha\ Ct_i\|^2 = \alpha^2\gamma_i(C) = \alpha^2\gamma_i(\delta C_i) \tag{40}$$

and for an arbitrary perturbation ΔC, the Schwartz inequality gives

$$\gamma_i(C+\Delta C) = a_i\|(C+\Delta C)t_i\|^2 \leq a_i(\|Ct_i\| + \|\Delta Ct_i\|)^2$$

$$= \gamma_i(C) + \gamma_i(\Delta C) + 2\sqrt{\gamma_i(C)\gamma_i(\Delta C)} \tag{41}$$

If ΔC turns the above inequality into equality, the maximum value of $\gamma_i(C+\Delta C)$ is obtained. In consequence, the variation $\Delta C = -\alpha\ \delta C_i$, for $\alpha \geq 0$ maximizes γ_i. Indeed, for $\Delta C = -\alpha\ \delta C_i$, from (38), (40) and (41) one obtains $(\alpha+1)^2\gamma_i(C) \leq (1+|\alpha|)^2\gamma_i(C)$.

The equality is obtained for $\alpha \geq 0$. For $\Delta C \in \mathcal{B}_c$ the maximal increase of $\gamma_i(C+\Delta C)$ is for $\alpha = \alpha_o = \rho \|C\| / \|\delta C_i\|$. Introducing $\Delta C = \alpha_o \delta C_i$ to the definition (36), one obtains $\Delta_{\gamma i} = \alpha_o(\alpha_o + 2)$. A proof for variations of B is similar.□

Theorem 7 shows that a variation ΔC co-linear with opposite sign to the cross-complement δC_i (or a variation ΔB co-linear with opposite sign to the cross-complement δB_i) causes the largest variation of the i-th Hankel singular value.

The factor $\Delta_{\gamma i}$ is compared with the variations of C. From Lemma A2 in the Appendix, $\|\delta C_i\| = \|Ct_i\| / \|t_i\|$ and from (38),

$$\Delta_{\gamma i} = \rho \beta_{oi}(\rho \beta_{oi} + 2) \tag{42a}$$

where $\beta_{oi} = \|C\| \|t_i\| / \|Ct_i\|$, and $\rho = \|\Delta C\| / \|C\|$. Since β_{oi} is large when compared to small Hankel singular values, therefore from (42a)

$$\Delta_{\gamma i} \gg \|\Delta C\| / \|C\|, \quad \text{and} \quad \Delta_{\gamma i} \approx \alpha_o^2, \tag{43}$$

Thus the amplification of the small Hankel singular value is much larger than the variation of C. For a balanced system, $T = I_n$, $t_i = e_i$, where e_i is the i-th column of the identity matrix I_n, then

$$\beta_{oi} = \|C_b\| / \|c_{bi}\|, \tag{42b}$$

where c_{bi} is the i-th column of C_b. From (42), it follows that the smaller value is the i-th column of C_b the larger value is the amplification of the i-th Hankel singular value.

It follows from (34) that by replacing C with B^T and t_i with s_i^T, the same results apply. Namely, the smaller the i-th row of B in the balanced representation, or the smaller is the cross-complement δB_i, the larger amplification of the i-th Hankel singular value is expected.

*Example   7.   A   four-degrees-of-freedom   (eight-state   variable) structure is investigated. Its triplet (A,B,C) is as follows

$$A = \begin{bmatrix} 0 & 0 & 0 & 0 & 1 & 0 & 0 & 0 \\ 0 & 0 & 0 & 0 & 0 & 1 & 0 & 0 \\ 0 & 0 & 0 & 0 & 0 & 0 & 1 & 0 \\ 0 & 0 & 0 & 0 & 0 & 0 & 0 & 1 \\ 1500 & -500 & 0 & 0 & 0.450 & -0.150 & 0 & 0 \\ -50 & 60 & -10 & 0 & -0.015 & 0.018 & -0.003 & 0 \\ 0 & -100 & 200 & -500 & 0 & -0.030 & 0.060 & -0.150 \\ 0 & 0 & -100 & 600 & 0 & 0 & -0.030 & 0.180 \end{bmatrix}$$

$$B^T = \begin{bmatrix} 0 & 0 & 0 & 0 & 0.5401 & -29.188 & 70.3528 & -0.9190 \\ 0 & 0 & 0 & 0 & 3.1384 & -70.401 & 10.4543 & -0.4474 \end{bmatrix},$$

$$C = \begin{bmatrix} 0 & 0 & 0 & 0 & -0.6411 & -1.5303 & -0.5371 & -0.9259 \\ 0 & 0 & 0 & 0 & -3.3794 & -3.3732 & -1.1707 & -0.6911 \end{bmatrix}.$$

The system is transformed into balanced representation (A_b, B_b, C_b), and its Hankel singular values are determined $\gamma_{1,2}^o = 506.26$, $\gamma_{3,4}^o = 261.42$, $\gamma_{5,6}^o = 6.16$, $\gamma_{7,8}^o = 0.618$. The plots of the system transfer functions from input 1 and 2 to output 2 are shown in Fig.17, dashed line, and its Hankel singular values (proportional to the resonance amplitudes) are seen in the plot. Resonance peaks corresponding to the small Hankel singular values are small (for frequencies 5.5416 and 26.4622).

Random changes ΔC_b of C_b are imposed, 100 samples of ΔC_b are generated, such that $\|\Delta C_b\| \le 0.1 \|C_b\|$. For each new value of $C_b + \Delta C_b$ the factor $\delta_{\gamma i} = (\gamma_i - \gamma_i^o)/\gamma_i^o$ is computed and shown in Fig.18. In this figure the small Hankel singular values increased significantly ($\gamma_{5,6}$ 9 times, and $\gamma_{7,8}$ 18 times, in average) while the large values changed insignificantly. The maximal amplification factor $\Delta_{\gamma i}$ is determined for small Hankel singular values. From (42b) $\beta_{o3}=25.61$, $\beta_{o4}=37.01$, and for $\rho=0.1$, one obtains from (42a) $\Delta_{\gamma 3}=11.68$ $\Delta_{\gamma 4}=21.10$. The largest actual amplification factors obtained from the simulations are $\delta_{\gamma 3}=10.2$ and $\delta_{\gamma 4}=19.8$. The plot of the transfer function for $C_b+\Delta C_b$ rather then C_b in the system representation is shown in Fig.17, solid line. In this figure the small resonance peaks increase significantly.

Example 8. A solar panel model, which consists of 12 modes (24 states) is considered. Its Hankel singular values are shown in Fig.19, solid line, and its transfer function in Fig.20, solid line. The last four frequencies are invisible in this plot, and the first one is weakly exposed. By random perturbation of sensor configuration one obtains new Hankel singular values, as in Fig.19, dashed line. A significant increase (of order 100) of small Hankel singular values is observed. The transfer function for the perturbed model is shown in Fig.20, dashed line. In this case, the high-frequency modes are now visible, and the visibility of the first mode is significantly improved.

6. Conclusions

In a manner similar to grammians, antigrammians, reflect controllability and observability properties of a system, but unlike grammians, they do exist for systems with integrators, making their reduction possible. In this paper a reduction algorithm for systems with integrators is derived. The algorithm is based on balancing of the antigrammians with the same computational effort as a regular balancing procedure.

Properties and procedures for sensor/actuator configuration are presented, such that a required degree of controllability and observability is acquired. The case of equally controlled and observed states (uniformly balanced systems) is given special attention. The assignment procedures and properties are specified for flexible structures, for which any positive definite grammian is always approximately assigned.

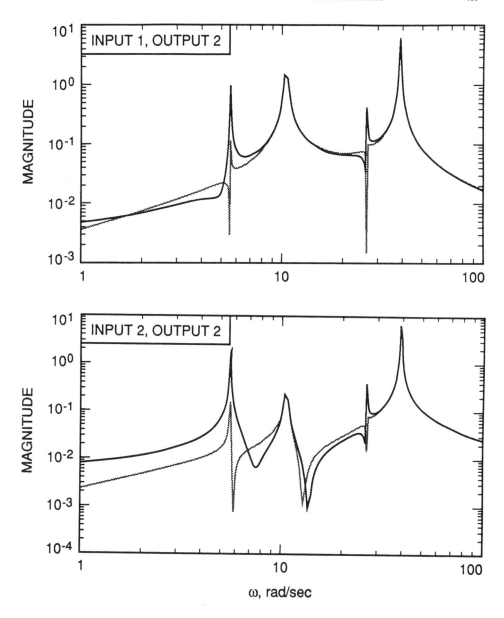

Fig.17. Magnitudes of transfer function for the original and perturbed system.

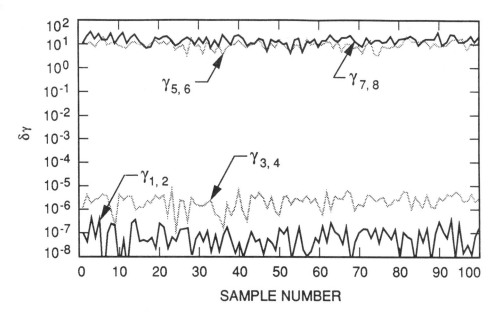

Fig.18. Amplification factors for randomly perturbed matrix C.

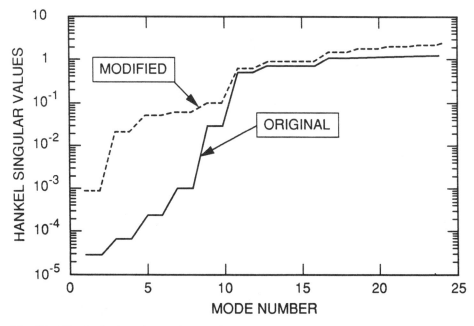

Fig.19. Hankel singular values of the solar panel.

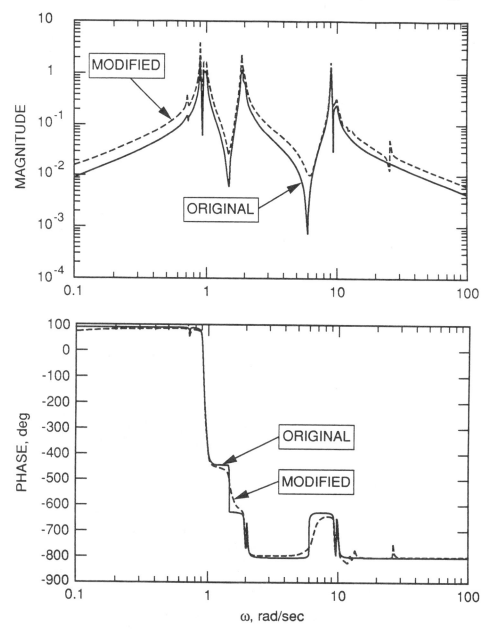

Fig.20. Magnitude and phase of the transfer function of the solar panel.

In many cases small Hankel singular values in structures (small or non-detectable resonances) are due to specific configuration of sensors and/or actuators. It has been shown that small Hankel singular values can be significantly amplified by variations of sensor and/or actuator configurations. This approach can be useful in system identification, particularly in system order determination from noisy data.

Acknowledgments.

This research was performed at the Jet Propulsion Laboratory, California Institute of Technology, under a contract with the National Aeronautics and Space Administration. Contributions from F.Y. Hadaegh, M.A. Mellstrom, and R.E. Scheid is greatly appreciated.

REFERENCES

1. K. Glover: "All Optimal Hankel Norm Approximations of Linear Multivariable Systems and Their L$^\infty$-Error Bounds." *International Journal of Control*, vol.39, 1984.

2. A.J. Laub, M.T. Heath, C.C. Paige, and R.C. Ward: "Computation of System Balancing Transformation and Other Applications of Simultaneous Diagonalization Algorithm." *IEEE Trans. on Automat. Control*, vol.32, No.2, 1987.

3. B.C. Moore: "Principal Component Analysis in Linear Systems: Controllability, Observability and Model Reduction," *IEEE Trans. on Automat. Control*, vol.26, No.1, Jan. 1981.

4. W. Gawronski, and T. Williams: "Model Reduction for Flexible Space Structures," *Journal of Guidance, Control, and Dynamics*, vol.14, no.1, 1991.

5. W. Gawronski, and J.-N. Juang: "Model Reduction for Flexible Structures," in: *Control and Dynamics Systems*, ed. C.T. Leondes, vol.36, Academic Press, New York, 1990.

6. L. Fortuna, A.Gallo, and G. Nunnar: "New Results Involving Open-Loop Balanced Realization Schemes." *Proc. 28th IEEE Conf. on Decision and Control*, Tampa, FL, Dec. 1989.

7. C.S. Hsu, and D. Hou: "Model Reduction of Unstable Linear Control Systems," *Third Int. Conference on Advances in Communication and Control Systems*, Victoria, Canada, 1991.

8. S. Weiland: "Balanced Representations and Approximations of Linear Systems." *Proc. 28th IEEE Conf. on Decision and Control*, Tampa, FL, Dec. 1989.

9. Y. Liu, and B.D.O. Anderson: "Singular Perturbation Approximation of Balanced Systems." *Proc. 28th Conf. on Decision and Control*, Tampa, FL, 1989.

10. R. Prakash, and S. Vittal Rao: "Model Reduction by Low Frequency Approximation of Internally Balanced Representation." *Proc. 28th Conf. on Decision and Control*, Tampa, FL, 1989.

11. E.A. Jonckheere: "Principal Component Analysis of Flexible Systems - Open Loop Case." *IEEE Trans. on Automat. Control*, vol.27, 1984.

12. G.F. Franklin, J.D. Powell and A. Emami-Naeini: *Feedback Control of Dynamic Systems*, Addison-Wesley, Reading, 1988.

13. M. Morari, and E. Zafiriou: *Robust Process Control*. Prentice Hall, Englewood Cliffs, 1989.

14. W. Gawronski, and J.A. Mellstrom: "Model Reduction for Systems with Integrators," *Third Int. Conference on Advances in Communication and Control Systems*, Victoria, Canada, 1991.

15. W. Gawronski, and J.A. Mellstrom: " Modeling and Simulation of the DSS-13 Antenna Control System," *JPL/TDA Progress Report*, vol.42-106, 1991.

16. W. Gawronski: "Model Reduction for Flexible Structures: Test Data Approach". *Journal of Guidance, Control, and Dynamics*, vol.14, 1991.

17. M.A. Wicks, and R.A. DeCarlo: "Gramian Assignment of the Lyapunov Equation," *IEEE Trans. on Automat. Control*, vol.35, 1990.

18. W. Gawronski, and F.Y. Hadaegh: "Balanced Input -Output Assignment," *Proc. 28th IEEE Conf. on Decision and Control*, Tampa, FL, 1989.

19. B.D.O. Anderson: "A System Theory Criterion for Positive Real Matrices." *J. SIAM Control*, vol.5, 1967.

20. W. Gawronski, F.Y. Hadaegh, and R.E. Scheid: "A Hankel Singular Value Approach for Identification of Low Level Dynamics in Flexible Structures," *Proc. 29th IEEE Conf. on Decision and Control*, Honolulu, Hawaii, 1990.

APPENDIX. Auxiliary Lemmas

For all real ω, and $p \le m$ let the $p \times m$ matrix $U_1(\omega)$ have p orthonormal rows, i.e. $U_1(\omega)U_1^*(\omega)=I_p$, and for $p \ge m$ let the $p \times m$ matrix $U_2(\omega)$ have m orthonormal columns; i.e., $U_2(\omega)U_2^*(\omega)=I_m$. Let $G(j\omega)=C(j\omega I-A)^{-1}B$ be a transfer function of the triplet (A,B,C), and $s=j\omega$ be not a pole of an element of $G(s)$, then

Lemma A1. If $rank(U_1(\omega))=rank(G(j\omega))=p$, for all real ω, and G satisfies the equation

$$GU_1+U_1^*G^*=\gamma^{-1}GG^* \qquad (A1)$$

or if $rank(U_2(\omega))=rank(G(j\omega))=m$, and G satisfies the equation

$$U_2G+G^*U_2^*=\gamma^{-1}G^*G \qquad (A2)$$

then the system is stable.

Proof. Introduce $G_1=GU_1$ and $G_2=U_2G$. It follows from (A1) that G_1 is positive real, and from (A2) that G_2 is positive real, hence both are stable. But (A,BU_1,C) is the state-space representation of G_1, and (A,B,U_2C) is the state-space representation of G_2. The representations G, G_1, and G_2 include stable matrix A, hence G, G_1, and G_2 are stable.□

Definition A1. A matrix δQ, $p \times q$, is a cross-complement of Q, $p \times q$, with respect to y, $q \times s$, if δQ is the smallest matrix (in terms of spectral norm) such that $(Q+\delta Q)y=0$. That is

$$\min_{(Q+\delta Q)y=0} \|\delta Q\|. \qquad (A3)$$

Lemma A2. A cross complement δQ of Q with respect to y has the following properties

$$\|\delta Q\| = \|\delta Q\, y\|/\|y\| = \|Q\, y\|/\|y\| \qquad (A4)$$

Proof. For the projection Q_0 of the rows of Q onto y^T

$$Q_0 = Q \, y \, y^T/(y^T y), \tag{A5}$$

and by substitution $\delta Q = -Q_0 + R$, the problem (A3) is restated as

$$\min_{Ry=0} \|-Q_0 + R\| \tag{A6}$$

Note that, from the definition of Q_0, $Ry=0$ implies $RQ_0^T = 0$. Then, by the properties of spectral norm

$$\min_{Ry=0} \|-Q_0 + R\| = \min_{Ry=0} \|-Q_0^T + R^T\| = \min_{Ry=0} \sup_{\|u\|=1} u^T(Q_0 Q_0^T + RR^T - Q_0 R^T - RQ_0^T)u$$

$$= \min_{Ry=0} \sup_{\|u\|=1} u^T(Q_0 Q_0^T + RR^T)u.$$

Since both $Q_0 Q_0^T$ and RR^T are positive semi-definite, it is necessary that $R=0$ in order to obtain the minimum, and thus $\delta Q = -Q_0 = -Q \, y \, y^T/(y^T y)$. From Definition A1 $Qy = -\delta Qy$, hence,

$$\delta Q = \delta Q \, y \, y^T/y^T y. \tag{A7}$$

From (A7) and Schwartz inequality $\|\delta Q\| \|y\|^2 = \|\delta Q \, y \, y^T\| \le \|\delta Q \, y\| \|y\|$, consequently

$$\|\delta Q\| \, \|y\| \le \|\delta Q \, y\| \tag{A8}$$

Note, however, that the Schwartz inequality holds

$$\|\delta Q\| \, \|y\| \ge \|\delta Q \, y\| \tag{A9}$$

hence, the first part of (A4) follows from (A8) and (A9), and (A5) and the first part of (A4) verifies the second part of (A4).□

Response-Only Measurement Techniques for the Determination of Aerospace System Structural Characteristics

An—Chen Lee

Juhn—Horng Chen

Department of Mechanical Engineering
National Chiao Tung University
Hsinchu, Taiwan, R.O.C.

I. Introduction

Experimental modal analysis has become an increasingly important engineering tool during the past 40 years in the aerospace, automotive, and machine tool industries. Modal parameters estimates obtained from experimental modal analysis are being used in the direct solution of vibration and/or acoustic problems, for correlation with output from finite element programs, and for prediction of changes in system dynamics due to structural changes. In all cases, the quality of the modal parameter estimates is of major concern.

Time series methods have been applied to the synthesis of structural systems excited by random forcing functions as well as to the identification of the natural frequencies and the damping ratio. The Autoregressive moving—average (ARMA) models have been used to estimate the characteristics of buildings being excited by wind force and the characteristics of the cutting process that participated with random cutting forces[1,2]. Recently, there has been a great deal of interest in determining modal parameters from measured response data

This chapter is based in part on the authors' paper in the Journal of Aircraft, Volume 27, Number 12, December 1990. Copyright American Institute of Aeronautics and Astronautics, Inc. © 1990, reprinted with permission.

taken on operating systems (e.g., turbulent flow over an airfoil, road inputs to automobiles, and environmental inputs to proposed large space structures). When random input data are not or cannot be measured, then only the available output data can be fitted to a time domain ARMA model. This estimation process always produces a minimum—phase system. The transfer functions defined this way by the ARMA model are successful in the estimation of magnitudes of the true transfer functions but do not give the correct phase information[3,4] except when the true system is minimum phase. In other words, the mode shape cannot be determined uniquely.

The objective of this paper is to solve the aforementioned problem and estimate the modal parameters when the input force is an unmeasured white noise sequence. The selection of the sampling interval for estimating the modal parameters is also considered. Emphasis is placed on the optimum design of uniform data sampling intervals when experimental constraints allow only a limited number of discrete time measurements of the output from the continuous system and the parameters of interest are natural frequencies, damping ratio, and time constant of the continuous system.

II. Statement of the Problem

Consider a structural system with a white noise input sequence $a(k)$ and response $x(k)$, $k=0,1,2,....,L$, then the only available set of data $x(k)$ can be modeled as an $\text{ARMA}(n,m)$ model of the form[5]

$$x(k) - \phi_1 \, x(k-1) - ... - \phi_n \, x(k-n)$$

$$= a(k) - \theta_1 \, a(k-1) - ... - \theta_m \, a(k-m) \tag{1}$$

where $\phi_1, \phi_2, ... , \phi_n$ are the autoregressive parameters
and $\theta_1, \theta_2, ... , \theta_m$ are the moving average parameters.
Applying z transform to each side of Eq. (1), we have

$$X(z) - \phi_1 \, X(z) \, z^{-1} - ... - \phi_n \, X(z) \, z^{-n}$$

$$= A(z) - \theta_1 \, A(z) \, z^{-1} - ... - \theta_m \, A(z) \, z^{-m}$$

$$H(z) \equiv \frac{\theta(z)}{\phi(z)} = \frac{X(z)}{A(z)} = \frac{1 - \sum\limits_{i=1}^{m} \theta_i z^{-i}}{1 - \sum\limits_{i=1}^{n} \phi_i z^{-i}} \tag{2}$$

where $H(z)$ is the transfer function, $A(z)$ and $X(z)$ are z transforms of the sequences $a(k)$ and $x(k)$. By using only the available set of the data $x(k)$, the modeling procedure is based on the minimization of the following function :

$$V = \sum_{k=1}^{L} [a(k)]^2 \tag{3}$$

By using the total square integral formula[6], the function V can be written as :

$$V = \sum_{k=1}^{L} [a(k)]^2 = \frac{1}{2\pi j} \oint |A(z)|^2 \frac{dz}{z} \tag{4}$$

where the contour of integration is the unit circle in z plane. By considering Eq. (2), this equation can also be expressed as

$$V = \sum_{k=1}^{L} [a(k)]^2 = \frac{1}{2\pi j} \oint \left| \frac{\phi(z)}{\theta(z)} X(z) \right|^2 \frac{dz}{z}$$

$$= \frac{1}{2\pi j} \oint \left| \frac{\phi(z)}{\theta(z)} \right|^2 |X(z)|^2 \frac{dz}{z} \tag{5}$$

Obviously, the function V does not involve any phase information, since it only involves the absolute value of polynomials $\phi(z)$, $\theta(z)$ and $X(z)$; i.e., the polynomials $\phi(z)$ and $\theta(z)$ which satisfy the condition V=minimum, are not uniquely determined. Therefore, even if we constrain the estimated $\phi(z)$ polynomial to be of minimum phase nature, which guarantees a stable model, we cannot decide which of the polynomials $\theta(z)$ with the same magnitude characteristics is the one that truly represents the actual physical system. The problem posed here is closely related to the so—called " stochastic realization," i.e., the problem of obtaining a model of a process $x(k)$, given its covariance function or second—order information[7]. In general, there exist many such models; however, with the inversibility condition, the only stable and stably invertible model is the (unique) innovations representation (IR)[8]. The inverse model is the whitening filter that produces a white noise process, the innovations $a(k)$, when driven by the observed data. If the process $x(k)$ is stationary, the problem of obtaining the IR essentially reduces to one of spectral factorization[9].

In conclusion, when input signals are unmeasurable, we can not estimate unique transfer functions. Therefore, we have to obtain more information such as velocity to constrain our solution. This condition will be discussed in next section.

III. Mathematical Formulation

Let a randomly excited structural system with n degree of freedom be expressed by

$$
\underset{(n \times n)(n \times 1)}{M \quad \ddot{X}} + \underset{(n \times n)(n \times 1)}{C \quad \dot{X}} + \underset{(n \times n)(n \times 1)}{K \quad X} = \underset{(n \times 1)}{f(t)} \tag{6}
$$

where M, C, and K represent the mass, damping, and stiffness matrix respectively, X is a displacement vector, and f is random excitation forces.

By using a new set of variables, q and $\dot{q}$, Eq. (6) can be rewritten as

$$
\dot{q} = \begin{bmatrix} 0 & I \\ -M^{-1}K & -M^{-1}C \end{bmatrix} q + \begin{bmatrix} 0 \\ M^{-1} \end{bmatrix} f(t)
$$

$$
(2n \times 1) \qquad (2n \times 2n) \qquad (2n \times 1) \quad (2n \times n) \ (n \times 1)
$$

$$= - \underset{(2n \times 2n)}{A} \quad \underset{(2n \times 1)}{q} + \underset{(2n \times 1)}{Z(t)} \tag{7}$$

where $I = n \times n$ identity matrix

$$q^t = [\, x_1, x_2, \cdots, x_n, \dot{x}_1, \dot{x}_2, \cdots, \dot{x}_n \,]$$

and x_i is the displacement component, $\dot{x}_i$ the velocity component, and q^t the transpose of q

The matrix Eq. (7) is the vector first—order differential equation with white noise as the forcing function and is called the continuous vector autoregressive model of first order, i.e., the continuous VAR(1) model. The device of expressing Eq. (6) in the form of Eq. (7) is derived by using the so—called state—space (or Markovian) representation of the relationship between input and output rather than the explicit form[10]. It may be viewed alternatively as a special case of the technique of writing certain types of non—Markov processes involving only finite stage dependence as vector Markov processes.

Although continuous time stochastic processes are frequently encountered in physical science and engineering, with the advent of digital computers, their discretely sampled observations have attracted interests for the purpose of analysis. This naturally leads to the corresponding discrete representation of those processes, i.e., the discrete vector first—order autoregressive models [the discrete VAR(1) models]. The relationship of the continuous VAR(1) and the discrete VAR(1) models to obtain the modal parameters will be discussed in the following subsection.

A. Discrete VAR(1) Model

The discrete VAR(1) model represents the vector first—order stochastic difference equation. It can be expressed as

$$\underset{(2n \times 1)}{p(k)} = \underset{(2n \times 2n)}{\phi} \quad \underset{(2n \times 1)}{p(k-1)} + \underset{(2n \times 1)}{a(k)} \tag{8}$$

where

$$\boldsymbol{p}^t(k) = [x_1(k),\ x_2(k),\ \cdots, x_n(k),\ \dot{x}_1(k),\ \dot{x}_2(k),\ \cdots,\ \dot{x}_n(k)]$$

$$= [\ p_1(k),\ p_2(k),\ ...,\ p_{2n}(k)\ ]$$

and ϕ is the discrete autoregressive matrix, $\boldsymbol{a}$ the discrete vector random forces, and

$$E[\ \boldsymbol{a}(k)\ \boldsymbol{a}^t(k{-}l)] = \delta_l\,\Gamma_a$$

$$\delta_l = \begin{bmatrix} 1 & \text{for } l = 0 \\ 0 & \text{for } l \neq 0 \end{bmatrix}$$

where δ_l is the Kronecker delta function, Γ_a the variance matrix of $\boldsymbol{a}(k)$, and E the expectation operator

The relationship between input signal $\boldsymbol{a}$ and output signal $\boldsymbol{p}$ is expressed by the Green function matrix $\boldsymbol{G}$:

$$\boldsymbol{p}(k) = \sum_{j=0}^{\infty} \boldsymbol{G}_j\ \boldsymbol{a}(k{-}j) = \sum_{j=-\infty}^{k} \boldsymbol{G}_{k-j}\ \boldsymbol{a}(j\ ) \tag{9}$$

where $\boldsymbol{G}_j$ is a $2n{\times}2n$ matrix.

Using Eq. (8) we have

$$\boldsymbol{p}(k) = (\boldsymbol{I} - \phi\ B)^{-1}\ \boldsymbol{a}(k) = \sum_{j=0}^{\infty} \phi^j\ \boldsymbol{a}(k{-}j) \tag{10}$$

where B is a backshift operator and has the property

$$B\ f(k) = f(k{-}1).$$

From Eqs. (9) and (10), we obtain the Green function matrix of the discrete VAR(1) model is

$$\boldsymbol{G}_j = \phi^j \tag{11}$$

The eigenvalue matrix (Λ) and eigenvector matrix (T) can be solved by

$$(I\,\lambda - \phi)\ T = 0 \tag{12}$$

and the matrix ϕ can be expressed by the eigenvalue matrix and the eigenvector matrix

$$\phi = T\,\Lambda\ T^{-1} \tag{13}$$

where

$$\Lambda = \begin{bmatrix} \lambda_1 & & 0 \\ & \lambda_2 & \\ & & \ddots \\ 0 & & \lambda_{2n} \end{bmatrix}$$

$$T = \begin{bmatrix} T_{1,1} & T_{1,2} & \cdots & T_{1,2n} \\ T_{2,1} & \cdots & & \\ \vdots & & & \\ T_{2n,1} & & \cdots & T_{2n,2n} \end{bmatrix} = [\,T_1,\ T_2,\ \cdots,\ T_{2n}\,]$$

$$T^{-1} = \begin{bmatrix} T^{1,1} & T^{1,2} & \cdots & T^{1,2n} \\ T^{2,1} & \cdots & & \\ \vdots & & & \\ T^{2n,1} & & \cdots & T^{2n,2n} \end{bmatrix} = \begin{bmatrix} (T^1)^t \\ (T^2)^t \\ \vdots \\ (T^{2n})^t \end{bmatrix}$$

$$T_i^{\,t} = [\ T_{1,i}\ \ T_{2,i}\ \cdots, T_{2n,i}\,]$$

$$(T^{\,i})^t = [\ T^{i,1},\ T^{i,2},\ ..., T^{i,2n}\,]$$

in which the matrix T^{-1} is the inverse matrix of the matrix T. Substituting Eq. (13) to Eq. (11), the following equation can be derived :

$$G_j = \phi^j = T \Lambda^j T^{-1} = \sum_{i=1}^{2n} g_i \lambda_i^j \tag{14}$$

where

$$g_i = T_i (T^i)^t$$

The state covariance matrix function, which indicates how the state variables are affected or related with other state variable at different number of lag k can be expressed as

$$\Gamma_k = E[p(l) \, p^t(l-k)] \tag{15}$$

Substitute Eq. (10) into Eq. (15) we obtain

$$\Gamma_k = \sum_{i=0}^{\infty} \phi^{i+k} \Gamma_a (\phi^i)^t$$

$$= \phi^k \sum_{i=0}^{\infty} \phi^i \Gamma_a (\phi^i)^t = \phi^k \Gamma_0 \tag{16}$$

From Eqs. (10) and (14), Eq. (16) can be rewritten as

$$\Gamma_k = \sum_{i=1}^{2n} \sum_{j=1}^{2n} [\, g_i \Gamma_a g_j^t \frac{1}{1-\lambda_i \lambda_j}] \, \lambda_i^k$$

$$= \sum_{i=1}^{2n} d_i \lambda_i^k \tag{17}$$

where

$$d_i = \sum_{j=1}^{2n} g_i \, \Gamma_a \, g_j^t \, \frac{1}{1-\lambda_i \lambda_j}$$

is a $2n \times 2n$ matrix.

B. Continuous VAR(1) Model

The continuous VAR(1) model represents the vector first—order stochastic differential equation. It can be expressed as

$$\underset{(2n \times 1)}{\dot{q}(t)} + \underset{(2n \times 2n)}{A} \underset{(2n \times 1)}{q(t)} = \underset{(2n \times 1)}{Z(t)} \tag{18}$$

where

$$q^t(t) = [\, x_1, \, x_2, \cdots, \, x_n, \, \dot{x}_1, \, \dot{x}_2, \cdots, \, \dot{x}_n \,]$$

$$= [\, q_1(t), \, q_2(t), \, ..., \, q_{2n}(t) \,]$$

$$E[Z(t)] = 0$$

$$E[Z(t) \, Z^t(t{-}v)] = \delta(v) \, \Gamma_z$$

where $\delta(v)$ is Dirac delta function and Γ_z the variance matrix of $Z(t)$. The Green function matrix $G(t)$ can be expressed as

$$q(t) = \int_0^\infty G(\tau) \, Z(t{-}\tau) \, d\tau \tag{19}$$

Solving Eq. (18), we have

$$q(t) = \int_0^\infty e^{-A\tau} \, Z(t{-}\tau) \, d\tau \tag{20}$$

From Eqs. (19) and (20), the Green function matrix of the continuous VAR(1) model is

$$G(t) = e^{-At} \tag{21}$$

Equation (21) is commonly called the state transition matrix in linear time—invariant systems.

The eigenvalue matrix (U), eigenvalue (u_1, u_2,..., u_{2n}), and the eigenvector matrix ($\mathscr{I}$) can be found by solving

$$(I u + A) \ \mathscr{I} = 0 \tag{22}$$

With the same procedure as that in Sec. III.A, the following equations can be derived :

$$G(t) = \exp(-At) = \exp[\mathscr{I} U (\mathscr{I}^{-1})^{t}]$$

$$= \mathscr{I} \exp(U t) \mathscr{I}^{-1} = \sum_{i=1}^{2n} \mathscr{g}_i \exp(u_i t) \tag{23}$$

where

$$\mathscr{g}_i = \mathscr{I}_i (\mathscr{I}^i)^{t}$$

$$U = \begin{bmatrix} u_1 & & 0 \\ & u_2 & \\ & & \ddots \\ 0 & & u_{2n} \end{bmatrix}$$

$$\mathscr{I} = \begin{bmatrix} \mathscr{I}_{1,1} & \mathscr{I}_{1,2} & \cdots & \mathscr{I}_{1,2n} \\ \mathscr{I}_{2,1} & \cdots & & \\ \vdots & & & \\ \mathscr{I}_{2n,1} & & \cdots & \mathscr{I}_{2n,2n} \end{bmatrix} = [\mathscr{I}_1, \mathscr{I}_2, ..., \mathscr{I}_{2n}]$$

$$\mathscr{T}^{-1} = \begin{bmatrix} \mathscr{T}^{1,1} & \mathscr{T}^{1,2} & \cdots & \mathscr{T}^{1,2n} \\ \mathscr{T}^{2,1} & \cdots & & \\ \vdots & & & \\ \mathscr{T}^{2n,1} & \cdots & & \mathscr{T}^{2n,2n} \end{bmatrix} = \begin{bmatrix} (\mathscr{T}^1)^t \\ (\mathscr{T}^2)^t \\ \vdots \\ (\mathscr{T}^{2n})^t \end{bmatrix}$$

$$\mathscr{T}_i^t = [\ \mathscr{T}_{1,i},\ \mathscr{T}_{2,i},\ \cdots,\ \mathscr{T}_{2n,i}\ ]\ ;$$

$$(\mathscr{T}^i)^t = [\ \mathscr{T}^{i,1},\ \mathscr{T}^{i,2},\ \cdots,\ \mathscr{T}^{i,2n}\ ]$$

Using Eq. (20) the state covariance matrix function of $q(t)$ is given by

$$\Gamma_q(v) = E[q(t+v)\ q^t(t)]$$

$$= \int_o^\infty \exp[-A(t+v)]\Gamma_z \exp[-A(t)]dt = \exp(-Av)\Gamma_q(0)$$

$$(24)$$

From Eqs. (20) and (23), Eq. (24) can be rewritten as

$$\Gamma_q(v) = E[q(t+v)\ q^t(t)]$$

$$= \sum_{i=1}^{2n} (\ \sum_{j=1}^{2n} \mathscr{T}_i \Gamma_z \mathscr{T}_j^t \frac{-1}{u_i + u_j}\ ) \exp(u_i v)$$

$$= \sum_{i=1}^{2n} d_i \exp(u_i v) \qquad (25)$$

where

$$d_i = \sum_{j=1}^{2n} \mathscr{T}_i \Gamma_z \mathscr{T}_j^t \frac{-1}{u_i + u_j}$$

C. Step–by–Step Procedure of the New Method

The way to find the values of the continuous parameters corresponding to the discrete parameters is through the covariance invariant principle[4], i.e., Γ_k in Eqs. (16) and (17) must be equal to the continuous covariance function $\Gamma(k\Delta)$ in Eqs. (24) and (25) by uniformly sampling. That is

$$\Gamma_k = \Gamma_q(k\Delta) = \phi^k \Gamma_0 = \sum_{i=1}^{2n} d_i \lambda_i^k$$

$$= \exp(-Ak\Delta)\Gamma_q(0) = \sum_{i=1}^{2n} d_i' \exp(u_i k\Delta) \tag{26}$$

With such treatment we have

$$\lambda_i = \exp(u_i \Delta) \tag{27}$$

$$\Gamma_0 = \Gamma_q(0) \tag{28}$$

$$\phi = T \Lambda T^{-1} = \mathcal{T} e^{U\Delta} \mathcal{T}^{-1} = \exp(-A\Delta) \tag{29}$$

and from Eqs. (27) and (29) we obtain

$$\mathbf{T} = \mathcal{T} \tag{30}$$

The modal parameters (i.e. natural frequency, damping ratio, and mode shape) can be obtained by following procedure:

1) The measured displacement and velocity are fitted into the discrete VAR(1) model. The matrix ϕ in Eqs. (8) could be obtained.

2) By the solving eigenvalue–eigenvector problem in Eq. (12), we could get the modal vectors (i.e., eigenvectors) and discrete eigenvalues.

3) Continuous eigenvalues u_i and A matrix could be found by using Eqs. (27) and (29); i.e., the stochastic model in Eq. (7) could be obtained.

4) The damping ratio ξ_i and natural frequency ω_i are solved by following equation[11] :

$$u_i, u_i^* = \omega_i \left(-\xi_i \pm j\sqrt{1 - \xi_i^2} \right) \tag{31}$$

where u_i^* is the complex conjugate of u_i

The uniqueness representation and parameter estimation of the discrete VAR(1) model are presented in Appendices A and B.

IV. Fisher Information Matrix with Respect to the Discrete Eigenvalues

In this section we shall develop the Fisher information matrix for a VAR(1) model and derive the variance—covariance matrix with respect to the parameters of interest.

Eq. (8) can also be expressed as

$$\left(I - \sum_{i=1}^{2n} g_i \lambda_i B \right) p(k) = a(k) \tag{32}$$

If the number of the observations N is large , then the likelihood function[12] can be approximately expressed as

$$f(\beta \mid \alpha) = [2\pi \det(\gamma_a)]^{-N/2} \exp \left\{ - \sum_{k=1}^{N} [\alpha^t(k) \gamma_a^{-1} \alpha(k)]/2 \right\}$$

where

$$\alpha^t(k) = [a_{n+1}(k), a_{n+2}(k), ..., a_{2n}(k)]$$

$$E[\alpha(k) \alpha^t(k-l)] = \delta_l \gamma_a$$

$$\beta^t = [\lambda_1, \lambda_2, ..., \lambda_{2n}]$$

$\det(\gamma_a)$ is the determinant of matrix γ_a and the log–likelihood function is

$$L = -(N/2) \log[2\pi \det(\gamma_a)]$$

$$- \sum_{k=1}^{N} \{ [\alpha^t(k) \, \gamma_a^{-1} \, \alpha(k)] / 2 \} \tag{33}$$

Then the $(2n \times 2n)$ matrix $I(\beta)$ which is called the Fisher information matrix for the parameter β is given by[13]

$$I_{ij}(\beta) = E[- \frac{\partial^2 L}{\partial \lambda_i \, \partial \lambda_j}] \tag{34}$$

For structural systems, however, the eigenvalue set always contains complex conjugates. Since $\partial L / \partial \lambda_j$ is complex when λ_j is complex, Eq. (34) can be generalized as

$$I_{ij}(\beta) = E[- \frac{\partial^2 L}{\partial \lambda_i \, \partial \lambda_j^*}] \tag{35}$$

and λ_j^* is the complex conjugate of the λ_j.

From Eqs. (33) and (35), we obtain the (i,j) element of the Fisher information matrix $I(\beta)$ as

$$I_{ij}(\beta) = -E[\frac{\partial^2 L}{\partial \lambda_i \, \partial \lambda_j^*}]$$

$$= \frac{1}{2} \sum_{k=1}^{N} E \left[\frac{\partial^2 (\boldsymbol{\alpha}^t(k) \, \gamma_a^{-1}(k) \boldsymbol{\alpha})}{\partial \lambda_i \, \partial \lambda_j^*} \right]$$

$$= \frac{1}{2} \sum_{k=1}^{N} \text{trace} \left\{ E \left[\frac{\partial^2 (\boldsymbol{\alpha}(k) \, \boldsymbol{\alpha}^t(k))}{\partial \lambda_i \, \partial \lambda_j^*} \right] \gamma_a^{-1} \right\}$$

$$= \frac{1}{2} \sum_{k=1}^{N} \text{trace} \left\{ \text{sub} \left[E \left(\frac{\partial^2 [\boldsymbol{a}(k) \, \boldsymbol{a}^t(k)]}{\partial \lambda_i \, \partial \lambda_j^*} \right) \right] \gamma_a^{-1} \right\}$$

$$(36)$$

where the relationship

$$\boldsymbol{a}^t(k) = [0, 0, ..., a_{n+1}(k), a_{n+2}(k), ..., a_{2n}(k)]$$

$$= [0, 0, ..., \boldsymbol{\alpha}^t(k)]$$

was used and the notation sub(...) denotes the lower—right corner $n \times n$ submatrix of the matrix (...). The derivative of $\boldsymbol{a}(k)$ can be found from Eq. (32) as

$$\frac{\partial \, \boldsymbol{a}(k)}{\partial \, \lambda_i} = \frac{\partial \left[(I - \sum_{l=1}^{2n} g_l \lambda_l B) \boldsymbol{p}(k) \right]}{\partial \, \lambda_i}$$

$$= -g_i B \, \boldsymbol{p}(k) = -g_i \frac{I - \phi B}{I - \phi B} \, \boldsymbol{p}(k-1)$$

$$= -g_i \left[I + (\phi B) + (\phi B)^2 + (\phi B)^3 + ... \right] \boldsymbol{a}(k-1)$$

$$= -g_i \left[a(k-1) + \phi \, a(k-2) + \phi^2 \, a(k-3) + \cdots \right]$$

(37)

Similarly

$$\frac{\partial \, a^t(k)}{\partial \, \lambda^*_j} = - \left[a^t(k-1) + a^t(k-2) \, \phi^{*t} \right.$$

$$\left. + a(k-3)(\phi^{*t})^2 + \cdots \right] (g^*_j)^t$$

(38)

Because

$$E \left[a(k-l) \, a^t(k-m) \right] = \begin{bmatrix} 0 & \text{when } l = m \\ \Gamma_a & \text{when } l \neq m \end{bmatrix}$$

and from Eqs. (37) and (38), we obtain

$$E \left[\sum_{k=1}^{N} \frac{\partial^2 \, [a(k) \quad a^t(k)]}{\partial \, \lambda_i \partial \, \lambda^*_j} \right]$$

$$= 2 \sum_{k=1}^{N} g_i [\Gamma_a + \phi \, \Gamma_a \phi^{*t}$$

$$+ \phi^2 \, \Gamma_a (\phi^{*t})^2 + \cdots](g^*_j)^t$$

$$= 2 N g_i [\gamma_a + \sum_{l=1}^{2n} \sum_{m=1}^{2n} \lambda_l \lambda^*_m \, g_l \Gamma_a \, (g^*_m)^t$$

$$+ \sum_{l=1}^{2n} \sum_{m=1}^{2n} \lambda^2_l (\lambda^*_m)^2 \, g_l \Gamma_a \, (g^*_m)^t + \cdots] \, (g^*_j)^t$$

$$= 2 N g_i [\sum_{l=1}^{2n} \sum_{m=1}^{2n} \frac{1}{1 - \lambda_l \lambda_m^*} g_l \Gamma_a (g_m^*)^t] (g_j^*)^t$$

(39)

where

$$\phi^i = \sum_{l=1}^{2n} g_l \lambda_l^i$$

Substituting Eq. (39) into Eq. (36), the (i,j) element of the Fisher information matrix $[I_{ij}(\beta)]$ can be obtained as

$$I_{ij}(\beta)=N \text{ trace}\{ \text{ sub}[g_i(\sum_{l=1}^{2n} \sum_{m=1}^{2n} \frac{1}{1 - \lambda_l \lambda_m^*}$$

$$\cdot g_l \Gamma_a (g_m^*)^t) (g_j^*)^t] \gamma_a^{-1} \}$$

$$= N \text{ trace}\{ \text{sub}[(g)_i R (g^*)_j^t] \gamma_a^{-1} \}$$

(40)

where

$$R = \sum_{l=1}^{2n} \sum_{m=1}^{2n} \frac{1}{1 - \lambda_l \lambda_m^*} (g)_l \Gamma_a (g^*)_m^t$$

V. Fisher Information Matrix with Respect to Natural Frequencies and Damping Ratios

In modal analysis application, the parameters of interest are natural frequencies and damping ratios. From Eqs. (27) and (31) we can find the transformation is

$$\lambda_k , \lambda_k^* = \exp(-(\xi_k \pm \sqrt{ 1- \xi_k^2} \text{ j}) \omega_k \Delta)$$

(41)

The Fisher information matrix for the transformed

parameters can be obtained by using the transformation (41) and Eq. (40). Define Y^t as

$$Y^t = [Y_1, Y_2, ..., Y_{2n}]$$

$$= [\omega_1, \xi_1, \omega_2, \xi_2, ..., \omega_n, \xi_n]$$

Then, by the chain rule

$$\frac{\partial a_{n+j}(k)}{\partial Y_i} = \sum_{l=1}^{2n} \frac{\partial a_{n+j}(k)}{\partial \lambda_l} \frac{\partial \lambda_l}{\partial Y_i}$$

$$(j = 1, 2, ..., n) \qquad (42)$$

By taking the expectation of

$$\left(\frac{\partial^2 a_{n+j}(k)}{\partial Y_i \, \partial Y_j^*} \right)$$

the Fisher information matrix for Y is given by

$$I(Y) = J_a^t \, I(\beta) \, J_a^* \qquad (43)$$

where J_a is the Jacobian matrix given by

$$
J_a = \begin{bmatrix}
\dfrac{\partial \lambda_1}{\partial Y_1} & \dfrac{\partial \lambda_1}{\partial Y_2} & \cdots & \dfrac{\partial \lambda_1}{\partial Y_{2n}} \\[2mm]
\dfrac{\partial \lambda_2}{\partial Y_1} & \ddots & & \\[2mm]
\vdots & & \ddots & \vdots \\[2mm]
\dfrac{\partial \lambda_{2n}}{\partial Y_1} & \cdots & & \dfrac{\partial \lambda_{2n}}{\partial Y_{2n}}
\end{bmatrix}
\tag{44}
$$

The elements of J_a can be found using Eq. (41), and the analytical derivatives are

$$
\frac{\partial \lambda_i}{\partial \omega_j} =
\begin{bmatrix}
-(\xi_i + \sqrt{1-\xi_i^2}\, j)\Delta\lambda_i & \text{for } i = j \\[3mm]
0 & \text{for } i \neq j
\end{bmatrix}
$$

$$
\frac{\partial \lambda_i}{\partial \xi_j} =
\begin{bmatrix}
-(1 - \dfrac{\xi_i}{\sqrt{1-\xi_i^2}}\, j)\, \omega_i \Delta \lambda_i & \text{for } i = j \\[3mm]
0 & \text{for } i \neq j
\end{bmatrix}
$$

Thus the Fisher information matrix with respect to ξ_n and ω_n can be found, and the variance–covariance matrix var(Y) for the transformed parameters can be obtained by inverting the Fisher information matrix, i.e.

$$
\text{var}(Y) = I^{-1}(Y)
\tag{45}
$$

If the variance–covariance matrix is solved, the standard errors for these parameters can then be obtained from the diagonal elements. The approximate 95% confidence limits of the parameters are[10]

$$
Y_i \pm 1.96 \sqrt{\text{var}(Y_i)}
$$

where $\sqrt{\text{var}(Y_i)}$ is the standard error obtained from the diagonal elements of the estimated variance—covariance matrix of Y.

VI. Choice of Optimal Sampling Interval

In principle, we can differentiate the particular elements of Eq. (45) with respect to the sampling interval and equate the result to zero to obtain the optimal sampling interval that minimizes the variance of the interested parameter. However, because of the complexity in the nonlinear algebraic equations involved, in general, there is no closed form solution except for first—order systems (see Appendix C). Consequently, one is compelled to obtain the solution numerically, and it is sometimes the case that a single sampling interval cannot cover the whole frequency range of interest. Therefore, for higher—order systems, the optimal sampling intervals might be respectively chosen for each mode in the system; i.e., one mode has priority at a time. The computational procedures are almost the same as that of first order systems.

The other alternative to the problem is to choose the optimal sampling interval for ξ and ω by reducing it to one parameter problem. The transformation is

$$q = w_1 \omega_1 + w_2 \xi_1 + w_3 \omega_2 + w_4 \xi_2 + \dots + w_{2n-1} \omega_n + w_{2n} \xi_n$$

$$= w^t Y \tag{46}$$

where w_{2i-1}, $w_{2i}(i=1, 2, \dots, n)$ are predetermined weighting factors based on the practical justification about relative importance of ω_i and ξ_i $(i=1, 2, \dots, n)$. The optimal sampling interval can then be defined as the sampling interval that yields minimum variance of q:

$$\text{var}(q) = w^t \text{var}(Y)\, w \tag{47}$$

In practical implementation, the exact values of the natural frequencies, and damping ratios can not be known a priori to obtain the optimum sampling interval. An iterative procedure is suggested by first choosing a reasonable initial guess of sampling interval based on physical reasoning and find the estimated modal parameters by the procedures presented in Sec. III.C, then acquire the corresponding Fisher information matrix with respect to those estimated $\hat{\omega}$ and $\hat{\xi}$ by using Eq. (43) and obtain the var($\hat{q}$) by Eqs. (46) and (47). In consequence, a new sampling interval can be obtained by minimizing the var($\hat{q}$) to refine the estimation accuracy.

VII. Numerical Examples

$$
\begin{bmatrix}
1 & 0 & 0 & 0 & 0 & 0 \\
0 & 1 & 0 & 0 & 0 & 0 \\
0 & 0 & 2 & 0 & 0 & 0 \\
0 & 0 & 0 & 3 & 0 & 0 \\
0 & 0 & 0 & 0 & 2.5 & 0 \\
0 & 0 & 0 & 0 & 0 & 2.8
\end{bmatrix}
\begin{bmatrix}
\ddot{x}_1(t) \\
\ddot{x}_2(t) \\
\ddot{x}_3(t) \\
\ddot{x}_4(t) \\
\ddot{x}_5(t) \\
\ddot{x}_6(t)
\end{bmatrix}
+
\begin{bmatrix}
0.4 & -.2 & 0 & 0 & 0 & 0 \\
-.2 & 0.5 & -.3 & 0 & 0 & 0 \\
0 & -.3 & 0.6 & -.3 & 0 & 0 \\
0 & 0 & -.3 & 0.8 & -.5 & 0 \\
0 & 0 & 0 & -.5 & 0.6 & -.1 \\
0 & 0 & 0 & 0 & -.1 & 0.3
\end{bmatrix}
\begin{bmatrix}
\dot{x}_1(t) \\
\dot{x}_2(t) \\
\dot{x}_3(t) \\
\dot{x}_4(t) \\
\dot{x}_5(t) \\
\dot{x}_6(t)
\end{bmatrix}
$$

$$
+
\begin{bmatrix}
43 & -40 & 0 & 0 & 0 & 0 \\
-40 & 54 & -14 & 0 & 0 & 0 \\
0 & -14 & 144 & -130 & 0 & 0 \\
0 & 0 & -130 & 210 & -80 & 0 \\
0 & 0 & 0 & -80 & 84 & -4 \\
0 & 0 & 0 & 0 & -4 & 6.5
\end{bmatrix}
\begin{bmatrix}
x_1(t) \\
x_2(t) \\
x_3(t) \\
x_4(t) \\
x_5(t) \\
x_6(t)
\end{bmatrix}
=
\begin{bmatrix}
f_1(t) \\
f_2(t) \\
f_3(t) \\
f_4(t) \\
f_5(t) \\
f_6(t)
\end{bmatrix}
\tag{48}
$$

The lumped–mass system described by Eq. (48) with linear springs and dampers was chosen to illustrate the applicability of this new method. The numerical example was executed on a VAX8800 digital computer and the computer package, Matrix$_x$[14], was used to

Table 1 Simulation results together with theoretical modal
parameters (eigenvalues, natural frequencies, and damping ratios)

sampling interval, s	eigenvalue	damping ratio	natural frequency, rad./s
1st Mode			
theoretical	−1.2142E−2±0.6170E 0j	1.9675E−2	0.6171E 0
0.030	−0.6740E−2±0.6218E 0j	1.0839E−2	0.6218E 0
0.243	−0.4703E−2±0.6188E 0j	0.7600E−2	0.6188E 0
2nd Mode			
theoretical	−5.3963E−2±1.6045E 0j	3.3613E−2	1.6054E 0
0.030	−7.0018E−2±1.6927E 0j	4.1330E−2	1.6942E 0
0.243	−7.6611E−2±1.6052E 0j	4.7673E−2	1.6070E 0
3rd Mode			
theoretical	−1.2068E−1±3.0128E 0j	4.0023E−2	3.0152E 0
0.030	−2.0303E−1±2.8950E 0j	6.9957E−2	2.9022E 0
0.243	−1.3027E−1±2.9829E 0j	4.3632E−2	2.9857E 0
4th mode			
theoretical	−1.5268E−1±6.6777E 0j	2.2858E−2	6.6795E 0
0.030	−2.4305E−1±6.5901E 0j	3.6855E−2	6.5946E 0
0.243	−1.2056E−1±6.6843E 0j	1.8034E−2	6.6853E 0
5th mode			
theoretical	−3.2326E−1±9.4955E 0j	3.4298E−2	9.4251E 0
0.030	−4.3795E−1±9.4569E 0j	4.6260E−2	9.4671E 0
0.243	−3.4946E−1±9.4395E 0j	3.6963E−2	9.4460E 0
6th Mode			
theoretical	−2.4420E−1±1.1372E 1j	2.1468E−2	1.1375E 01
0.030	−2.6060E−1±1.1326E 1j	2.3003E−2	1.1329E 01
0.243	−2.3960E−1±1.1400E 1j	2.1013E−2	1.1402E 01

Table 2 The simulation results together with theoretical
modal parameters (modal vectors)

sampling interval, s	modal vector

1st Mode

theoretical (1.0, 1.07E 0—2.51E—3j, 1.22E 0—6.96E—3j
1.23E 0—7.32E—3j, 1.23E 0—7.15E—3j, 9.07E—1+4.64E—3j)

0.03 (1.0, 1.06E 0—1.43E—3j, 1.25E 0—1.40E—2j
1.26E 0—1.30E—2j, 1.26E 0—1.17E—2j. 9.12E—1+4.55E—2j)

0.243 (1.0, 1.07E 0—3.27E—3j, 1.22E 0—8.40E—3j
1.23E 0—8.92E—3j, 1.23E 0—8.46E—3j, 9.19E—1+8.25E—3j)

2nd Mode

theoretical (1.0, 1.01E 0—3.61E—3j, 8.54E—1—6.36E—3j
8.04E—1—3.74E—3j, 6.44E—1+4.81E—3j,—3.59E 0+9.98E—2j)

0.03 (1.0, 1.01E 0+2.59E—3j, 7.50E—1+1.26E—2j
6.96E—1+3.64E—2j, 5.52E—1+6.80E—2j,—3.07E 0+8.10E—1j)

0.243 (1.0, 1.01E 0+1.80E—3j, 8.41E 0—1.65E—2j
7.85E—1—1.32E—2j, 6.30E—1—5.19E—3j,—3.55E 0+4.27E—2j)

3rd Mode

theoretical (1.0, 8.48E—1+8.21E—4j,—1.38E—1—1.04E—2j
—2.24E—1—5.47E—3j, —2.89E—1—1.91E—3j, 6.09E—2—5.49E—4j)

0.03 (1.0, 8.30E—1—3.08E—2j,—1.59E—1—8.86E—2j
—2.49E—1—6.84E—2j, —3.47E—1—4.75E—2j, 2.21E—1—5.59E—2j)

0.243 (1.0, 8.52E—1—1.51E—4j,—1.35E—1—4.00E—3j
—2.23E—1+9.13E—3j, —2.90E—1+1.27E—2j, 3.11E—2+1.28E—2j)

4th Mode

theoretical (1.0, —4.02E—2—1.72E—2j,—2.85E 0—3.21E—1j
—1.19E 0—1.54E—1j, 3.48E 0+4.37E—1j,—1.20E—1+1.18E—3j)

0.03 (1.0, —5.45E—2—7.89E—2j,—2.74E 0—1.08E 0j
—1.12E 0—6.23E—1j, 3.27E 0+1.26E 0j,—5.60E—1—1.97E—1j)

0.243 (1.0, —2.36E—2—1.62E—2j,—2.70E 0—3.25E—1j
—1.17E 0—1.61E—1j, 3.27E 0+3.83E—1j,—3.24E—2+8.15E—2j)

5th Mode

theoretical (1.0, —1.15E 0+4.09E—3j,—7.98E—3+9.90E—3j
1.25E—1—2.56E—2j, —7.24E—2+1.41E—2j, 1.16E—3—4.45E—4j)

0.03 (1.0, —1.17E 0—7.07E—2j, 1.06E—1—2.30E—2j
1.27E—1+3.85E—3j, —1.32E—1+4.20E—2j, 8.91E—2+4.20E—2j)

0.243 (1.0, —1.16E 0—1.95E—2j, 6.97E—3—3.27E—3j
1.30E—1—2.13E—2j, —6.99E—2—2.19E—2j,—3.58E—3+4.83E—3j)

Table 2 The simulation results together with theoretical
modal parameters (modal vectors) (continued)

sampling interval, s	modal vector
6th Mode	
theoretical	(1.0, $-2.16E$ $0-9.76E-2$j, 8.14E $0+2.71E$ 0j
	$-6.98E$ $0-2.35E$ 0j, 2.36E $0+6.88E-1$j,$-2.86E-2-1.10E-3$j)
0.03	(1.0, $-1.54E$ $0-1.83E-1$j, 6.04E $0+2.39E$ 0j
	$-5.04E$ $0-2.24E$ 0j, 1.46E $0+6.77E-1$j, 2.69E$-1+1.44E-1$j)
0.243	(1.0, $-2.37E$ $0-5.88E-2$j,10.43E $0+3.29E$ 0j
	$-9.05E$ $0-2.87E$ 0j, 3.09E $0+7.87E-1$j,$-3.57E-3+3.98E-2$j)

simulation the system. The band—limited white noise was given to the
system as excitation force to generate response. The number of
observations was chosen to be 1000 for each variable. Although it is
not possible to build a generator of perfect white noise, as long as the
flat range of a practical generator extends beyond the frequency
response of the system being considered, the "nonwhiteness" will not
present any difficulty. The variance vector of exciting random force

was $[1, 1, 0.25, 0.11, 0.16, 0.13]^t$, and the sampling interval was 0.03 s.
From the results shown in Tables 1 and 2, the estimated modal
parameters are quite closed to the theoretical ones. The reasonableness
of the proposed method was demonstrated. For the optimal sampling
interval, three cases are considered: 1) The sixth mode is of interest;

i.e., the weighting vector (w) is $[0, 0, 0, 0, 0, 0, 0, 0, 0, 0, 0.5, 0.5]^t$. 2)
The even—weighting of the six natural frequencies is considered; i.e.,
the weighting vector (w) is $[0.167, 0, 0.167, 0, 0.167, 0, 0.167, 0, 0.167,$

$0, 0.167, 0]^t$. 3) The even—weighting of the six damping ratios is
considered; i.e., the weighting vector (w) is $[0, 0.167, 0, 0.167, 0, 0.167,$

$0, 0.167, 0, 0.167, 0, 0.167]^t$. The initial guess of sampling interval was
chosen 0.03 s. for those three cases. In case 1, the final sampling
interval was obtained to be 0.243 s after 3 iterations. The results are
shown in Tables 1 and 2 which indicate that the estimation accuracy
of the sixth mode is improved. The variance of q vs. the product of
sampling interval and the highest natural frequency of the system is
shown in Fig. 1 for each iteration. For the cases 2 and 3, the typical
plots are shown in Figs. 2 and 3. The numerical scheme converged
rapidly after few iterations and final results are 0.25 and 0.27 s. for

cases 2 and 3 respectively.

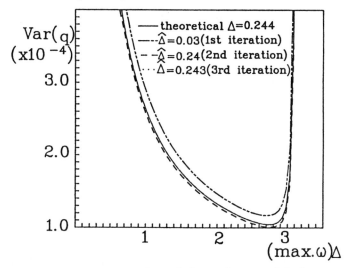

Fig.1 Typical plot of var(q) vs. (max. ω)Δ for case 1.

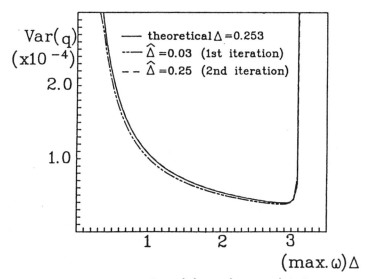

Fig.2 Typical plot of var(q) vs. (max. ω)Δ for case 2.

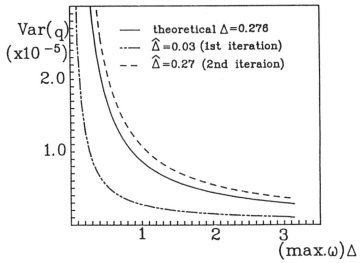

Fig.3 Typical plot of var(q) vs. (max. ω)Δ for case 3.

VIII. Concluding Remarks

This chapter presented an approach to overcome the difficulty of nonuniqueness of mode shape in modal analysis when random input data are not or cannot be measured. Also, a work regarding the selection of optimal sampling interval is discussed. The additional velocity information that in consequence can be used in setting up the state covariance matrix at the points of measurement is the key point of this approach. The selection of the optimal sampling interval is based on on the fact that the length of the confidence interval (or variance) of an estimated parameter is a measure of the accuracy of estimation; hence, for a vector of parameters, the optimization criterion is to minimize the particular elements of the covariance matrix of the estimated continuous parameter vector as a function of the sampling interval. One drawback to the method is that the full states must be measured in order to back out the state transition matrix. This cannot be accomplished in practice, and the method should be modified to include unmeasured states.

Appendix A: Uniqueness of the Discrete VAR(1) Model

Equation (8) can be rewritten as following

$$p_i(k) = \sum_{j=1}^{2n} \phi_{ij} \, p_j(k-1) + a_i(k)$$

$$(i=1, 2, ..., 2n) \qquad \qquad \text{(A1)}$$

$$[p_i(k) - \phi_{ii} \, p_i(k-1)] = \sum_{\substack{j=1 \\ j \neq i}}^{2n} \phi_{ij} \, p_j(k-1) + a_i(k)$$

$$(i=1, 2, ..., 2n) \qquad \qquad \text{(A2)}$$

Applying z transform to each side of Eq. (A2)

$$A_i(z) = \phi_{ii}(z) \, p_i(z) - \sum_{\substack{j=1 \\ j \neq i}}^{2n} \phi_{ij}(z) \, p_j(z)$$

$$(i=1, 2, ..., 2n) \qquad \qquad \text{(A3)}$$

where

$$\phi_{ii}(z) = 1 - \phi_{ii} z^{-1}$$

$$\phi_{ij}(z) = \phi_{ij} z^{-1} \qquad \qquad \text{for } i \neq j$$

and $A_i(z)$ and $p_i(z)$ are z transform of $a_i(k)$ and $p_i(k)$. The function V_i can be expressed as

$$V_i = \sum_{k=1}^{L} [a_i(k)]^2 = \frac{1}{2\pi j} \oint \left| \phi_{ii}(z) p_i(z) \right.$$

$$\left. - \sum_{\substack{j=1 \\ j \neq i}}^{2n} \phi_{ij}(z) \, p_j(z) \right|^2 \frac{dz}{z}$$

$$= \frac{1}{2\pi j} \oint (| \sum_{j=1}^{2n} \phi_{ij}(z) \, p_j(z) |^2$$

$$+ \sum_{\substack{k=1 \\ k \neq i}}^{2n} \sum_{\substack{j=1 \\ j \neq i,k}}^{2n} \phi_{ij}(z) \, p_j(z) \phi_{ik}(-z) \, p_k(-z)$$

$$- \phi_{ii}(z) p_i(z) \sum_{\substack{j=1 \\ j \neq i}}^{2n} \phi_{ij}(-z) \, p_j(-z)$$

$$- \phi_{ii}(-z) p_j(-z) \sum_{\substack{j=1 \\ j \neq i}}^{2n} \phi_{ij}(z) \, p_j(z)) \frac{dz}{z}$$

$$(i = 1, 2, ..., 2n) \qquad \text{(A4)}$$

From this relation we see that the phase of polynomials ϕ_{ij} $(i,j=1, 2, ..., 2n)$ are involved. This fact implies that the modal parameters of the discrete VAR(1) can be determined uniquely.

Appendix B: Parameter Estimation of the Discrete VAR(1) Model

To estimate the parameter matrix ϕ the sum of the cost function V_i $(i=1, 2, ..., 2n)$ must be minimized.

Taking transpose to each side of Eq (8) we obtain

$$\boldsymbol{p}^t(k) = \boldsymbol{p}^t(k-1) \, \boldsymbol{\phi}^t + \boldsymbol{a}^t(k) \qquad \text{(B1)}$$

Now suppose a set of $L+1$ measurements are obtained from the system response at times $t_0, t_1, ..., t_L$. We can then arrange the equations in the following form :

$$C = D \, \phi^t + \epsilon \qquad \text{(B2)}$$

where

$$C = \begin{bmatrix} \boldsymbol{p}^t(1) \\ \boldsymbol{p}^t(2) \\ \vdots \\ \boldsymbol{p}^t(L) \end{bmatrix}, \quad D = \begin{bmatrix} \boldsymbol{p}^t(0) \\ \boldsymbol{p}^t(1) \\ \vdots \\ \boldsymbol{p}^t(L-1) \end{bmatrix}, \quad \text{and} \quad \epsilon = \begin{bmatrix} \boldsymbol{a}^t(1) \\ \boldsymbol{a}^t(2) \\ \vdots \\ \boldsymbol{a}^t(L) \end{bmatrix}$$

The sum of the cost function can be written as

$$V = \sum_{i=1}^{2n} \sum_{k=1}^{L} a_i^2(k) = \text{trace}[\ \epsilon^t \epsilon\ ]$$

$$= \text{trace} [(C - \phi\, D^t)^t (\ C - \phi\, D^t)\ ]$$

$$= \text{trace} [\ C^t\, C - \phi\, D^t\, C - C^t\, D\, \phi^t + \phi\, D^t D\, \phi^t]$$

$$\text{(B3)}$$

Differentiate V with respect to the parameter matrix ϕ and equates the result to a zero matrix to determine the condition of the estimate $\hat{\phi}$ that minimizes V. Thus we obtain

$$\frac{\partial V}{\partial \phi}\bigg|_{\phi = \hat{\phi}} = 0 - C^t\, D - C^t D + 2\, \phi\, D^t D = 0 \qquad \text{(B4)}$$

where the following relations are used.

$$\frac{\partial(\ \text{trace}\, [A\ B\ A^t])}{\partial A} = 2\, A\, B$$

when B is symmetric

$$\frac{\partial(\text{trace}\, [A\ B\ ])}{\partial A} = B^t$$

Then the parameter matrix ϕ can be estimated in the following form:

$$\hat{\phi} = (C^t D) (D^t D)^{-1} \tag{B5}$$

It is noted that the parameter matrix estimated by the proceeding method is a kind of linear least—squares method that is equivalent to the maximum likelihood approach under an imposed random input condition.

Appendix C: First Order Processes

The first order process

$$(D + \alpha) X = Z \tag{C1}$$

has the discrete representation

$$X(k) = \phi X(k-1) + a(k) \tag{C2}$$

where

$$\phi = e^{-\alpha \Delta} = e^{-\Delta/\tau} = \lambda \tag{C3}$$

and $X(k)$, ϕ, and $a(k)$ are scalars.

From Eqs. (40) and (C3), the 1×1 Fisher information matrix for λ is

$$I(\lambda) = \frac{N}{1 - \phi^2} \tag{C4}$$

From Eq. (44), the Jacobian matrix is

$$J_a = \frac{\partial \phi}{\partial \tau} = \frac{\Delta \phi}{\tau^2} \tag{C5}$$

Therefore, the Fisher information matrix for τ is

$$I(\tau) = \frac{N(\Delta \phi)^2}{\tau^4 (1-\phi^2)} \tag{C6}$$

According to Eq. (45), the variance of τ is

$$\text{var}(\tau) = \frac{\tau^4(1-\phi^2)}{N(\Delta\phi)^2} \tag{C7}$$

The optimal sampling interval (Δ) is the one which yields the smallest var(τ). By taking the partial derivative of var(τ) with respect to Δ and setting it to zero, the optimal Δ can be derived as

$$\frac{\partial \ \text{var}(\tau)}{\partial \ \Delta}$$

$$= \frac{2\ \tau}{N}\ (\ \frac{\tau}{\Delta}\ )^3\ [(\ \frac{\Delta}{\tau} - 1)\ \exp(2\Delta/\tau) + 1]$$

$$= 0 \tag{C8}$$

When N and τ are constant

$$(\ \frac{\Delta}{\tau} - 1)\ \exp(2\Delta/\tau) + 1 = 0 \tag{C9}$$

It is concluded that the optimal Δ for a first order dynamic system is

$$\Delta = 0.7968\ \tau \tag{C10}$$

and

$$\phi = \exp(-0.7968) = 0.45$$

where τ is the time constant of the continuous system. In other words, if the data observed were the realization from the first-order continuous stochastic system, one can consider the sampling interval to be adequate if the value of the estimated parameter ϕ of the discrete model is in the neighborhood of 0.45.

References

1.   F.A. Burney, S.M. Pandit and S.M. Wu, "A Stochastic Approach to Characterization of Machine Tool System Dynamics under Actual Working Conditions," *Transactions of the ASME, Series B* **98**(2), pp. 614–619(1976).

2. W. Gersch and R. Liu, "Time Series Method for the Synthesis of Random Vibration System," *Journal of Applied Mechanics* **43**(1), pp. 159–165(1976).

3. K. F. Eman and K. J. Kim, "Modal Analysis of Machine Tool Structures Based on Experimental Data," *ASME Journal of Engineering for Industry* **105**(4), pp. 282–287(1983).

4. J. Perl and L. L. Scharf, "Covariance–Invariant Digital Filtering," *IEEE Transactions on Acoustics, Speech, and Signal Processing* **ASSP–25**(2), pp. 143–151(1977).

5. S. M. Pandit, and S. M. Wu, *Time Series and System Analysis with Application*, Wiley, New York, 1983.

6. E. I. Jury, *Theory and Application of the z–Transform Method*, Wiley, New York, 1964.

7. H. Akaike, "Stochastic Theory of Minimal Realization," *IEEE Transactions on Automatic Control* **AC–19**(6), pp. 667–673(1974).

8. Dickinson, T. Kailath, and M. Morf, "Canonical Matrix Fraction and State–Space Descriptions for Deterministic and Stochastic Linear Systems," *IEEE Transactions on Automatic Control* **AC–19**(6), pp. 656–667(1974).

9. M. Morf, D. T. Lee, J. R. Nickolls and A. Vieira, "A Classification of Algorithms for ARMA Models and Ladder Realizations," *Proceedings of the IEEE International Conference on ASSP*, Inst. of Electrical and Electronics Engineers , New York, pp. 13–19(1977).

10. H. Akaike, "Markovian Representation of Stochastic Processes and Its Application to the Analysis of Autoregressive Moving Average Processes," *Annals of the Institute Statistical Math* **26**, pp. 363–387(1974).

11. D. J. Ewin, *Modal Testing : Theory and Practice*, Wiley, New York, 1984.

12. George E. P. Box, and Gwilym M. Jenkins, *Time Series Analysis : forecasting and Control*, Holden–Day, San Francisco, CA, 1976.

13. R. A. Fisher, *Statistical Method an Scientific Inference*, Oliver & Boyd, Edinburgh, Scotland, UK, 1956.

14. *MATRIX$_x$, Graphical Model Building and Simulation*, Integrated Systems, Inc. Palo Alto, CA, May 1986.

Krylov Vector Methods for Model Reduction and Control of Flexible Structures

Tzu-Jeng Su
Spacecraft Dynamics Branch
NASA Langley Research Center
Hampton, Virginia 23665

Roy R. Craig, Jr.
Aerospace Engineering and Engineering Mechanics
The University of Texas at Austin
Austin, Texas 78712

I. INTRODUCTION

Although all real structures are, in fact, distributed-parameter systems, it is usually inevitable for the structures to be modeled as discrete systems for the purpose of design and analysis. The finite element method is the discretization approaches that is most frequently used. The finite element model of a large or geometrically-complicated structure may attain tens of thousands of degrees of freedom, which is neither computationally economical nor "Riccati-solvable" for dynamic simulation and control design applications. Therefore, model order reduction plays an indispensable role in the dynamic analysis and control design of large structures.

Usually, model reduction of a structural dynamics system is performed by the Rayleigh-Ritz method, which transforms the system equation into a smaller order by using a projection subspace. It is indisputable that the choice of projection subspace is important to the accuracy of the reduced model. The eigen-subspace, or the normal mode subspace, is frequently used for projection because it has a clear physical meaning and because it preserves the system's natural frequencies. However, with regard to the accuracy of system response, numerical experience has shown that preservation of the natural frequencies is usually not the first concern. Other than normal modes, there are other static modes, e.g., constraint modes, attachment modes, inertia-relief modes, which are frequently used in component mode synthesis [1]. In this chapter, Krylov vectors, which can be considered as static modes, are used for model reduction.

There has been a lot of research concerning the convergence and efficiency of Krylov vectors in application to eigenvalue analysis and to the structural dynamics model reduction problem [2–6]. The major purpose of this chapter is to discuss the possible application of Krylov vectors to controller design of flexible structures. Since a controller design problem includes actuators and measurements, the structural dynamics system studied here is described by a second-order matrix differential equation together with an output measurement equation. The Krylov model reduction algorithm developed here is related to parameter-matching methods for model reduction. Parameter-matching constitutes a class of efficient methods for model order reduction of general linear systems [7–12]. A parameter-matching method constructs a reduced-order model that matches a certain number of system parameters, for instance, low-frequency moments and/or high-frequency moments. All of the previous parameter-matching methods deal with systems described either by transfer functions or by the first-order state-space form. To apply the existing parameter-matching methods to a structural dynamics system, it is necessary to put the system equation into first-order form. There is some disadvantage to this approach, namely, the symmetry and physical meaning of the system matrices is destroyed. Therefore, the model reduction algorithm developed here is based on a second-order formulation.

The major application of the proposed model reduction method is to the control of flexible structures. A basic topic in the control of flexible structures is how to reduce the spillover of control energy, which is a direct result of model reduction. Spillover of control energy from the controlled subsystem into the residual subsystem usually degrades the controller performance and sometimes may lead to stability problems. It will be shown that there are three types of spillover: *control spillover, observation spillover,* and *dynamic spillover.* The transformed system equation in Krylov coordinates turns out to have dynamic spillover but no control or observation spillover, which is the major difference between this method and the normal mode method.

This chapter is organized as follows. In Section II, first the parameter-matching method for general linear, time-invariant systems is briefly reviewed; then the Krylov model reduction algorithm based on the concept of parameter-matching is developed for structural dynamics systems described by matrix differential equations in second-order form. Undamped and damped systems are discussed separately. In Section III, the characteristics of spillover are depicted and a Krylov formulation for control of flexible structures is derived. In Section IV, a model reduction example and a flexible structure control example are used to illustrate the efficacy of the Krylov method. This chapter is concluded in Section V.

II. MODEL REDUCTION

A. GENERAL LINEAR SYSTEMS

Consider an n-th order, linear, time-invariant system described by

$$
\begin{aligned}
\dot{z} &= Az + Bu & z &\in R^n, \ u \in R^l \\
y &= Cz & y &\in R^m
\end{aligned}
\tag{1}
$$

for which the transfer function $G(s) = C(sI - A)^{-1}B$ can be formally expanded in a Laurent series around $s = \infty$ as

$$
G(s) = \sum_{i=0}^{\infty} CA^i B s^{-i-1}
\tag{2}
$$

If the system has no pole at the origin, then the Taylor series expansion of $G(s)$ around $s = 0$ yields

$$
G(s) = \sum_{i=0}^{\infty} -CA^{-i-1}B s^i
\tag{3}
$$

From Eq. (2), we get a set of system parameters $\{CA^iB \mid i = 0, 1, \ldots\}$, which are called *Markov parameters* [9,11], or *high-frequency moments* [12]. From Eq. (3), we get another set of system parameters $\{CA^{-i}B \mid i = 1, 2, \ldots\}$, which are called *time moments* [9,11], or *low-frequency moments* [12]. These two sets of parameters constitute pieces of system data for the triple (A, B, C). They provide a database for system identification (for instance, the ERA method [13]). In the model reduction area, one approach, called the *parameter-matching method,* seeks to construct a reduced-order model such that it matches a certain number of parameters of the full-order system. Villemagne and Skelton provide in [12] a toolbox for producing parameter-matching reduced-order models. The reduced-order model is obtained by an oblique projection approach and is described by

$$\dot{z}_R = A_R z_R + B_R u \qquad z_R \in R^r$$
$$y = C_R z_R \tag{4}$$

where $r < n$, $A_R = TAR$, $B_R = TB$, $C_R = CR$, and $TR = I_r$. T and R are the left and right projection matrices. It is shown in [12] that if T and R are chosen such that $span\{T\} = span\{(A^T)^{-p}C^T, (A^T)^{-p+1}C^T, \ldots, (A^T)^q C^T\}$ and $span\{R\} = span\{A^{-s}B, A^{-s+1}B, \ldots, A^t B\}$ with $p, q, s, t \geq 0$ and $p+q=s+t$, then the reduced-order model matches $p + s$ low-frequency moments and $q + t$ high-frequency moments. That is, $C_R A_R^i B_R = CA^i B$, for $i = -p-s, \ldots, q+t$.

B. UNDAMPED STRUCTURAL DYNAMICS SYSTEMS

An undamped structural dynamics system can be described by the input-output form

$$M\ddot{x} + Kx = Pu$$
$$y = Vx + W\dot{x} \tag{5}$$

where $x \in R^n$ is the displacement vector, $u \in R^l$ is the input force vector, $y \in R^m$ is the output measurement vector, M and K are the system mass and stiffness matrices; P is the force distribution matrix; and V and W are the displacement and velocity sensor distribution matrices. In most practical cases, we can assume that l and m are much smaller than n.

Applying the Fourier Transform to Eq. (5a) yields the frequency response solution $X(\omega) = (K - \omega^2 M)^{-1} PU(\omega)$, with $X(\omega)$ and $U(\omega)$ the Fourier transforms of x and u. If the system is assumed to have no rigid-body motion, then a Taylor expansion of the frequency response around $\omega = 0$ is possible

$$
\begin{aligned}
X(\omega) &= (I - \omega^2 K^{-1} M)^{-1} K^{-1} PU(\omega) \\
&= \sum_{i=0}^{\infty} \omega^{2i} (K^{-1} M)^i K^{-1} PU(\omega)
\end{aligned}
\tag{6}
$$

Combining Eq. (5b) and Eq. (6), the system output frequency response can be expressed as

$$
Y(\omega) = \sum_{i=0}^{\infty} [V(K^{-1} M)^i K^{-1} P + j\omega W (K^{-1} M)^i K^{-1} P] \omega^{2i} U(\omega)
\tag{7}
$$

In the above expressions, $V(K^{-1} M)^i K^{-1} P$ and $W(K^{-1} M)^i K^{-1} P$ play roles similar to that of low-frequency moments in the first-order state-space formulation. Therefore, we have the following definition.

Definition 1 *The low-frequency moments of an undamped structural dynamics system described by Eqs. (5) are $V(K^{-1} M)^i K^{-1} P$ and $W(K^{-1} M)^i K^{-1} P$, for $i = 0, 1, 2, \ldots$.*

Model reduction of structural dynamics systems is usually done by using the Rayleigh-Ritz method. The reduced system equation takes the form

$$
\begin{aligned}
\bar{M}\ddot{\bar{x}} + \bar{K}\bar{x} &= \bar{P}u \quad \bar{x} \in R^r \\
y &= \bar{V}\bar{x} + \bar{W}\dot{\bar{x}}
\end{aligned}
\tag{8}
$$

with the reduced system coordinate $\bar{x}$ and the original system coordinate x related by

$$
x = L\bar{x}
\tag{9}
$$

The reduced system matrices and the original system matrices are related by

$$
\bar{M} = L^T M L \ , \quad \bar{K} = L^T K L \ , \quad \bar{P} = L^T P \ , \quad \bar{V} = VL \ , \quad \bar{W} = WL
\tag{10}
$$

L is called the *projection matrix*. The choice of the projection matrix governs the accuracy of the reduced system. The following theorem indicates that if L is formed by a set of Krylov vectors, then the reduced system has a parameter-matching property.

Theorem 1 *If span$\{L\}$=span$\{L_P \; L_V \; L_W\}$ with*

$$L_P = \left[\; K^{-1}P \;\; (K^{-1}M)K^{-1}P \;\; \cdots \;\; (K^{-1}M)^p K^{-1}P \; \right]$$
$$L_V = \left[\; K^{-1}V^T \;\; (K^{-1}M)K^{-1}V^T \;\; \cdots \;\; (K^{-1}M)^q K^{-1}V^T \; \right]$$
$$L_W = \left[\; K^{-1}W^T \;\; (K^{-1}M)K^{-1}W^T \;\; \cdots \;\; (K^{-1}M)^s K^{-1}W^T \; \right]$$

for p, q, $s \geq 0$, then the reduced system matches the low frequency moments $V(K^{-1}M)^i K^{-1}P$ for $i = 0$, 1, $\ldots$, $p + q + 1$ and $W(K^{-1}M)^i K^{-1}P$, for $i = 0$, 1, 2, $\ldots$, $p + s + 1$.

Proof: The proof is recursive. First, we have the property: if a vector v is such that $K^{-1}v$ is contained in L, i.e., $K^{-1}v = L\alpha$, then $L(L^T K L)^{-1}L^T v = L(L^T K L)^{-1}L^T K(K^{-1}v) = L(L^T K L)^{-1}L^T K L\alpha = L\alpha = K^{-1}v$. By using this property, it can be shown that

$$L[(L^T K L)^{-1}(L^T M L)]^i (L^T K L)^{-1}L^T P = (K^{-1}M)^i K^{-1}P, \quad i = 0, \, 1, \, \cdots, \, p$$
$$V L[(L^T K L)^{-1}(L^T M L)]^i (L^T K L)^{-1}L^T = V(K^{-1}M)^i K^{-1}, \quad i = 0, \, 1, \, \cdots, \, q$$
$$W L[(L^T K L)^{-1}(L^T M L)]^i (L^T K L)^{-1}L^T = W(K^{-1}M)^i K^{-1}, \quad i = 0, \, 1, \, \cdots, \, s$$

Therefore,

$$\bar{V}(\bar{K}^{-1}\bar{M})^i \bar{K}^{-1}\bar{P} = V L[(L^T K L)^{-1}(L^T M L)]^i (L^T K L)^{-1}L^T P$$
$$= V(K^{-1}M)^i K^{-1}P \quad \text{for } i = 0, \, 1, \, \cdots, \, p + q + 1$$

$$\bar{W}(\bar{K}^{-1}\bar{M})^i \bar{K}^{-1}\bar{P} = W L[(L^T K L)^{-1}(L^T M L)]^i (L^T K L)^{-1}L^T P$$
$$= W(K^{-1}M)^i K^{-1}P \quad \text{for } i = 0, \, 1, \, \cdots, \, p + s + 1 \quad \blacksquare$$

The vectors contained in L_P are frequently referred to in the literature as *Krylov vectors*, or *Krylov modes*, which can be generated by a simple iteration formula

$$\begin{aligned} Q_1 &= K^{-1}P \\ Q_{i+1} &= K^{-1}MQ_i \end{aligned} \tag{11}$$

Krylov modes can be considered as static modes and have a clear physical interpretation [14]. The first vector $K^{-1}P$ is the *system's static deflection due to the force distribution P*. The vector Q_{i+1} can be interpreted as the *static* *deflection produced by the inertia force associated with the* Q_i. If only the dynamic response simulation is concerned, we usually choose $L = L_P$ [2–5]. In this case, the reduced model matches $p + 1$ low-frequency moments. As to the vectors in L_V and L_W, a physical interpretation such as the *static deflection due to sensor distribution* may be inadequate. However, from an input-output point of view, L_V, L_W, and L_P are equally important as far as parameter-matching of the reduced-order model is concerned. Since the major concern here is to construct a reduced-order model for control application, where the input-output property is important, the following block-Lanczos algorithm for generating the projection matrix L is proposed.

Algorithm 1 (Undamped Krylov Model Reduction Algorithm)

(1) *Starting block of vectors*:
    (a) $Q_0 = 0$
    (b) $R_0 = K^{-1}\tilde{P}$, $\tilde{P} =$ *linearly-independent portion of* $[P \ V^T \ W^T]$
    (c) $R_0^T K R_0 = U_0 \Sigma_0 U_0^T$ (*singular-value decomposition*)
    (d) $Q_1 = R_0 U_0 \Sigma_0^{-\frac{1}{2}}$ (*normalization*)

(2) *For $j = 1, 2, \ldots, k-1$, repeat*:
    (e) $\bar{R}_j = K^{-1} M Q_j$
    (f) $R_j = \bar{R}_j - Q_j A_j - Q_{j-1} B_j$ (*orthogonalization*)
        $A_j = Q_j^T K \bar{R}_j$, $B_j = U_{j-1} \Sigma_{j-1}^{\frac{1}{2}}$
    (g) $R_j^T K R_j = U_j \Sigma_j U_j^T$ (*singular-value decomposition*)
    (h) $Q_{j+1} = R_j B_{j+1}^{-T} = R_j U_j \Sigma_j^{-\frac{1}{2}}$ (*normalization*)

(3) *Form the k-block projection matrix* $L = \begin{bmatrix} Q_1 & Q_2 & \cdots & Q_k \end{bmatrix}$.

The above algorithm is a modification of the Lanczos algorithm in Ref. [3]. The major difference between the above algorithm and the Lanczos algorithm in Ref. [3] is that the K matrix instead of the M matrix is used as the normalization weighting matrix. By using K as the normalization weighting matrix, the generated Lanczos vectors normalize the K matrix and tridiagonalize the M matrix. Therefore, the standard Ritz method can be employed to transform the system's dynamic equation into the following form

$$L^T M L \ddot{\bar{x}} + L^T K L \bar{x} = L^T P u$$

where the transformed mass matrix $L^T M L$ is block tridiagonal. On the other hand, the Lanczos algorithm in Ref. [3] generates a set of Lanczos vectors that orthonormalize M and tridiagonalize $MK^{-1}M$. Therefore, the following transformation of the system equation has to be employed in order to take advantage of the tridiagonal form.

$$L^T M K^{-1} M L \ddot{\bar{x}} + L^T M K^{-1} K L \bar{x} = L^T M K^{-1} P u$$

The transformation employed is not a standard Ritz method.

The vectors generated by Algorithm 1 lie in the Krylov subspace, because they are generated by the Krylov iteration formula in Step (e). The algorithm is called a Lanczos algorithm because of the three-term orthogonalization scheme used in Step (f). The only difference between the column vectors of the L matrix in Algorithm 1 and the column vectors of the L matrix in Theorem 1 is that the former ones are orthonormalized with respect to the K matrix. But, they still span the same subspace.

A few remarks can be made about the numerical details of Algorithm 1 and the property of the reduced-order model.

Remark 1. Loss of orthogonality among the current vectors and those vectors generated two iterations before may occur due to the tendency of the vectors to converge to eigenvectors, or due to the limited number of significant figures used by the computers. Although the orthogonalization scheme in Algorithm 1 poses as a three-term recursion, in practice complete reorthogonalization [15] or selective reorthogonalization [16] is necessary to prevent the loss of orthogonality.

Remark 2. The algorithm is a block-vector algorithm instead of a single-vector algorithm. In order to produce a reduced-order model with parameter-matching property, a block algorithm is absolutely necessary for the multi-input/multi-output case. Another advantage of using a block Lanczos algorithm is that the system's repeated (or closely-spaced) natural frequencies and their corresponding modes can be reproduced or approximated by the reduced-model. This is the same as the advantage of using subspace iteration instead of single-vector iteration to solve for eigenvalues and eigenvectors.

Remark 3. It is possible that after many iterations the R_j matrix generated at

step (f) may not have full column rank. If this happens, then only the nonzero
singular-value portion of R_j needs to be retained to obtain the new block Q_{j+1}
at step (h). Therefore, it is possible that the size of the block becomes smaller
during the iteration process. If the algorithm proceeds to generate as many
vectors as possible, then, in general, n vectors can be generated provided that
orthogonalization is performed successfully. If it terminates before n vectors
are generated, then the full-order system is not a minimum realization. For
structures, this occurs only when the actuator distribution and/or the sensor
distribution is in the subspace spanned by some normal modes. In this case, an
exact reduced-order model can be obtained by projecting the full-order system
onto the L subspace.

Remark 4. According to Theorem 1, it can easily be shown that if the
projection matrix L generated by Algorithm 1 is employed to perform model
reduction, then the reduced-order model matches the low-frequency moments
$V(K^{-1}M)^i K^{-1}P$ and $W(K^{-1}M)^i K^{-1}P$, for $i = 0,\ 1,\ 2,\ \ldots,\ 2k - 1$. Therefore, k, the number of blocks included in the projection matrix, determines the
number of the low-frequency moments that are matched by the reduced model.
Since k also determines the dimension of the projection matrix L, the order of
the Krylov reduced-model increases as the value of k increases. For the purposes of dynamic response simulation and control design, it is usually desirable
to keep the order of the reduced model small. Therefore, there is a trade-off between the order and the accuracy of the Krylov reduced model, which happens
to all the existing model reduction methods. Reference [3] uses a participation
factor as the truncation criterion. However, the concept of participation factor
is valid only for system equations in normal coordinates, and, therefore, is not
quite appropriate for Krylov model reduction. As will be discussed in Remark 6,
the system equation in Krylov coordinates represents the structure of a tandem
system. Hence, it is recommended here to use the norm of the off-diagonal
submatrices in the transformed mass and stiffness matrices as the truncation
criterion.

Remark 5. Instead of low-frequency moments, it is possible to construct a
reduced-order model that matches the high-frequency moments of the full-order
structure. The only changes that need to be made in the algorithm are to replace

$K^{-1}M$ at Step (e) by $M^{-1}K$ and to replace $K^{-1}\tilde{P}$ at step (b) by $M^{-1}\tilde{P}$. By doing so, the reduced-order model obtained approximates the higher frequency range of the full-order system. However, since the lower-frequency modes usually have more significant contribution to the response than the higher modes, matching the low-frequency moments is recommended.

Remark 6. Due to the choice of starting vectors, K-orthogonalization, and three-term recurrence, the transformed system equation in the Lanczos coordinates has a mass matrix in block-tridiagonal form, a stiffness matrix equal to the identity matrix, and force distribution and measurement distribution matrices with nonzero elements only in the first block. The form of the transformed system equation is

$$
\begin{bmatrix}
\times & \times & & & & & & \\
\times & \times & \times & & & & & \\
& \times & \times & \cdot & & & & \\
& & \cdot & \cdot & \cdot & & & \\
& & & \cdot & \cdot & \cdot & & \\
& & & & \cdot & \cdot & \times & \\
& & & & & \times & \times &
\end{bmatrix}
\ddot{\bar{x}} + \bar{x} =
\begin{Bmatrix}
\times \\
0 \\
0 \\
\cdot \\
\cdot \\
\cdot \\
0
\end{Bmatrix} u
\tag{12}
$$

$$
y = [\, \times \quad 0 \quad 0 \quad \cdots \quad 0\,]\bar{x} + [\, \times \quad 0 \quad 0 \quad \cdots \quad 0\,]\dot{\bar{x}}
$$

where $\times$ denotes the location of nonzero elements. This special form reflects the structure of a tandem system (Fig. 1), in which only subsystem S_1 is directly

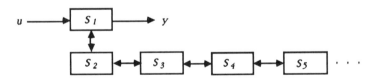

Figure 1: Structure of a tandem system.

controlled and measured while the remaining subsystems, S_i, $i = 2, 3, \ldots,$ are excited through chained dynamic coupling. If the off-diagonal elements, which represent the coupling between subsystems in Fig. 1, are small, then this special form serves as a truncation basis for model reduction. It will be shown in Section III that this special form of the transformed system equation has a major advantage in control applications.

C. DAMPED STRUCTURAL DYNAMICS SYSTEMS

A damped structural dynamics system with input-output configuration is described by

$$M\ddot{x} + D\dot{x} + Kx = Pu$$
$$y = Vx + W\dot{x} \tag{13}$$

Here the damping matrix D can be general damping, except that it is assumed to be symmetric. The output frequency response can be formally represented by a Taylor series

$$
\begin{aligned}
Y(\omega) &= (V + j\omega W)(-\omega^2 M + j\omega D + K)^{-1} P U(\omega) \\
&= (V + j\omega W)(I + j\omega K^{-1}D - \omega^2 K^{-1}M)^{-1} K^{-1} P U(\omega) \\
&= \sum_{i=0}^{\infty} (V + j\omega W)(\omega^2 K^{-1}M - j\omega K^{-1}D)^i K^{-1} P U(\omega) \\
&= \big\{ V K^{-1}P + j\omega(W - V K^{-1}D)K^{-1}P + \omega^2[V K^{-1}M \\
&\quad + W K^{-1}D - V(K^{-1}D)^2]K^{-1}P + \cdots \big\} U(\omega)
\end{aligned}
\tag{14}
$$

It is seen that the low-frequency moments defined in Definition 1 for an undamped system do not constitute a complete basis for the representation of the output frequency response of a damped system. The low-frequency moments of a damped system are defined as the coefficient matrices in the above expansion series. To arrive at an algorithm for constructing a reduced-order model that matches low-frequency moments, it is easier to start from the first-order formulation. The first-order differential equation equivalent to Eq. (13) can be expressed as

$$
\begin{bmatrix} D & M \\ M & 0 \end{bmatrix} \begin{Bmatrix} \dot{x} \\ \ddot{x} \end{Bmatrix} + \begin{bmatrix} K & 0 \\ 0 & -M \end{bmatrix} \begin{Bmatrix} x \\ \dot{x} \end{Bmatrix} = \begin{Bmatrix} P \\ 0 \end{Bmatrix} u
$$
$$
y = [V \ W] \begin{Bmatrix} x \\ \dot{x} \end{Bmatrix}
\tag{15}
$$

or

$$
\hat{M}\dot{z} + \hat{K}z = \hat{P}u
$$
$$
y = \hat{V}z
\tag{16}
$$

with

$$\hat{M} = \begin{bmatrix} D & M \\ M & 0 \end{bmatrix} , \quad \hat{K} = \begin{bmatrix} K & 0 \\ 0 & -M \end{bmatrix} , \quad \hat{P} = \begin{Bmatrix} P \\ 0 \end{Bmatrix} , \quad \hat{V} = \begin{bmatrix} V & W \end{bmatrix} \quad (17)$$

Or, in standard state-space form, the system equation becomes

$$\begin{aligned} \dot{z} &= -\hat{M}^{-1}\hat{K}z + \hat{M}^{-1}\hat{P}u \\ y &= \hat{V}z \end{aligned} \tag{18}$$

Recalling that $CA^{-i}B$ is the low-frequency moment for a system described by Eq. (1), we can express the low-frequency moments for the above system by

$$T_i = \hat{V}(-\hat{M}^{-1}\hat{K})^{-i}\hat{M}^{-1}\hat{P} = (-1)^i\hat{V}(\hat{K}^{-1}\hat{M})^{i-1}\hat{K}^{-1}\hat{P}, \quad i = 1, 2, \cdots \quad (19)$$

It can be shown that $\hat{V}(\hat{K}^{-1}\hat{M})^i\hat{K}^{-1}\hat{P}$ is equal to the coefficient matrix associated with the $(j\omega)^{i-1}$ term in Eq. (14). This is not a coincidence, since the definition of low-frequency moments is based on the Taylor series expansion of the output frequency response, which is unique whether the formulation is of first-order or second-order. Note the similarity between Eq. (19) and the low-frequency moments of undamped systems defined in Definition 1. Because of this similarity, we may call $\hat{M}$ the *generalized mass matrix*, $\hat{K}$ the *generalized stiffness matrix*, and $\hat{P}$ the *generalized force distribution matrix* of the system.

At the first sight of the similarity between the expression of the low-frequency moments of an undamped system and that of a damped system, it may appear that one can simply use Algorithm 1, but replace M, K, and P by $\hat{M}$, $\hat{K}$, and $\hat{P}$, and use $\hat{K}^{-1}\hat{P}$ as the starting block of vectors, to generate a set of Lanczos vectors for the damped system. If this can be done, then, of course, the reduced system equation would also be in the first-order state-space form. However, there are two reasons that this may not be the best approach. First of all, the $\hat{K}$ matrix is negative definite, while the Lanczos algorithm requires a positive-definite $\hat{K}$ matrix to perform normalization. Even though this difficulty can be overcome by modification of the algorithm, for instance, by employing the shifting method in Ref. [17], there still exists another drawback. That is, there is no guarantee that the reduced-order model is stable. In other words, there is no guarantee that all the eigenvalues of the pair $(-\hat{L}^T\hat{K}\hat{L}, \hat{L}^T\hat{M}\hat{L})$ will have negative real parts for any arbitrary projection matrix $\hat{L}$, even though all of the eigenvalues of the full-order system $(-\hat{K}, \hat{M})$ have negative real parts. This is, in fact, the major drawback of a first-order formulation approach. For this reason, this study is devoted to seeking a projection matrix that can be employed to perform model reduction to the second-order equation and can produce a reduced-order model with the moment-matching property.

Before establishing an algorithm for generating the projection matrix, a Krylov iteration procedure can be developed by using Eq. (19). Substitution of Eq. (17) into Eq. (19) gives

$$T_i = (-1)^i \begin{bmatrix} V & W \end{bmatrix} \begin{bmatrix} -K^{-1}D & -K^{-1}M \\ I & 0 \end{bmatrix}^{i-1} \begin{Bmatrix} K^{-1}P \\ 0 \end{Bmatrix}$$

This expression suggests the following iteration formula

$$\begin{Bmatrix} Q^d_{j+1} \\ Q^v_{j+1} \end{Bmatrix} = \begin{bmatrix} -K^{-1}D & -K^{-1}M \\ I & 0 \end{bmatrix} \begin{Bmatrix} Q^d_j \\ Q^v_j \end{Bmatrix} \tag{20}$$

or, equivalently,

$$\begin{aligned} Q^d_{j+1} &= -K^{-1}DQ^d_j - K^{-1}MQ^v_j = -K^{-1}DQ^d_j - K^{-1}MQ^d_{j-1} \\ Q^v_{j+1} &= Q^d_j \end{aligned} \tag{21}$$

Superscripts d and v denote displacement and velocity portions of the vector, respectively. The matrix formed by the generated vector sequence has the form

$$\begin{bmatrix} Q^d_1 & Q^d_2 & Q^d_3 & \cdots \\ Q^v_1 & Q^d_1 & Q^d_2 & \cdots \end{bmatrix}$$

Based on this form, the following theorem concerning the choice of projection matrix and the moment-matching property is established.

Theorem 2 *Let*

$$L_{\hat{P}} = \begin{bmatrix} Q^d_1 & Q^d_2 & Q^d_3 & \cdots & Q^d_p \\ 0 & Q^d_1 & Q^d_2 & \cdots & Q^d_{p-1} \end{bmatrix}$$

be the sequence of vectors generated by Eq. (21) with $\hat{K}^{-1}\hat{P}$ the starting block of vectors, i.e., $Q^d_1 = K^{-1}P$, $Q^v_1 = 0$, and let

$$L_{\hat{V}} = \begin{bmatrix} P^d_1 & P^d_2 & P^d_3 & \cdots & P^d_q \\ P^v_1 & P^d_1 & P^d_2 & \cdots & P^d_{q-1} \end{bmatrix}$$

be the subspace of vectors generated by Eq. (21) with $\hat{K}^{-1}\hat{V}$ the starting block of vectors, i.e., $P_1^d = K^{-1}V^T$, $P_1^v = -M^{-1}W^T$. If the projection matrix L is chosen such that $span\{L\} = span\{ Q_1^d \ \cdots \ Q_p^d \ P_1^d \ \cdots \ P_q^d \ P_1^v \}$ and is used for the model reduction of the damped structural dynamics system described by Eq. (13), then the reduced-order model matches the system parameters $\hat{V}(\hat{K}^{-1}\hat{M})^i\hat{K}^{-1}\hat{P}$, for $i = 0, 1, \ldots, p+q-1$.

Proof: The proof is similar to that of Theorem 1. First, the reduced system equation is described by

$$L^T M L\ddot{\bar{x}} + L^T D L\dot{\bar{x}} + L^T K L\bar{x} = L^T P u$$
$$y = V L\bar{x} + W L\dot{\bar{x}}$$

$$(22)$$

with

$$x = L\bar{x}$$

If we define

$$\hat{L} = \begin{bmatrix} L & 0 \\ 0 & L \end{bmatrix}$$

then the preceding equation can be rewritten into a first-order form like Eq. (16).

$$\hat{L}^T \hat{M} \hat{L}\dot{\bar{z}} + \hat{L}^T \hat{K} \hat{L}\bar{z} = \hat{L}^T \hat{P} u$$
$$y = \hat{V}\hat{L}\bar{z}$$

$$(23)$$

for which the low frequency moments, according to Eqs. (19) and (16), are

$$(\hat{V}\hat{L})[(\hat{L}^T\hat{K}\hat{L})^{-1}(\hat{L}^T\hat{M}\hat{L})]^{i-1}(\hat{L}^T\hat{K}\hat{L})^{-1}\hat{L}^T\hat{P}, \quad i = 1, 2, \cdots$$

Then, by using a procedure similar to the proof of Theorem 1, the moment-matching property of the reduced-order model can be proved.

∎

The $L_{\hat{P}}$ matrix and $L_{\hat{V}}$ matrix in Theorem 2 are the *generalized controllability matrix* and the *generalized observability matrix* of the system described by Eq. (18). For the purpose of response simulation only, one can choose the projection matrix L to be the d-portion of the $L_{\hat{P}}$ matrix. For control applications, as will be shown in Section III, there is an advantage of having a reduced-order model without control and observation spillover. Therefore, the following algorithm is proposed for the model reduction of damped structural dynamics systems.

Algorithm 2 (Damped Krylov Model Reduction Algorithm)

(1) *Starting blocks of vectors:*

$$Q_0^d = Q_0^v = 0$$
$$R_0^d = K^{-1}\tilde{P}$$
$$\tilde{P} = linearly\text{-}independent\ portion\ of\ \left[\begin{array}{cccc} P & V^T & W^T & (M^{-1}W^T) \end{array} \right]$$
$$R_0^v = -M^{-1}W^T$$
$$(R_0^d)^T K R_0 = U_0 \Sigma_0 U_0^T \quad (singular\text{-}value\ decomposition)$$
$$Q_1^d = R_0^d U_0 \Sigma_0^{-\frac{1}{2}} \quad (Q_1^d\ normalized\ w.r.t.\ K)$$
$$Q_1^v = R_0^v U_0 \Sigma_0^{-\frac{1}{2}} \quad (Q_1^v\ normalized\ using\ same\ scaling\ of\ Q_1^d)$$

(2) *For $j = 1, 2, \ldots, k-1$, repeat:*

$$\left. \begin{array}{l} R_j^d = -K^{-1}DQ_j^d - K^{-1}MQ_j^v \\ R_j^v = Q_j^d \end{array} \right\} \ (new\ vector)$$

For $i = 1, 2, \ldots, j$ repeat:

$$\left. \begin{array}{l} \alpha_i = (Q_i^d)^T K R_j^d \\ R_j^d = R_j^d - Q_i^d \alpha_i \\ R_j^v = R_j^v - Q_i^v \alpha_i \end{array} \right\} \ (orthogonalization)$$

end;

$$\left. \begin{array}{l} (R_j^d)^T K R_j^d = U_j \Sigma_j U_j^T \\ Q_{j+1}^d = R_j^d U_j \Sigma_j^{-\frac{1}{2}} \\ Q_{j+1}^v = R_j^v U_j \Sigma_j^{-\frac{1}{2}} \end{array} \right\} \ (normalization)$$

end;

(3) *Form the k-block projection matrix* $L = \left[\begin{array}{ccccc} Q_1^v & Q_1^d & Q_2^d & \cdots & Q_k^d \end{array} \right].$

In the above algorithm, the displacement portion and the velocity portion of the vectors are normalized and orthogonalized by using the same scaling, so that the vectors generated span the same subspace as those generated by Eq. (21). In other words, when the new vector $\left\{ \begin{array}{c} R_j^d \\ R_j^v \end{array} \right\}$ is to be orthogonalized with respect to the previous vectors, the displacement portion and the velocity portion cannot be orthogonalized separately.

A few remarks may be made about the model reduction of damped structural dynamics systems and the advantage of using Algorithm 2.

Remark 7. In Theorem 2, the starting vectors are $Q_1^d = K^{-1}P$ and $P_1^d = K^{-1}V^T$. But, for the starting vectors of Algorithm 2, not only P and V^T but also W^T and $(M^{-1}W^T)$ are included in $\tilde{P}$. The purpose of such choice is to produce an L matrix that is K-normalized and a transformed system equation with force distribution and sensor distribution matrices having nonzero elements only in the first block. By including W^T and $M^{-1}W^T$ in the starting block of vectors $\tilde{P}$, the size of the block can be quite large, if there are many inputs and many outputs. In practice, it is always possible to choose a collocated actuator/sensor design, which means that the column vectors of P, V^T, and W^T are not completely linearly-independent, so that the size of $\tilde{P}$ is kept small.

Remark 8. Unlike the undamped case, the transformed mass and damping matrices produced by Algorithm 2 do not have block tridiagonal form. In general, they are full matrices. Therefore, the transformed system does not have the tandem structure shown in Fig. 1. However, numerical experience so far indicates that the off-diagonal coupling in the transformed mass and damping matrices is usually small.

Remark 9. An alternative approach to create a model reduction basis for damped structural dynamics systems, as proposed in Ref. [3], is to ignore the effect of the damping matrix and simply use the undamped Krylov model reduction algorithm (Algorithm 1). For the case of Rayleigh damping ($D = \alpha M + \beta K$), this approach also produces a transformed damping matrix in the attractive block tridiagonal form. However, this approach is not recommended because, from the parameter-matching point of view, the undamped Krylov vectors can not produce a reduced-order model with parameter-matching property. In Section IV, an example will be used to show that for damped systems, a reduced-order model based on undamped Krylov vectors is far less accurate than a reduced-order model based on damped Krylov vectors.

Remark 10. Another alternative method for the model reduction of damped structural dynamics systems is to use the unsymmetric Lanczos algorithms in Refs. [17] and [18]. The unsymmetric algorithm is based on a first-order formulation of the system dynamic equations. Because the system matrix is unsymmetric, it requires the formation of two sets of vectors, the right Lanczos vectors and the left Lanczos vectors, and it suffers more numerical difficulties than the symmetric algorithm. The major disadvantage of using an unsymmetric algorithm is that the reduced system is not assured to be stable. It is indicated in both Refs. [17] and [18] that the unsymmetric Lanczos algorithm may result in reduced-order models with unstable poles.

Remark 11. The reduced-order model produced by Algorithm 2 is guaranteed to be stable as long as the damping matrix is positive definite. This can be proved by using the Lyapunov Theorem [19], which says that, if the system is controllable and if there exists a positive definite function $\mathcal{U}$, which is a function of the system states, such that $\dot{\mathcal{U}}$ is negative definite, then the system is asymptotically stable. Let us define the Lyapunov function $\mathcal{U}$ of the reduced system to be the system's total energy

$$\mathcal{U} = \frac{1}{2}\left[\dot{\bar{x}}^T(L^TML)\dot{\bar{x}} + \bar{x}^T(L^TKL)\bar{x}\right]$$

Then, since D is positive definite,

$$\dot{\mathcal{U}} = \dot{\bar{x}}^T(L^TML\ddot{\bar{x}} + L^TKL\bar{x}) = -\dot{\bar{x}}^TL^TDL\dot{\bar{x}} \leq 0$$

$\dot{\mathcal{U}} = 0$ is valid only for the case when $L\dot{\bar{x}} = 0$, which is possible only when $\dot{\bar{x}} = 0$, since the column vectors in L are linearly-independent. Except for the case that the system is completely at rest, the zero state condition $\dot{\bar{x}} = 0$ can occur only at some discrete instants of time, but not over any finite interval of time, no matter how small. Therefore, the system's total energy $\mathcal{U}$ decreases with time and eventually approaches zero. This means that the reduced system is asymptotically stable. This stability property is an advantage over the unsymmetric Lanczos algorithm.

D. SYSTEMS WITH RIGID-BODY MOTION

For systems having rigid-body modes, the algorithms developed in the previous sections do not apply because of the nonexistence of K^{-1}. To deal with this case, one can separate the rigid-body components and the flexible components of the system's motion and perform Krylov model reduction to the flexible components only. To accomplish this, the system equation must be transformed to a subspace corresponding to the flexible components.

Let $\Phi_r \in R^{n \times n_r}$ be the rigid-body mode matrix which satisfies

$$K\Phi_r = 0$$

and which is normalized with respect to the mass matrix

$$\Phi_r^T M \Phi_r = I$$

Construct a matrix $\Phi_e \in R^{n \times (n-n_r)}$ such that $[\Phi_r \ \Phi_e]$ has full column rank. Then, the displacement vector x can be decomposed into two portions

$$x = \begin{bmatrix} \Phi_r & \Phi_e \end{bmatrix} \begin{Bmatrix} r \\ e \end{Bmatrix} \tag{24}$$

The transformed system equation in the new coordinates becomes

$$\begin{bmatrix} I & \Phi_r^T M \Phi_e \\ \Phi_e^T M \Phi_r & \Phi_e^T M \Phi_e \end{bmatrix} \begin{Bmatrix} \ddot{r} \\ \ddot{e} \end{Bmatrix} + \begin{bmatrix} \Phi_r^T D \Phi_r & \Phi_r^T D \Phi_e \\ \Phi_e^T D \Phi_r & \Phi_e^T D \Phi_e \end{bmatrix} \begin{Bmatrix} \dot{r} \\ \dot{e} \end{Bmatrix}$$

$$+ \begin{bmatrix} 0 & 0 \\ 0 & \Phi_e^T K \Phi_e \end{bmatrix} \begin{Bmatrix} r \\ e \end{Bmatrix} = \begin{Bmatrix} \Phi_r^T P \\ \Phi_e^T P \end{Bmatrix} u \tag{25}$$

$$y = [V\Phi_r \ V\Phi_e] \begin{Bmatrix} r \\ e \end{Bmatrix} + [W\Phi_r \ W\Phi_e] \begin{Bmatrix} \dot{r} \\ \dot{e} \end{Bmatrix}$$

To decouple the rigid-body components and the flexible components in the above equation, Φ_e must be chosen such that $\Phi_e^T M \Phi_r = 0$ and $\Phi_e^T D \Phi_r = 0$. For general damping matrix, there might not exist a Φ_e that satisfies the above conditions. For this case, a first-order state-space formulation similar to the one in Ref. [20] must be used. Here it is assumed that the damping matrix D is either a proportional damping matrix in the form

$$D = M \sum_i a_i [M^{-1} K]^i$$

or in the form

$$D = \sum_i \left(\frac{2\zeta_i \omega_i}{\phi_i^T M \phi_i} \right) (M \phi_i)(M \phi_i)^T$$

or satisfies $D\Phi_r = 0$. This type of assumption is frequently used in structural dynamics analyses [21]. With this assumption, Φ_e can be chosen as

$$\Phi_e = (I - \Phi_r \Phi_r^T M)S = \mathcal{P}^T S \tag{26}$$

where $\mathcal{P} = I - M\Phi_r \Phi_r^T$ is an $n \times n$ projection matrix and S is an arbitrary $n \times (n - n_r)$ matrix with full column rank.

The decoupled system equation in the new coordinates becomes

$$
\begin{bmatrix} I & 0 \\ 0 & \Phi_e^T M \Phi_e \end{bmatrix} \begin{Bmatrix} \ddot{r} \\ \ddot{e} \end{Bmatrix} + \begin{bmatrix} 0 & 0 \\ 0 & \Phi_e^T D \Phi_e \end{bmatrix} \begin{Bmatrix} \dot{r} \\ \dot{e} \end{Bmatrix}
$$

$$
+ \begin{bmatrix} 0 & 0 \\ 0 & \Phi_e^T K \Phi_e \end{bmatrix} \begin{Bmatrix} r \\ e \end{Bmatrix} = \begin{Bmatrix} \Phi_r^T P \\ \Phi_e^T P \end{Bmatrix} u \qquad (27)
$$

$$
y = [V\Phi_r \ \ V\Phi_e] \begin{Bmatrix} r \\ e \end{Bmatrix} + [W\Phi_r \ \ W\Phi_e] \begin{Bmatrix} \dot{r} \\ \dot{e} \end{Bmatrix}
$$

The Krylov model reduction method can be applied to obtain a reduced-order model for the e-portion of the above equation

$$
\Phi_e^T M \Phi_e \ddot{e} + \Phi_e^T D \Phi_e \dot{e} + \Phi_e^T K \Phi_e e = \Phi_e^T P u
$$
$$
y = V\Phi_e e + W\Phi_e \dot{e} \qquad\qquad (28)
$$

III. CONTROL OF FLEXIBLE STRUCTURES

Control of flexible structures has emerged as an important research topic since space technology has made the construction of large space structures possible. A major difficulty in the control of flexible structures comes from the fact that flexible structures are distributed-parameter systems. Although there exists distributed control theory, it is not applicable to structural control problems because most real structures have very complicated geometry so that only in rare cases can one write down the partial differential equations of motion for the systems. Therefore, in most cases model reduction is inevitably introduced at the very beginning stage when the structure is modelled as a discrete system by use of the finite element method. In addition to modelling, the limitations of computer memory and speed also make model reduction a necessity. Usually the finite element model of a real structure has hundreds of thousands of degrees of freedom. Such a model is neither computationally economical nor "Riccati-solvable" for use in the control design. Therefore, a further model reduction of the mathematical model is indispensable.

An immediate effect of model reduction to the control design is the spillover of control energy. Spillover effects come from the coupling between the modeled and unmodeled dynamics of the system. When a controller design that is based on the reduced model is applied to control the full-order structure, the unmodeled dynamics of the system will be excited due to the spillover of control energy. This effect will degrade the controller performance and may cause instability in the closed-loop system. Therefore, one major issue of structural control is to construct a reduced-order model that can reduce the spillover effect. In this section, characteristics of spillover will be discussed first, and then, a Krylov formulation for control of flexible structures will be proposed.

A. CHARACTERISTICS OF SPILLOVER

Let a structural dynamics system be partitioned into two subsystems, a controlled subsystem and a residual subsystem, and be described by

$$
\begin{bmatrix} M_c & M_{cr} \\ M_{rc} & M_r \end{bmatrix} \begin{Bmatrix} \ddot{x}_c \\ \ddot{x}_r \end{Bmatrix} + \begin{bmatrix} D_c & D_{cr} \\ D_{rc} & D_r \end{bmatrix} \begin{Bmatrix} \dot{x}_c \\ \dot{x}_r \end{Bmatrix} + \begin{bmatrix} K_c & K_{cr} \\ K_{rc} & K_r \end{bmatrix} \begin{Bmatrix} x_c \\ x_r \end{Bmatrix} = \begin{Bmatrix} P_c \\ P_r \end{Bmatrix} u
$$

$$
y = \begin{bmatrix} V_c & V_r \end{bmatrix} \begin{Bmatrix} x_c \\ x_r \end{Bmatrix} + \begin{bmatrix} W_c & W_r \end{bmatrix} \begin{Bmatrix} \dot{x}_c \\ \dot{x}_r \end{Bmatrix}
$$

$$(29)$$

where subscripts c and r denote controlled subsystem and residual subsystem, respectively. Here the x_c and x_r vectors can be in physical coordinates or some generalized coordinates. Although there is interaction between the c-subsystem and the r-subsystem (in fact, there are three types of "interaction" which will be discussed later), these two subsystems are considered as being separate from each other at the control design stage. The residual subsystem is assumed to be a stable system with sufficient passive damping and, therefore, it does not require feedback control to introduce active damping. The controlled subsystem, however, can be unstable (for instance, slewing or pointing control problem must include the rigid-body motion), or neutrally stable (for instance, if there are undamped modes), or stable but with very slight passive damping. Therefore, the controlled subsystem needs feedback control to achieve the mission of pointing or vibration suppression. The controlled subsystem, on which the control design is based, is described by

$$
M_c \ddot{x}_c + D_c \dot{x}_c + K_c x_c = P_c u
$$

$$
y_c = V_c x_c + W_c \dot{x}_c
$$

$$(30)$$

A general dynamic output feedback controller for the controlled subsystem has the form

$$\dot{q} = Eq + Fy_c$$
$$u = Gq \tag{31}$$

in which the elements of the controller system matrices E, F, and G are the design parameters. Note that, although the actual measurement available is y, the signal used in the controller design is y_c, which is the output of the controlled subsystem. This controller can be designed by using any existing control approach, for instance, the linear quadratic optimal control theory. For the case of full-state feedback, the size of the q vector is twice that of the x_c vector.

Now, when the controller in Eq. (31) is applied to control the full-order system, Eq. (29), the closed-loop dynamics can be described by the following first-order equation:

$$
\begin{bmatrix}
I & 0 & 0 & \vdots & 0 & 0 \\
0 & 0 & I & \vdots & 0 & 0 \\
0 & M_c & D_c & \vdots & M_{cr} & D_{cr} \\
\cdots & \cdots & \cdots & & \cdots & \cdots \\
0 & 0 & 0 & \vdots & 0 & I \\
0 & M_{rc} & D_{rc} & \vdots & M_r & D_r
\end{bmatrix}
\begin{Bmatrix}
\dot{q} \\
\ddot{x}_c \\
\dot{x}_c \\
\cdots \\
\ddot{x}_r \\
\dot{x}_r
\end{Bmatrix}
=
\begin{bmatrix}
E & FW_c & FV_c & \vdots & FW_r & FV_r \\
0 & I & 0 & \vdots & 0 & 0 \\
P_cG & 0 & -K_c & \vdots & 0 & -K_{cr} \\
\cdots & \cdots & \cdots & & \cdots & \cdots \\
0 & 0 & 0 & \vdots & I & 0 \\
P_rG & 0 & -K_{rc} & \vdots & 0 & -K_r
\end{bmatrix}
\begin{Bmatrix}
q \\
x_c \\
x_c \\
\cdots \\
\dot{x}_r \\
x_r
\end{Bmatrix}
\tag{32}
$$

From this expression, we see that although the controller is intentionally designed to control only the c-subsystem and leave the r-subsystem undisturbed, there are three channels through which the control energy is "spilled over" into the residual subsystem. These three spillover factors are depicted in Fig. 2 and can be clearly identified in the system equation (29). In Ref. [22], the P_r submatrix is called the *control spillover* factor, through which the control input u goes directly into the r-subsystem; and the V_r and W_r submatrices are called the *observation spillover* terms, through which the measurement y is contaminated by the output signal coming from the r-subsystem. The third factor, recently referred to as the *dynamic spillover* in Ref. [23], is due to the coupling submatrices M_{cr}, D_{cr}, and K_{cr}. All three of these types of spillover have a direct effect on the performance and stability of the closed-loop system.

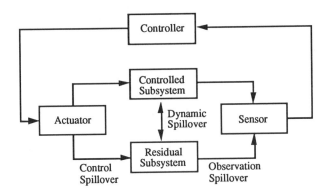

Figure 2: Characteristics of spillover.

Suppose that there exists a coordinate transformation such that the afore-mentioned three spillover factors are eliminated. Then, the poles of the closed-loop system would simply be the union of the eigenvalues of the two diagonal submatrices in Eq. (32), that is, the union of the poles of the closed-loop c-subsystem and the poles of the undisturbed r-subsystem. In this case, the closed-loop stability is guaranteed. However, such a situation occurs only when the full-order system is not a minimum realization, or, in other words, only when it is not completely controllable and/or not completely observable. For instance, if the actuator and sensor distributions are in a subspace spanned by some normal modes, then those corresponding modes can be controlled and measured independently while leaving the other modes undisturbed. This ac-tuator/sensor configuration requires an infinite number of discrete actuators and sensors or distributed actuators and sensors, which is not practical in real implementations. Therefore, in most practical cases, there always exists some spillover. The closed-loop poles, then, will be perturbed from the designed lo-cations by the nonzero off-diagonal submatrices in Eq. (32).

For example, if model reduction and control design are based on a normal-mode reduced-model, with the assumption that damping is proportional (i.e., $D_{cr} = D_{rc} = 0$), then the closed-loop system equation is described by

$$
\begin{bmatrix}
I & 0 & 0 & \vdots & 0 & 0 \\
0 & 0 & I & \vdots & 0 & 0 \\
0 & I & C_c & \vdots & 0 & 0 \\
\cdots & \cdots & \cdots & & \cdots & \cdots \\
0 & 0 & 0 & \vdots & 0 & I \\
0 & 0 & 0 & \vdots & I & 0
\end{bmatrix}
\begin{Bmatrix}
\dot{q} \\
\ddot{x}_c \\
\dot{x}_c \\
\cdots \\
\ddot{x}_r \\
\dot{x}_r
\end{Bmatrix}
=
\begin{bmatrix}
E & FW_c & FV_c & \vdots & FW_r & FV_r \\
0 & I & 0 & \vdots & 0 & 0 \\
P_cG & 0 & -\Omega_c^2 & \vdots & 0 & 0 \\
\cdots & \cdots & \cdots & & \cdots & \cdots \\
0 & 0 & 0 & \vdots & I & 0 \\
P_rG & 0 & 0 & \vdots & 0 & -\Omega_r^2
\end{bmatrix}
\begin{Bmatrix}
q \\
\dot{x}_c \\
x_c \\
\cdots \\
\dot{x}_r \\
x_r
\end{Bmatrix}
$$

In the above expression, it is seen that P_rG, the control spillover term, as well as FV_r and FW_r, the observation spillover terms, are responsible for the closed-loop system pole perturbation. Balas showed in Ref. [22] that the combined effect of control and observation spillover usually degrades the performance of the controller and sometimes can destabilize the closed-loop system. This can be effectively explained by the fact that large perturbation may move some of the closed-loop system poles to the right half of the complex plane.

Recently, Yam proposed a flexible system model reduction approach based upon actuator and sensor influence functions [23]. The transformed system equation is free from control and observation spillover but has dynamic coupling. The advantage is that, unlike the normal mode formulation, the design parameter F and G matrices do not enter into the perturbation, though the closed-loop system is still coupled by the dynamic spillover terms M_{cr}, D_{cr}, and K_{cr}. In this case, a perturbation technique can be used effectively to study and estimate the closed-loop stability without the design parameters complicating the analysis. However, the transformation matrix used in Ref. [23] has no physical significance and seems somewhat artificial. There is also lack of certainty about the "smallness" of the dynamic spillover terms. In the following section, it will be shown that a Krylov-based reduced model can eliminate both control spillover and observation spillover, and thus, it has some characteristics similar to the formulation in Ref. [23].

B. KRYLOV METHODS FOR CONTROL OF FLEXIBLE STRUCTURES

As derived in Section II, the transformed system equation for a damped structural dynamics system in the Krylov coordinates has the following form

$$
\begin{bmatrix} M_c & M_{cr} \\ M_{rc} & M_r \end{bmatrix} \begin{Bmatrix} \ddot{x}_c \\ \ddot{x}_r \end{Bmatrix} + \begin{bmatrix} D_c & D_{cr} \\ D_{rc} & D_r \end{bmatrix} \begin{Bmatrix} \dot{x}_c \\ \dot{x}_r \end{Bmatrix} + \begin{bmatrix} I & 0 \\ 0 & I \end{bmatrix} \begin{Bmatrix} x_c \\ x_r \end{Bmatrix} = \begin{bmatrix} P_c \\ 0 \end{bmatrix} u
$$

$$
y = \begin{bmatrix} V_c & 0 \end{bmatrix} \begin{Bmatrix} x_c \\ x_r \end{Bmatrix} + \begin{bmatrix} W_c & 0 \end{bmatrix} \begin{Bmatrix} \dot{x}_c \\ \dot{x}_r \end{Bmatrix}
$$

(33)

where

$$
P_c = \begin{Bmatrix} P_1 \\ 0 \\ \vdots \\ 0 \end{Bmatrix}, \qquad V_c = \begin{bmatrix} V_1 & 0 & \cdots & 0 \end{bmatrix}, \qquad W_c = \begin{bmatrix} W_1 & 0 & \cdots & 0 \end{bmatrix}
$$

It is seen that there is no control or observation spillover term in the transformed system equation.

Combining Eq. (33) with the controller system equation (31) gives the closed-loop system equation in the form

$$
\begin{bmatrix} I & 0 & 0 & \vdots & 0 & 0 \\ 0 & 0 & I & \vdots & 0 & 0 \\ 0 & M_c & D_c & \vdots & M_{cr} & D_{cr} \\ \cdots & \cdots & \cdots & & \cdots & \cdots \\ 0 & 0 & 0 & \vdots & 0 & I \\ 0 & M_{rc} & D_{rc} & \vdots & M_r & D_r \end{bmatrix} \begin{Bmatrix} \dot{q} \\ \ddot{x}_c \\ \dot{x}_c \\ \cdots \\ \ddot{x}_r \\ \dot{x}_r \end{Bmatrix} = \begin{bmatrix} E & FW_c & FV_c & \vdots & 0 & 0 \\ 0 & I & 0 & \vdots & 0 & 0 \\ P_c G & 0 & -I & \vdots & 0 & 0 \\ \cdots & \cdots & \cdots & & \cdots & \cdots \\ 0 & 0 & 0 & \vdots & I & 0 \\ 0 & 0 & 0 & \vdots & 0 & -I \end{bmatrix} \begin{Bmatrix} q \\ \dot{x}_c \\ x_c \\ \cdots \\ \dot{x}_r \\ x_r \end{Bmatrix}
$$

(34)

This is called the *Krylov form of the closed-loop system equation*, in which the controller is designed based on the Krylov reduced model. The Krylov form of the closed-loop system equation, like the formulation in Ref. [23], is coupled only by the dynamic spillover terms. If the dynamic coupling terms are small compared to the magnitude of the diagonal submatrices, then, the perturbation of the closed-loop poles is small and the closed-loop system is less likely to be unstable. Therefore, the norm of the off-diagonal submatrices in Eq. (33a) can be used as a truncation basis for Krylov model reduction. Reference [24] has an example showing that the dynamic spillover terms in the Krylov formulation are small.

IV. EXAMPLES

A. MODEL REDUCTION EXAMPLE

The example considered is a plane truss structure similar to the one in Ref. [3] but on a smaller scale (see Fig. 3). The structure has 48 degrees of freedom and has a force actuator at f and a displacement sensor at d. The structure geometry is constructed to provide closely-spaced natural frequencies. The damping matrix is proportional such that modes 1–5 have a 3% damping ratio, modes 6–10 have a 5% damping ratio, and the remaining higher modes have successively higher damping.

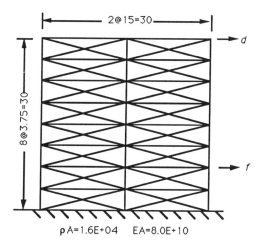

Figure 3: Details of plane truss structure for model reduction example.

The structure is reduced to eight degrees of freedom by using eight Krylov vectors generated by Algorithm 2. In this case, the reduced-order model matches the low-frequency moments $\hat{V}(\hat{K}^{-1}\hat{M})^i\hat{K}^{-1}\hat{P}$, for $i = 1, 2, \ldots, 7$, of the full-order structure. Another reduced-order model obtained by using eight "un-damped Krylov vectors" generated by Algorithm 1 is also examined. Figures 4 to 6 compare the accuracy of the impulse response of the normal mode reduced model, "damped Krylov reduced model," and "undamped Krylov reduced model." It is seen that for this example the normal mode reduced model and the damped Krylov reduced model have about the same accuracy in predicting the system's impulse response. The undamped Krylov reduced model, however, does not approximate the system's impulse response very well. The eigenvalues of the full-order system and the two Krylov reduced-order models are compared in Table 1. It shows that both the undamped and the damped Krylov reduced models approximate the lower frequency range of the full-order system. This example illustrates that for the purpose of response simulation, a reduced model that simply preserves the lower natural frequencies of the system is not sufficient.

Table 1. Eigenvalues of full-order system and reduced-order models of model reduction example.

Full-Order System		Damped Krylov		Undamped Krylov	
real	imag	real	imag	real	imag
−5.8057E−1	1.9343E+1	−5.8057E−1	1.9343E+1	−5.8057E−1	1.9343E+1
−1.7695E+0	5.8958E+1	−1.7695E+0	5.8958E+1	−1.7695E+0	5.8958E+1
−3.0517E+0	1.0167E+2	−3.2945E+0	1.0948E+2	−3.2464E+0	1.0598E+2
−3.2728E+0	1.0904E+2	−4.5641E+0	1.3440E+2	−3.2736E+0	1.0905E+2
−4.2244E+0	1.4075E+2	−8.0259E+0	1.6603E+2	−9.3053E+0	1.8596E+2
−8.1867E+0	1.6353E+2	−9.1026E+0	1.9223E+2	−1.2117E+1	2.3019E+2
−9.2901E+0	1.8557E+2	−1.5183E+1	2.5975E+2	−1.9084E+1	2.9124E+2
−1.0217E+1	2.0409E+2	−1.6359E+1	2.6979E+2	−2.4060E+1	3.2684E+2
−1.0249E+1	2.0474E+2				
−1.1216E+1	2.2403E+2				
−1.2396E+1	2.3550E+2				
−1.3306E+1	2.4396E+2				
−1.3628E+1	2.4689E+2				
−1.3695E+1	2.4750E+2				
−1.3884E+1	2.4919E+2				
−1.4079E+1	2.5093E+2				

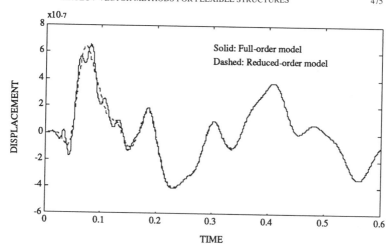

Figure 4: Impulse response; eight normal modes and exact solution.

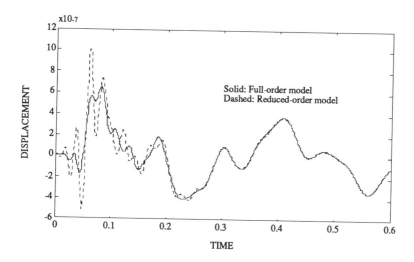

Figure 5: Impulse response; eight undamped Krylov modes and exact solution.

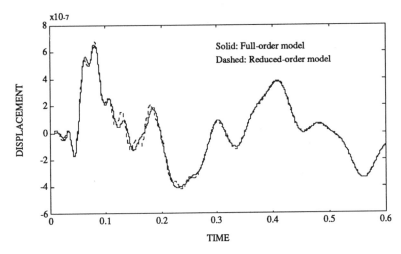

Figure 6: Impulse response; eight damped Krylov modes and exact solution.

B. EXAMPLE OF CONTROL OF A FLEXIBLE STRUCTURE

The example considered is a 20 degree-of-freedom lightly-damped plane truss structure as shown in Fig. 7. The damping matrix is proportional such that modes 1–5 have 0.1% damping ratio, modes 6–10 have 0.2% damping ratio, and the remaining higher modes have successively higher damping up to about 0.5%. A force actuator is located at f and a displacement sensor is located at d. It is assumed that at the actuator location there is a zero-mean white-noise disturbance with intensity 10^{-3}. Also, the sensor is assumed to be contaminated by a zero-mean white noise with intensity 10^{-12}. Usually, in the LQG design, the intensities of input disturbance and measurement noise are used as adjustable parameters for tuning the gains and pole locations of the estimator. Here they are chosen arbitrarily just to compare the LQG designs based on different reduced-order models.

The structure is reduced to low-order models by using either Krylov modes or normal modes: five Krylov reduced models with order 2, 4, 6, 8, and 10, respectively, and five normal mode reduced models with order 2, 4, 6, 8, and

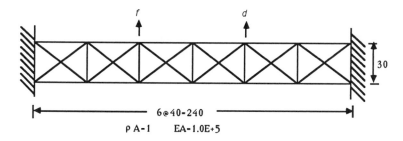

Figure 7: Details of plane truss structure for control example.

10, respectively. Based on each reduced-order model, an LQG control design is carried out to minimize the performance index

$$J = \frac{1}{2} \lim_{t \to \infty} E[\dot{x}_c^T M_c \dot{x}_c + x_c^T K_c x_c + \rho u^T u]$$

in which the first two terms represent the total energy of the reduced system and the third term represents the control cost. A positive scalar ρ is used to adjust the relative weighting of the regulation cost and control cost penalties. Overall controller authority, actuator mean-square force levels and controller bandwidth are all inversely proportional to ρ. The value of ρ was varied from 0.05 to 500 to study the closed-loop stability and controller performance. All of the calculations and designs were performed by using the CTRL-C package run on a VAX computer.

The results are summarized by Table 2 and Fig. 8, in which K2 stands for the controller designed based on the 2nd-order Krylov reduced model, N2 stands for the controller designed based on the 2nd-order normal mode reduced model, and so on. From Table 2, it is seen that controllers designed using normal mode reduced models are more likely to cause closed-loop instability than controllers designed using Krylov reduced models, especially when the controller bandwidth is large (small ρ) and the order of the reduced model is low. The performance of the controllers is presented in Fig. 8, which shows the regulation cost $J_e = \lim_{t \to \infty} E[\dot{x}^T M \dot{x} + x^T K x]$ as a function of the control cost $J_e = \lim_{t \to \infty} E[u^T u]$ (obtained by varying the value of ρ). It is seen that the K6 controller has a performance very close to that of the full-order optimal LQG controller. The N6 controller, which fails to yield stable design for $\rho < 5.0$, has a performance not very much better than that of the K4 controller. Performance curves of K8, K10, N8, and N10 are all very close to the optimal one and, hence, are not shown in Fig 8.

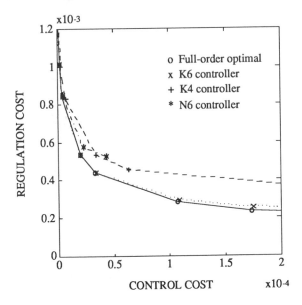

Figure 8: Performance plot.

Table 2. Stability of the controllers.

Controller	ρ=.05	0.1	0.5	1.0	5.0	10.0	50.0	100.0	500.0
K2	U	U	U	U	S	S	S	S	S
K4	U	S	S	S	S	S	S	S	S
K6	S	S	S	S	S	S	S	S	S
K8	U	S	S	S	S	S	S	S	S
K10	S	S	S	S	S	S	S	S	S
N2	U	U	U	U	U	U	S	S	S
N4	U	U	U	U	U	U	S	S	S
N6	U	U	U	U	S	S	S	S	S
N8	S	S	S	S	S	S	S	S	S
N10	S	S	S	S	S	S	S	S	S

S: the closed-loop system is stable.   U: unstable

V. CONCLUSIONS

In this study, Krylov vectors and the concept of parameter-matching are combined together to develop model-reduction algorithms for structural dynamics systems. Although all of the other existing parameter-matching methods are based on either the first-order state-space description or the transfer function of the system, the method presented here is derived for a structural dynamics system described by a second-order matrix differential equation. The advantage of using a second-order formulation is that the stability of the reduced-order model is guaranteed. The Krylov reduced models are shown to have a promising aspect in the application to control of flexible structures. The formulation based on Krylov vectors can eliminate control and observation spillovers while leaving only the dynamic spillover terms to be considered. Two examples: one model-order reduction example and one flexible structure control example are provided to show the efficacy of the Krylov method.

VI. ACKNOWLEDGEMENTS

This work was conducted under contract NAS9-17254 and NAS9-357 with the NASA-Lyndon B. Johnson Space Center. The interest of Mr. Rodney Rocha and Mr. John Sunkel is gratefully acknowledged.

VII. REFERENCES

1. R. R. Craig, Jr., "A Review of Time-Domain and Frequency-Domain Component Modes Synthesis Methods," *Combined Experimental/Analytical Modeling of Dynamic Structural Systems*, AMD-Vol. 67, ASME, NY, pp. 1–30 (1985).

2. E. L. Wilson, M. W. Yuan, and J. M. Dickens, "Dynamic Analysis by Direct Superposition of Ritz Vectors," *Earthquake Eng. & Struc. Dyn.*, **10**, pp. 813–82 (1982).

3. B. Nour-Omid and R. W. Clough, "Dynamic Analysis of Structures Using Lanczos Co-ordinates," *Earthquake Eng. & Struc. Dyn.*, **12**, pp. 565–577 (1984).

4. R. R. Craig, Jr. and A. L. Hale, "The Block-Krylov Component Synthesis Method for Structural Model Reduction," *AIAA J. Guidance, Control, and Dynamics*, 11, pp. 562–570 (1988).

5. R. R. Arnold, R. L. Citerley, M. Chargin, and D. Galant, "Application of Ritz Vectors for Dynamic Analysis of Large Structures," *Computers and Structures*, 21, pp. 461–467 (1985).

6. T. J. Su and R. R. Craig, Jr., "Model Reduction and Control of Flexible Structures Using Krylov Vectors," *AIAA J. Guidance, Control, and Dynamics*, 14, pp. 260–267 (1991).

7. T. J. Su and R. R. Craig, Jr., "Controller Reduction By Preserving Impulse Response Energy," AIAA Paper 89-3432, *AIAA Guidance, Navigation and Control Conference*, Boston, MA, pp. 55–64 (1989).

8. C. F. Chen, "Model Reduction of Multivariable Control System by Means of Matrix Continued Fraction," *Int. J. Control*, 20, pp. 225–238 (1974).

9. J. Hickin and N. K. Sinha, "Model Reduction for Linear Multivariable Systems," *IEEE Trans. Automat. Contr.*, AC-25, pp. 1121–1127 (1980).

10. J. S. Lai and J. C. Hung, "Practical Model Reduction Methods," *IEEE Trans. Industrial Electronics*, 34, pp. 70–77 (1987).

11. L. S. Shieh and Y. J. Wei, "A Mixed Method for Multivariable System Reduction," *IEEE Trans. Automat. Contr.*, AC-20, pp. 429–432 (1975).

12. C. D. Villemagne and R. E. Skelton, "Model Reduction Using a Projection Formulation," *IEEE Trans. Automat. Contr.*, AC-46, pp. 2141–2169 (1987).

13. J. N. Juang and R. S. Pappa, "An Eigensystem Realization Algorithm for Modal Parameter Identification and Modal Reduction," *J. Guidance*, 8, pp. 620–627 (1985).

14. R. R. Craig, Jr., T. J. Su, and H. M. Kim, "Use of Lanczos Vectors in Structural Dynamics," *Proc. International Conference: 'Spacecraft Structures and Mechanical Testing'*, Noordwijk, The Netherlands, pp. 187–192 (1988).

15. C. C. Paige, "Practical Use of the Symmetric Lanczos Process with Re-orthogonalization," *BIT*, **10**, pp. 183–195 (1970).

16. B. N. Parlett and D. S. Scott, "The Lanczos Algorithm with Selective Orthogonalization," *Math. Comp.*, **33**, pp. 217–238 (1979).

17. B. Nour-Omid and M. E. Regelbrugge, "Lanczos Method for Dynamic Analysis of Damped Structural Systems," *Earthquake Eng. & Struc. Dyn.*, **18**, pp. 1091–1104 (1989).

18. H. M. Kim and R. R. Craig, Jr., "Structural Dynamics Analysis Using An Unsymmetric Block Lanczos Algorithm," *Int. J. Numer. Methods Eng.*, **26**, pp. 2305–2318 (1988).

19. W. L. Brogan, *Modern Control Theory*, 2nd Edition, Prentice-Hall, Inc., Englewood Cliffs, NJ (1985).

20. R. R. Craig, Jr., T. J. Su, and Z. Ni, "State-Variable Models of Structures Having Rigid-Body Modes," *AIAA J. Guidance, Control, and Dynamics*, **13**, pp. 1157–1160 (1990).

21. R. R. Craig, Jr., *Structural Dynamics - An Introduction to Computer Methods*, John Wiley & Sons, Inc. NY, 1981.

22. M. J. Balas, "Feedback Control of Flexible Systems," *IEEE Trans. Automat. Contr.*, **AC-23**, pp. 673–679 (1978).

23. Y. Yam, T. L. Johnson, and J. H. Lang, "Flexible System Model Reduction and Control System Design Based Upon Actuator and Sensor Influence Functions," *IEEE Trans. Automat. Contr.*, **AC-32**, pp. 573–582 (1988).

24. T. J. Su and R. R. Craig, Jr., "Model Reduction and Controller Design of Flexible Structures Using Krylov Subspaces," *30th AIAA/ASME/ASCE/ AHS Structures, Structural Dynamics and Materials Conference*, Mobile, AL, pp.691-700 (1989).

MANEUVERING TARGET TRACKING :
IMAGING AND NON-IMAGING SENSORS

Michel Mariton

Signal and Image Processing Laboratory
MATRA MS2i
BP 235 - 38 boulevard Paul Cezanne
78052 St Quentin en Yvelines Cedex
FRANCE

David D. Sworder

Department of AMES
University of California at San Diego
La Jolla, CA 92093-0411
USofA

1. INTRODUCTION

As the term is commonly interpreted, tracking refers to estimating the current state of a target from a spatiotemporal observation [1]. Prediction of future target location can be made on the basis of this estimate. Tracking an agile target is a difficult task, even with the new generation of electro-optical (EO) sensors. There is a well developed theory of estimation and prediction, quite useful in benign environments, but it is difficult to adapt the results to situations involving rapid change. The models upon which the algorithms are based frequently do not exhibit the clearly differentiated maneuver regimes typical of a hostile encounter. While the algorithms can be formally modified to include maneuver detection and equalization, unacceptable performance often results. In this report, recent results are presented towards the development of new tracking algorithms that would incorporate explicit models of the maneuvering/non maneuvering phases of the encounter. It is believed that the imaging capacity of the new generation of EO sensors makes it possible to reconfigure the tracker bandwidth following a fast and reliable maneuver detection and identification based on a detailed analysis of the target image.

The rest of the paper is organized as follows. First a discussion of tracking systems is proposed to emphasize the need for more accurate maneuver models. Second tracking with non imaging sensors is discussed and it is shown through an air defense scenario that multiple model filters can significantly improve performance over conventional trackers. However there is an inherent limitation to point measurements provided by non imaging sensors and the third part of the paper discusses the role of emerging imaging sensors. Finally simulations with realistic infrared images are presented to show the feasibility of image based target maneuver detection with currently available sensors.

2. A DISCUSSION OF TRACKING SYSTEMS

Synthesis of model-based tracking algorithms begins with the equations of motion of the target. The motion dynamics are conventionally rendered in terms of a Linear Gauss Markov (LGM) model

$$dx_t = Ax_t dt + dw_t \tag{1}$$

in which $\{x_t\}$ is the target state process including position, velocity, etc., and $\{\frac{dw_t}{dt}\}$ is a wide band (white) process of intensity W, ($Wdt = (dw_t)(dw_t)'$), selected to introduce uncertainty into the path of the target. The initial conditions in Eq.(1) are Gaussian and independent of $\{w_t\}$.

It is customary to assume that the tracker avails itself of sensors producing noisy measurements of the target states :

$$dy_t = Dx_t dt + dn_x \tag{2}$$

where $\{n_x\}$ is a Brownian motion process with intensity $R_x > 0$, independent of both $\{w_t\}$ and the initial conditions on Eq.(1). In applications, Eq.(2) is frequently a localization of a nonlinear measurement link.

The solution to the estimation problem described in Eqs.(1) and (2) can be phrased as follows. Denoting by $\mathcal{Y}_t$ the information pattern (filtration) generated by $\{y_t\}$, the conditional mean of the encounter state $(\hat{x}_t = E\{x_t|\mathcal{Y}_t\}$ is given by the (Extended) Kalman Filter (EKF) [2]

$$d\hat{x}_t = A\hat{x}_t dt + P_{xx}D'R_x^{-1}dv_t \qquad (3)$$

where $dv_t = dy - D\hat{x}dt$ (the innovations process), and P_{xx} is the error covariance matrix.

Equation (3) has an intuitive appeal which motivates its use in situations more general than that captured by the LGM model. The increment in $\hat{x}_t$ is expressed as a sum of a drift and a correction. The drift is in the direction of the mean increment $(E\{d\hat{x}_t|\mathcal{Y}_t\} = A\hat{x}_t dt)$, and the correction is proportional to the innovations process $(d\hat{x}_t - E\{d\hat{x}_t|\mathcal{Y}_t\} = P_{xx}D'R_x^{-1}dv_x)$. The former term has an obvious rationale ; i.e., in the absence of new data, extrapolate from the most recent estimate. The latter is more engaging. The innovations gain, $P_{xx}D'R_x^{-1}$, is inversely proportional to the measurement noise intensity (R_x^{-1}), and directly proportional to the signal amplification (D) ; implying that as sensor quality improves, the estimate becomes more sensitive to new data. The innovations gain is also proportional to the uncertainty in the estimate (P_{xx}). It is only this factor that is not explicit in the model of the observation link, and indeed P_{xx} is determined jointly by the signal dynamics and the observation fidelity.

Note that when P_{xx} is small, the innovations process receives little note, and the estimate propagates forward along the field of the unexcited system. As the uncertainty in the state estimate increases, new information is accorded increasing value ; i.e., as the estimator becomes less sure of the true state, it is more willing to modify its prior estimate in response to new data. It can be shown that $\{P_{xx}\}$ is given by the solution to a matrix stochastic differential equation (see [3], (18.19))

$$dP_{xx} = [AP_{xx} + (AP_{xx})' + W - P_{xx}D'R_x^{-1}DP_{xx}]dt$$
$$\qquad (4)$$
$$+ \sum_k \pi_{xx}(x_k)(D'R_x^{-1}dv_t)_k$$

where $\pi_{xx}(x_k) = E\{\tilde{x}\tilde{x}'\tilde{x}_k|\mathcal{Y}_t\}$ and where $\tilde{x} = x - \hat{x}$ is the estimation error. The right hand side of Eq.(5) is composed of readily interpretable terms. The error variance is amplified by the system dynamics $((AP_{xx} +$

$P_{xx}A')dt$), and is reduced by the observation process $(P_{xx}D'R_x^{-1}DP_{xx}dt)$. As the quality of the sensors improves, the error variance decreases. The equation of evolution of the error variance process is responsive to the influence of the exogenous state disturbance as well. As the intensity of this excitation, W, increases, the increment in $\{P_{xx}\}$, is increased proportionately. Observe that Eq.(1) that W has another interpretation. It is the intensity of the $\mathcal{Y}_t$-predictable quadratic variation of $\{x_t\}$; i.e.,

$$W dt = d < x, x >_t \tag{5}$$

The final term in Eq.(5) is anomalous, and warrants discussion. The error covariance is a random process which responds to the observation through a term containing the third conditional error moment as a factor.

It is known that $\{\tilde{x}_t\}$ is a Gaussian process, and as a consequence, this central moment is zero for all k and for all t ; i.e., $\pi_{xx}(x_k)(D'R_x^{-1}dv_x)_k \equiv 0$. Thus, the last term vanishes in Eq.(5), and the familiar Riccati ordinary differential equation results,

$$\dot{P}_{xx} = AP_{xx} + P_{xx}A' + W - P_{xx}D'R_x^{-1}DP_{xx} \tag{6}$$

The nonstochastic nature of the reduced form of Eq.(5) implies that the observation gain in the state estimator is independent of the sample path of the observation. This makes the estimation algorithm "precomputable", an attractive attribute in many applications.

The EKF algorithm is quite adept at following motions in unhurried tracking environments such as those which occur at long range. Target motions are small relative to the field-of-view (FOV), and so, any reasonable tracking algorithm will maintain the target within the tracking window. At shorter ranges the problem becomes more complex. Perceived target motions become increasingly volatile, and a representative description of the maneuver dynamics should be incorporated into the motion model. This is commonly done by including acceleration states in the target dynamics. A conscientious representation would differentiate between distinguishable constituents of the acceleration process. On time scales of interest, the acceleration has a natural decomposition into continuous and discontinuous parts. The former is produced as a combination of a variety of parasitic effects, individually small, but nonnegligible in aggregate, and is well modeled by a white noise process. The latter is produced by operator induced motions. As the pilot of the target vehicle seeks to avoid being tracked, and to make his future location difficult to predict, he may select a timed sequence of accelerations to create a complex motion pattern. The LGM algorithm, Eqs.(3)-(6), finds such motions difficult to follow because the accelerations are outside the family of sample paths generated from $\{w_t\}$. This leads to larger than expected errors in tracking, and unacceptable errors in prediction. Jinking maneuvers thus achieve precisely the evasive intent of the pilot.

A more realistic motion model would display the maneuver acceleration explicity. Suppose that the maneuver acceleration takes on values in a finite set, say $\{a_1...a_s\}$, and let $\{\alpha_t\}$ be a maneuver indicator ; i.e., let $\alpha_t = e_i$ if the maneuver is the ith[1]. Then $\{\alpha_t\}$ evolves on $\Re^s$, and the motion equation can be written as :

$$dx_t = Ax_t dt + dw_t + \rho' d\alpha_t \qquad (7)$$

where $\{w_t\}$ is Brownian motion, and ρ links jumps in acceleration to changes of $\{\alpha_t\}$. If there were no maneuver, say $\alpha_t \equiv \alpha_0$, Eq.(7) would be identical to Eq.(1). When there is a change in maneuver acceleration, $\Delta\alpha_t \neq 0$, and this is reflected in Eq.(7).

In contrast to the earlier motion model, Eq.(7) does not have the LGM structure which yields so readily to the Kalman filter formalism. Target location is seen to be the sum of a part due to $\{w_t\}$ –gaussian– and a part due to $\{\alpha_t\}$ –certainly nongaussian. In order to use the easily implemented filter given in Eqs.(3)-(6), the dynamics represented by Eq.(7) must be wedged into the LGM framework. This has been done in various ways. Under reasonable assumptions on the dynamic character of $\{\alpha_t\}$, the power spectral density of the maneuver acceleration can be expressed as a rational fraction in ω^2, and factored so as to produce a "shaping filter" with the same output spectrum. This shaping filter and a fictitious noise excitation can be integrated into Eq.(7) by increasing the dimensions of the state space ; thus, replacing $\{\rho'\alpha_t\}$ with a Gaussian surrogate. In [4], for example, a second order shaping filter is used to model roll maneuvers. The parameters of the filter are explicit functions of the frequency with which acceleration changes take place. Indeed, in the indicated reference, the target excitation model contains only terms associated with the shaping filter.

Because the LGM model coarsely describes the maneuver, it must be carefully tuned to the application [5]. This is difficult to do, especially when the amplitudes of the maneuver accelerations are significant, but the frequency low. The intensity of the fictitious white noise must be

---

[1]Denote the unit vector in the ith direction in $\Re^s$ by e_i.

small to match the energy in the maneuvers, and the time constants must be extended if the acceleration spectrum is to be reproduced. The small size of the associated block of W leads to the computation of a small error covariance, and this translates into long time constant in the filter. Although this improves performance in nonmaneuvering regimes, response to a maneuver tends to be retarded. This deficiency is mimicked in the velocity estimate, and is compounded in any predictor based upon velocity extrapolation.

To illustrate this, consider a target moving in the plane, and being tracked with a range-bearing sensor. At time t = 20 seconds the target turns for 20 seconds before returning to constant velocity flight. In the first case, the acceleration sojourns are of 10 sec. duration, and in the second, a single acceleration lasts for the full 20 sec. The parameters of the EKF depend upon the mean sojourn time in each acceleration regime, and on the frequency of maneuvers. Using a first order shaping filter to approximate the maneuvers, Figs. 1 and 2 show sample functions of the turning rate (piecewise constant) along with estimates as generated by two EKFs each tuned to the maneuver type; the first with a time constant of 20 seconds and high gain, and the second with a time constant of 40 seconds and low gain. Although lateral acceleration is two integrals removed from the position measurement, the estimate of acceleration matches the maneuver path reasonably well. The short time constants do, however, create considerable fluctuation during periods of constant acceleration. The second filter generates much smoother estimates, but by the time the acceleration is recognized the target has already returned to its quiescent mode.

The conventional way to avoid the large errors indicated in Fig. 2. is to simply increase the noise intensity W (see [2]). This can be done either in the components of W associated with the shaping filter excitation, or more directly by increasing the position-velocity subblock of W. In either case, an artificial increase in computed P_{xx} results. Although this procedure has been rationalized as representing various modeling errors, its fundamental attraction is the fact that it increases the gain of the EKF, and thus creates a primitive conservatism in so far as the innovations are accorded increased weight. The higher gain causes the EKF to respond more expeditiously, but as seen in Fig. 1., also increases the volatility of the estimate during intervals of constant velocity flight. This latter effect is unacceptable in high accuracy tracking applications where increased velocity errors are magnified by the predictor.

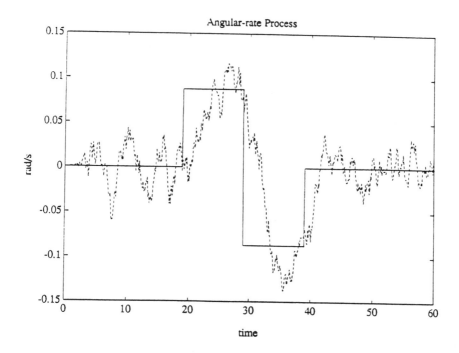

Figure 1: Turn rate estimate using a high gain EKF

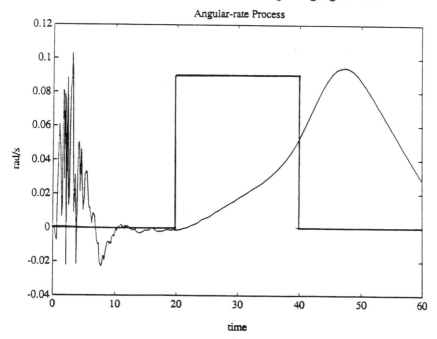

Figure 2: Turn rate estimate using a low gain EKF

3. POINT MASS MANEUVERING TARGET TRACKING

The difficulties alluded to above arise in part because the shaping filter approach so coarsely approximates the actual behavior of the maneuver process in order to embed it within the LGM framework. It would be far better to use the model given in Eq.(7) directly. It more carefully distinguishes the alternative acceleration hypotheses, and it is reasonable to suppose that the performance of an estimator retaining the individuated structure would be superior. Although Eq.(7) does not generate Gaussian sample paths, $\{x_t\}$ is conditionally Gaussian given $\{\alpha_t\}$. This fact suggests a multiple hypothesis approach to estimation ; i.e., find the conditional expectation of x_t given both $\mathcal{Y}_t$ and $\{x_t\}$, and then average over the acceleration paths. This approach uses multiple target models, and averages –or mixes– their outputs. Since there are infinitely many maneuver paths, an extreme pruning of hypotheses must be performed if the number of active alternatives is not to become overwhelming. The use of multiple models was reported in [6] and recent interest is focused on hybrid filters where a regime process describes the model jumps [7, 8, 9]

Within this class of solutions, the interacting multiple models (IMM) algorithm originally proposed by Blom [10] is a suboptimal hybrid filter that was shown to achieve the most interesting compromise between performance and complexity. This algorithm is based on the hybrid systems framework used throughout the present report (see [11, 12] for a general introduction to hybrid systems). The basic interest of hybrid models for tracking algorithms is that the occurrence of target maneuvers can be explicitly included in the kinematic equations through regime jumps. In the presence of clutter, the IMM has to be complemented to take into account the uncertainty of measurements origin and here the probabilistic data association (PDA) logic [13, 14] is chosen as an efficient solution for that aspect. Implementation issues are discussed in [10, 7, 14] with issues of filter balancing explored in [9].

When the sensors deliver point mass measurements, i.e. a non linear transform of the target position (range, azimuth, ...), the maneuver (which is physically an acceleration change) can only be detected and accomodated through a careful design of the position, velocity and acceleration chain of estimates. It is shown here that accurate models of the various kinematic regimes are then needed to obtain an improved performance : when general models are used to describe the maneuvering periods, the IMM behavior is not satisfactory, in that the innovations associated to the different models do not discriminate between the corresponding target maneuvering regimes. A more satisfactory behavior of the IMM algorithm can however be obtained by selecting accurate target motion models.

The derivation of the IMMPDA tracker (coupling the IMM algorithm and the PDA processing of multiple detections) can be found in [14] and the discussion here is limited to the choice of models to be used within the IMMPDA mechanization. As any model-based algorithm the IMM algorithm cannot perform better than the models on which it relies and in this paragraph three different sets of models to describe target maneuver scenarios are discussed. To each of these sets corresponds a different dynamic behavior and performance of the algorithm.

- Classical models : most multiple models tracking studies have used the augmentation of the target state vector has a means to capture maneuvers. The one dimensional models are then position and velocity in the absence of maneuvers and position velocity acceleration when a maneuver occurs, with kinematic relations as the state dynamics.

- Models with accurate constant rate turn equations : the exact kinematics of a mobile turning with a known constant angular rate ω in a plane leads to a state vector in $\Re^6$ with

$$x = (X_1, X_2, X_3, \dot{X}_1, \dot{X}_2, \dot{X}_3)'$$

and

$$\begin{cases} \ddot{X}_1 = -\dot{X}_2\omega \\ \ddot{X}_2 = \dot{X}_1\omega \\ \ddot{X}_3 = 0 \end{cases}$$

For a sampling period $T = 1$ the discrete-time dynamics matrix is

$$A = e^F = \begin{pmatrix} 1 & 0 & 0 & \frac{\sin(\omega)}{\omega} & \frac{\cos(\omega)-1}{\omega} & 0 \\ 0 & 1 & 0 & \frac{1-\cos(\omega)}{\omega} & \frac{\sin(\omega)}{\omega} & 0 \\ 0 & 0 & 1 & 0 & 0 & 1 \\ 0 & 0 & 0 & \cos(\omega) & -\sin(\omega) & 0 \\ 0 & 0 & 0 & \sin(\omega) & \cos(\omega) & 0 \\ 0 & 0 & 0 & 0 & 0 & 1 \end{pmatrix}$$

For $\omega > 0$ the above model describes a counterclockwise turn and its natural counterpart with $\omega < 0$ for a clockwise turn is also used. These models assume that the rotation rate of the target is known, as it is the case for a civilian aircraft because of flight rules constraints. For a military aircraft this assumption is less obvious, but pilots in combat or attack situations tend to fly at the limit of their flight envelope with maximum load factor and the corresponding approximately known ω.

- The case of unknown turning rates : the generalisation of the above models to the case where the turning rate is not known requires the addition of ω to the state vector

$$x = (X_1, X_2, X_3, \dot{X}_1, \dot{X}_2, \dot{X}_3, \omega)'$$

The continuous-time kinematics of the turn are then

$$\dot{x}_t = f(x_t)$$

with

$$f(x_t) = (\dot{X}_1, \dot{X}_2, \dot{X}_3, -\omega\dot{X}_2, \omega\dot{X}_1, 0, 0)'$$

and the sampled version becomes, up to second order in T,

$$x_{k+1} = g(x_k)$$

for

$$g(x) = \begin{pmatrix} X_1 + T\dot{X}_1 - \frac{T^2}{2}\dot{X}_2\omega \\ X_2 + T\dot{X}_2 + \frac{T^2}{2}\dot{X}_1\omega \\ X_3 + T\dot{X}_3 \\ \dot{X}_1 - T\dot{X}_2\omega - \frac{T^2}{2}\dot{X}_1\omega^2 \\ \dot{X}_2 + T\dot{X}_1\omega - \frac{T^2}{2}\dot{X}_2\omega^2 \\ \dot{X}_3 \\ \omega \end{pmatrix}$$

The IMM with the above sets of models was simulated in a Passive Air Defense scenario where the azimuth and elevation angles measured by two distributed infra-red sensors are fused to track the target in 3D. The false alarms are generated from a gaussian distribution centered on the true target position (decoys). The number of false alarms is uniformly distributed.

The different models were tested using the same trajectory, shown in Fig. 3., where the target initiates three circular turns at $t = 23s$, $t = 60s$ and $t = 90s$ with rectilinear segments connecting them. For comparison purposes, a single model Kalman tracker with a position and velocity state vector was also implemented.

- Using classical models : the error obtained is not significantly reduced compared to that obtainable using the single model filter. Both trackers show peak errors around 160m. The errors associated with the different models are comparable both on quiescent and maneuvering segments. This is contrary to the natural behavior of the IMM dynamics and the explanation is that the models are not discriminating enough. The tuning of the regime transition matrix (ie a priori information) is then important to recover the correct ordering of the regime probabilities, because the a posteriori information conveyed by the innovations conditioned on regime hypothesis are not contrasted enough.

- Using exact turn kinematics : the idea is to improve the above results by more carefully modeling the maneuvers. The three probabilities displayed on Fig. 4. show that the regimes are correctly ordered and it is seen on Fig. 5. that the dynamics of these probabilities are now significantly driven by the a posteriori information contained in the innovations : the errors of a given model are smallest during periods where it is found most likely by the IMM probabilities.

- With unknown turning rate : the second set of models provides a satisfactory performance. However, it depends on the hypothesis that the turning rate is known. When this assumption is too unrealistic ω is estimated through non linear state dynamics. Using a transition matrix with large off diagonal entries a different behavior of the IMM algorithm is obtained where no single regime is allowed to raise significantly above the others and where excellent estimates are produced through "regime mixing". This provides the best performance as shown on Fig. 6. where maneuvers are now hardly noticeable (the peak error is around 60m - compared to 160m with the single model filter - and only slightly above the errors during quiescent segments).

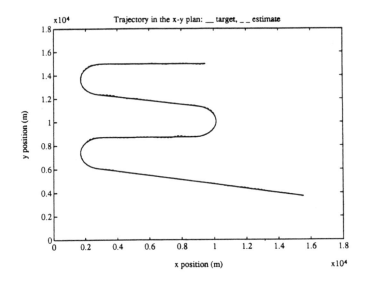

Figure 3: True and estimated trajectories (the estimate is that of the IMM with known turning rate)

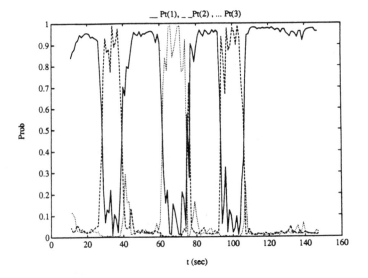

Figure 4: Regime probabilities for the IMM with known turning rate

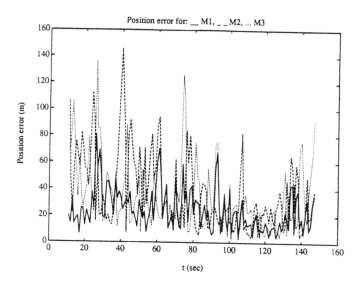

Figure 5: Model errors for the IMM with known turning rate

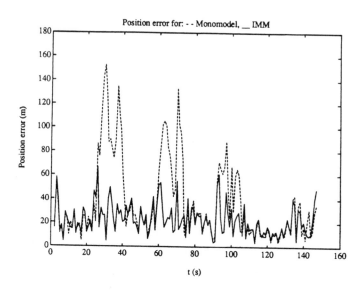

Figure 6: Position errors for IMM with unknown turning rate and the single model filter

4. IMAGE AUGMENTATION OF AN EKF

The above results concern tracking on the basis of a center-of-reflection representation of both target motion and its measurement. The IMM algorithm currently provides a satisfactory solution to maneuvering target tracking when no imaging sensor is available and it is believed that without the incorporation of image specific improvements it will be difficult to go beyond the performance of this class of solutions [9, 14]. Although Eq.(1) is frequently thought of as an intrinsic representation of the target, and Eq.(2) a product of this representation, in tracking the reverse is true. The target has many attributes, but only those related to measurable quantities are included in the motion dynamics. From this perspective, it is evident that the sensor suite provides the framework within which the motion model is phrased ; e.g., Eqs.(1),(2) are natural when the measurements are point-location properties. In some cases it is possible to augment the conventional sensor suite with an imager which creates a sequence of "pictures" of the target. Not only must the estimation architecture be modified to accommodate these new measurements, but so too must the encounter model be changed. In this Section, a study of the advantages which accrue to image enhancement is presented. It is shown that an image-adaptive EKF can be deduced with good quiescent performance and expeditious response to a maneuver. This is achieved by having a low filter gain in normal operation ; thus providing high noise rejection. Alternatively, during transient intervals, the gain is increased to reduce the response time of the filter. In the next Section the advantages of imaging sensors are further illustrated by presenting an image-based maneuver detector performing on realistic synthetic images.

Center-of-reflection tracking algorithms tend to produce unsatisfactory results in rapidly changing environments. The reason is the existence of a physical limit where the maneuver, which is an acceleration increment, is reflected in the measurements of a non imaging position sensor through two time integrations. Position related measurements can thus only be slow indicators of the maneuver, no matter what sophisticated processing is used. Any measurement which would give a more expeditious indication of a maneuver would reduce the lags inherent in current implementations. Indeed, it was proposed in one case that the tracked mobile transmit a signal to the tracker whenever it turned, thus permitting the estimator to adjust its coefficients to the changing dynamic regime [15]. Significantly improved performance was achieved with this novel data link, but the referenced application requires cooperation between target and tracker, as in Air Traffic Control. By contrast, a non cooperative target will select its motion pattern to exploit limitations in the sensor-processor architecture in an antagonistic environment. With imaging sensors, a change in target attitude is instantaneously reflected by a change at the pixels' level. The fact that the dominant portion of the acceleration is most often lateral acceleration and is modified by a change of attitude then makes it possible to use the processing of target attributes extracted from the image as fast maneuver indicators.

In [8, 16] a new tracking approach was proposed to better exploit image specific features of the sensors. An image-based measurement ; e.g., from a FLIR, was used to augment the conventional measurement architecture. For example, the center-of-area of the ostensible target pixels can be found from a single image, and a bearing angle deduced therefrom. Over time, a temporal sequence of bearing measurements is generated upon which a classical estimator/predictor can be based. This architecture, though it uses image data, differs in no essential way from that based upon a radar or similar device.

Mapping the output of an array of detectors to a single point certainly reduces the data rates in the subsequent blocks of the tracker, but it neglects much potentially useful information. At a suitable range, image information is detailed enough that spatial attributes of the target can be discerned, and used to augment or modify established estimation algorithms. These new attributes ; e.g., target type, target orientation with respect to the sensor, are only recognizable from properties of target extent ; e.g., shape, aspect ratio, etc.

Measurements of these supplementary features have a fundamentally different character than those associated with standard point-location states. Either, the nature of the thing being measured is discrete (e.g. target type), or else the image explication algorithm categorizes the feature into discrete categories (e.g. target orientation is placed within a fixed number of angular bins). In either case, the output of the image processor takes the form of a sequence of symbols from a fixed alphabet ; i.e., a spatiotemporal image sequence is converted to a temporal sequence of symbols, individual elements of which are identified with a particular status.

This unorthodox information pattern has important implications concerning the tracking architecture. Again a maneuver is produced by a sudden change in acceleration and, although the maneuver creates an immediate change in the curvature of the target path, a significant time may elapse before position deviates enough from the nonmaneuvering path to be recognizable in noise. During this post inception interval, errors in velocity estimation and position prediction build to large levels. Furthermore, these errors may remain surprisingly large after the maneuver ends. This latter effect is little discussed, but is an intrinsic property of the EKF. The time constant of the EKF is directly proportional to P_{xx}. During nonmaneuvering motion, P_{xx} is appropriately small, and the EKF does not chase phantom motions "seen" in ν_t. Subsequent to maneuver initiation, tracking error must necessarily increase as target motion deviates from that predicted on the basis of Eq.(1). Although P_{xx} remains small, ν_t increases to such a degree that sizable corrections in the state estimate take place. During this interval some components of $\{\hat{x}_t\}$ lag well behind $\{x_t\}$, particularly those components measured only indirectly ; e.g., velocity, turn rate, etc. When the maneuver ends, the target returns to constant velocity motion, and the assumptions which underlie the EKF are again true. It would reasonably be expected that the filter would perform quite well in this regime ; i.e., the realized tracking error would return to the size compatible with P_{xx}. While ultimately this is true, its realization is far slower than one would expect.

To give an example of this behavior, consider an agile target moving in the plane at constant velocity. It executes a near 180° turn and returns to constant velocity flight. The motion equation of the target can be written in the form

$$dx_t = Ax_t dt + dw_t + a_t C(k \times v_t^*)dt \qquad (8)$$

where x_t is a 4-vector containing the vehicle position (components 1 and 2) and velocity, v_t (components 3 and 4), v_t^* is a unit vector along v_t, k is a vertical unit vector, and C links the maneuver acceleration to an increment in velocity. The A matrix relates the states to their derivatives. The first two terms retain their identity from Eq.(7). The final term delineates the influence of a maneuver acceleration. It has sense and magnitude given by a_t, and its direction is perpendicular to the velocity. Suppose that there is a range-bearing sensor located at the origin with sample frequency 10 measurements/sec and standard errors 5 meters in range, and 0.25° in bearing, independent in time and type. Figure 7. shows the target path ("truth curve") along with an estimate of position generated by a conventional EKF[2]. The EKF neglects the last term in Eq.(8) and linearizes the range-bearing measurement about the state estimate. This provides the basic nonimaging estimate of target location.

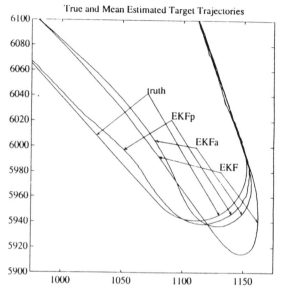

Figure 7: Tracking performance of the filters (m)

The sample path of the target shown in the figure is propagated with no wide band noise ; i.e., $W = 0$. If this fact were properly included in the calculation of P_{xx} the EKF would diverge. This follows from the fact that the solution to Eq.(6) converges to zero when $W = 0$, and Eq.(3) diverges with zero gain. The filter can be "stabilized" by including a pseudo-noise as recommended in [2]. The EKF shown in Fig. 7. is designed as if there were a small wideband acceleration in each velocity component of intensity : $(dw)_i^2 = (0.1)dt$. The bogus acceleration stabilizes the EKF, but with large error both during and after a maneuver.

As would be expected, position error grows quite rapidly during the initial phases of the maneuver, and returns lethargically to its quiescent level after the maneuver is complete. The slow decay of the tracking error stems from the unresponsiveness of P_{xx} to target motion. To see why this is so, note that large residuals are produced by a maneuver, and the state of the EKF is moved into an inappropriate region of the state space. This effect is most pronounced in the ancillary states. The

---

[2]The figures in this Section are based on simulations of R.G. Hutchins.

time history of the velocity errors of the EKF is shown in Fig. 8. for the indicated scenario. After the maneuver ends, the accumulated velocity errors are eliminated by the EKF, but the duration is long because the time constant of the EKF is large. This example illustrates a deficiency which also arises in estimators which compensate for maneuvers by first identifying a change in acceleration, usually from the position residuals, and then adding the estimated acceleration to the null-maneuver filter. Such an approach does not account for the deweighting of old residuals which must take place if the obsolete data is not to have undue influence on current estimates.

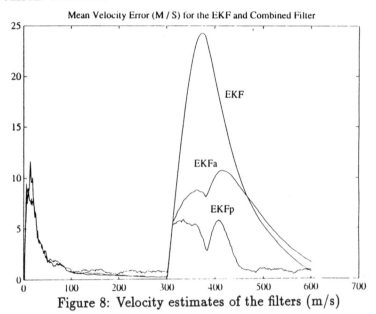

Figure 8: Velocity estimates of the filters (m/s)

It was suggested in [16] and [17] that the estimation architecture be modified to avail itself of a parallel image explication path. To illustrate the impact of an imager, consider the same tracker configuration along with the addition of a collocated imager. The scene created by the imager is interpreted by an image processor which interprets the raw signal from the detector array, and classifies target orientation within a fixed set of L alternative angular bins. Observe that the signature of an orientation hypothesis is not manifest in an individual pixel, but it is rather contained in the gestalt of the image. Suppose that corresponding to the ith orientation hypothesis, the image processor transmits a symbol u_i. Symbols will be assumed to arrive randomly at the tracker at the frame rate λ symbols/second and with exponential interarrival times.

If the symbols were infallible indications of condition, the tracker's task would be eased. Unfortunately, there are errors in image creation (clutter or obscuration), and in image interpretation (misidentifying target edges). It will be supposed that the fidelity in processing a particular image can be quantified in terms of a discernibility matrix $P = [P_{ij}]$ where P_{ij} is the probability that symbol i will be received if j is the target orientation at time of image creation. Let ρ_t be an indicator of the true orientation bin. Then

$$P_{ij} = \wp(u_i | \rho = e_j) \tag{9}$$

The image portion of the observation process can be written in terms of the L-vector counting process $\{z_t\}$ where the event $\Delta z_i = 1$ signifies the occurence of the ith symbol at time t.

Let $(\Omega, F, \wp)$ and filtration $\{\mathcal{F}_t\}$ be the basic probability space upon which the encounter is defined. Denote by $\{\mathcal{Z}_t\}$ the subfiltration generated by $\{z_t\}$. Note that $E\{dz_i | \mathcal{F}_t\} = \sum_k \lambda P_{ik} \rho_k dt$; i.e., the mean rate of receipt of the ith orientation symbol is proportional to the frame rate and to P_{ik} if $\rho_t = e_k$. If the discernibility matrix were the identity matrix, and if $\rho_t \equiv e_i$ then $E\{dz_i | \mathcal{F}_t\} = \lambda$ and $E\{dz_j | \mathcal{F}_t\} = 0$ for $j \neq i$; i.e., the status of the orientation condition would be reconfirmed roughly every λ^{-1} seconds. In applications, badly cluttered images may be received and misinterpreted, and this will create ambiguity in the tracker. Does a change in the received symbol truly signify a change in orientation, or is it merely an erroneous transmission, or an erroneous interpretation of the transmission ? The weight attached to a specific observation depends upon the perceived fidelity of the source, the clutter in the transmission and the temporal dynamics of the encounter. All of these influences are simultaneously weighed by the tracker in arriving at conclusions regarding the current status of the encounter.

The image-observation sequence can be written as an s-vector with drift $E\{dz | \mathcal{F}_t\} = \lambda P \rho dt$. Define

$$h' = \lambda P \tag{10}$$

with $h_{ij} = \lambda P_{ij}$. The image-observation $\{z_t\}$ has two different decompositions

$$dz = h' \hat{\rho}_t dt + (dz - h' \hat{\rho}_t dt) = h' \rho_t dt + (dz - h' \rho_t dt) \tag{11}$$

where

$$(dz - h'\rho_t dt) \quad \text{is} \quad \text{an} \quad \mathcal{F}_t - \text{mg}$$

and   (12)

$$(dz - h'\hat{\rho}_t dt) \quad \text{is} \quad \text{an} \quad \mathcal{Z}_t - \text{mg}.$$

Denote the martingales in Eq.(13) by

$$dz - h'\rho_t dt = dn_\rho \tag{13}$$

and

$$dz - h'\hat{\rho}_t dt = dv_\rho \tag{14}$$

where $\{dv_\rho\}$ is the image-innovations process, and $\{dn_\rho\}$ is the image-observation noise. For convenience let $\phi = \alpha \otimes \rho$. With a slight abuse of notation, Eq.(14) can be written

$$dz_t = h'\phi_t dt + dn_\rho \tag{15}$$

This equation has a form that is identical with Eq.(2) although it is written in $\Re^M$ with $M = L + s$. However, the $\{dn_\rho\}$ process has properties which distinguish it from the like term in that equation. While both are $\mathcal{F}_t$-martingales, unlike $\{n_x\}$, the quadratic variation of the noise in Eq.(15) is random

$$(dn_\rho)dn_\rho' = dz dz' \tag{16}$$

While in Eq.(2), the observation noise is Brownian motion, in Eq.(15) the structure of the noise is more complex. It can be shown that

$$E\{(dn_\rho)dn_\rho'|\mathcal{F}_t\} = diag(h'\phi_t)dt \tag{17}$$

$$E\{(dn_\rho)dn_\rho'|\mathcal{Z}_t\} = diag(h'\hat{\phi}_t)dt = R_\phi dt \tag{18}$$

Note that R_ϕ is the $\mathcal{Z}_t$-intensity of an effective observation noise. The system will be assumed to be such that $R_\phi > 0$, which is achieved if $P_{ik} > 0$ for some k and every i, and it is perforce diagonal.

Equations (2) and (15) provide complementary observations of the same object. Each observation generates an associated filtration, $\mathcal{Y}_t$ for

Eq.(2) and $\mathcal{Z}_t$ for Eq.(15). In the tracker, the two processes must be merged to form the estimate. Let $g_t = (y_t, z_t)$ and $G_t = \mathcal{Y}_t \vee \mathcal{Z}_t$.

Tracking entails the synthesis of an algorithm which generates the G_t-predictable mean of $\{x_t\}$. Although the observation processes are cogenerators of $\{G_t\}$, they have distinguishable roles in estimation. Current design practice maintains the identity of the two data paths in the tracking architecture. The motivation for this parallelism lies in the fundamentally different nature of the target qualities measured by the point-location sensor on the one hand, and the imager on the other. The attributes captured in Eq.(15) are frequently only weakly linked to $\{\mathcal{Y}_t\}$; e.g., when direct measurements exist, target orientation is indirectly and peripherally related to measurements of target location. When $\{\mathcal{Y}_t\}$ is the only available data link, motion inferences must necessarily be made on its basis. However, the path generating $\{\mathcal{Z}_t\}$ provides, in principle, the means for reducing the detection delays inherent in conventional maneuver detection algorithms.

A lateral acceleration is manifested first in changes in angular orientation, and thus a turn can be determined more expeditiously from image data than it can from the trajectory in the plane. Indeed, if the range-bearing data is noisy, turn rate estimates can be made exclusively from image data with little loss in accuracy ;

$$\hat{\phi}_t = E\{\phi_t | G_t\} \cong E\{\phi_t | \mathcal{Z}_t\} \tag{19}$$

That is, there is little loss in fidelity if evaluation of the maneuver hypothesis is based exclusively upon the image-observation. For the case where $\{a_t\}$ is a Markov process, the maneuver estimation equations are described in detail in [8], and take the form :

$$\hat{a}_t = \sum_i a_i \hat{\alpha}_t$$

subject to

$$\hat{\alpha}_t = (I_s \otimes 1'_L)\hat{\phi}_t \tag{20}$$

and

$$d\hat{\phi}_t = Q'\hat{\phi}_t dt$$
$$+ (diag(\hat{\phi}_t) - \hat{\phi}_t\hat{\phi}'_t)(1'_s \otimes I_L)\Lambda' diag(\Lambda(1_s \otimes I'_L)\hat{\phi}_t)^{-1} dz_t \tag{21}$$

where Q is dependent on the maneuver and orientation dynamics of the engagement, $\Lambda = \lambda P, I$ is the identity matrix, and 1 is a vector of ones. The augmented EKF (EKFa) discussed in [17] simply adds the mean maneuver acceleration $\{\hat{a}_t\}$ to the standard EKF.

Figure 7. shows the path generated by EKFa for the simple planar maneuver. Compared with the EKF, tracking is much improved with EKFa during the turn, but large errors remain after the maneuver ends. This is due in large part to the long time constant of both filters resulting from the small value of W. The maximum position error is reduced by adding the estimate of acceleration from 88 m. to 52 m (see Fig. 9.) but the decay is slow. A similar behavior is seen in the velocity trajectory.

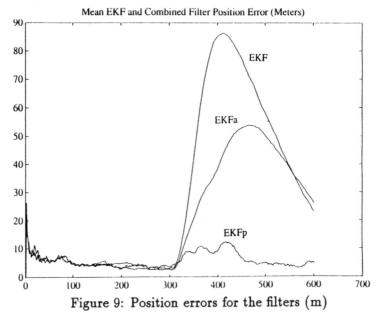

Figure 9: Position errors for the filters (m)

When the maneuver ends, the acceleration-augmented filter behaves much like the EKF. Although this behavior is perhaps surprising because in the nonmaneuvering regime, the LGM model correctly describes target motion, this slow decay in the error is due to the failure of either filter to properly deweight the large residuals generated during the maneuver. This deficiency could have been avoided if the time constant of the EKF were decreased after changes in $\{a_t\}$. Although various approaches to doing this have been proposed the proper implementation is controversial. At one level, adding pseudonoise through W reduce the filter time constant, and improves maneuver tracking, but makes nominal operation more volatile. Another option was proposed by Bekir in [18]. He suggested that $\{a_t\}$ be estimated, and the elements of the W matrix be augmented proportionately "to the amount of acceleration along each axis". In this way, "large or unnecessary" W augmentation is avoided during nominal operation, and is reserved for times of suspected maneuvers.

It is actually the uncertainty in maneuver estimation that induces a need to change the EKF dynamics, and the requisite information is readily available from $\{\widehat{\alpha}_t\}$. The maneuver variance is given by :

$$Var a_t = E\{(a_t - \hat{a}_t)^2 | \mathcal{Z}_t\}^2 = \sum_i a_i^2 \widehat{\alpha}_i - (\sum_i a_i \widehat{\alpha}_i)^2 \qquad (22)$$

Equation (22) has an interesting behavior pattern. When the maneuver is resolvable from the image sequence ($\widehat{\alpha}_t \cong e_i$), the acceleration uncertainty is small.

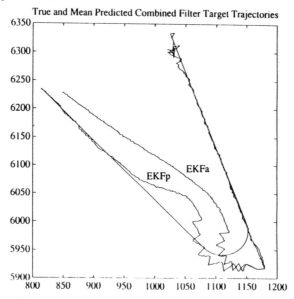

Figure 10: Prediction errors for the filters (m)

Alternatively, as the a posteriori probabilities of the maneuver hypotheses become more diffuse, Var $\{a_t\}$ grows as well. This is precisely the sort of situational adaptation required in the application. The process noise intensity W can be augmented proportionally to acceleration uncertainty (rather than acceleration magnitude), transformed into the cartesian coordinate system [19]. The corresponding value of P_{xx} is the solution to Eq.(6) with the increased W. Through this simple artifice, a maneuver adaptive tracker is created. Not only is the mean acceleration added as a bias, but the tracking time constant is modified as well. Denote the corresponding filter by EKFp. Tracking is much improved with covariance augmentation (Fig. 7.). Not only is the maximum velocity error reduced with this filter, but the decay to quiescent performance is much faster (Fig. 8.).

Accurate velocity estimates are of great importance in prediction because the effect of artifacts in the residuals tends to accumulate in velocity without any direct measurement to disabuse the filter of erroneous inferences. Covariance augmentation reduces velocity errors significantly, and covariance adaptation speeds the final velocity correction –note that EKFa still has a significant position bias at the end of the sample engagement. For this scenario, the peak 5-second prediction error in position generated by the EKF is 190 meters, with 25 meters remaining after 300 sec. The EKFa reduces the peak to 110 meters, but has about the same residual error after 300 sec. The adaptive EKFp has a 75 meters peak, but only an 8 meters prediction error after 150 sec. (see Fig. 10.).

There is of course some penalty to be paid for the improved performance during and after a maneuver. The EKF is derived on the basis of a model best tuned to the nonmaneuvering target, and consequently it wastes no effort on tracking putative maneuvers. This complacency leads to excellent nominal prediction performance (5 meters error). The adaptive filter "sees" more maneuvers than actually occur. Nevertheless the improved performance during a turn suggests that this modification will be useful in volatile encounters.

5. IMAGE BASED MANEUVER DETECTION

To further illustrate the impact of imaging sensors on tracking systems, the feasibility of image based target maneuver detection is demonstrated in this Section. A distinctive feature here is that compared to previous reports (e.g.[8, 20, 21, 22, 23]) images are actually generated, processed to extract attributes and the attributes are post-processed to detect maneuvers. Realistic synthetic images are simulated in the infrared spectral band where a CAD facet model of the target provides the scene geometry and the radiometries are computed through atmospheric trasmission from a temperature knowledge base. The target maneuvers are incorporated in the simulation with accurate flight dynamics.

Detecting a maneuver using image attributes is of course a special case of change detection in signals and systems and many approaches are available as witnessed by the survey [24] or the monograph [25]. The proposed detector is however specific to the imaging sensor in that it processes image attributes that are intrinsically integers (e.g. the number of pixels) without undue approximation by real numbers. Contrary to the analysis in the previous sections, the filter is derived in discrete time to espouse a digital image processor implementation.

The evolution of the target attributes is built from the superposition of two trends :

- the benign evolution of, say, the apparent size of the target due to the continuous evolution of the sensor to target range. Assuming that the range is known this influence can easily be described by a simple parametric model of the image formation process. After estimating the parameters of this model on a set of test images, the attributes can be normalized to compensate for the evolution that is not of direct interest.

- a more discontinuous component signalling a change of the aspect of the target ; e.g. a sudden change of roll angle at the onset of a maneuver. This is of course the interesting evolution to design a maneuver detection algorithm.

Figures 11 and 12 presents the evolution of two target attributes (the target width and number of pixels) under the no maneuver and maneuver hypothesis (the compensation for the range influence was obtained through a least square fit of a second order parametric model).

Assuming that there are M possible regimes, the indicator $\phi_t \in \Re^M$ is described as above by a Markov chain with

$$q_{ij} = \wp\{\phi_{t+1j} = 1 | \phi_{ti} = 1\}$$

The incorporation of the a priori information contained in $Q = (q_{ij})$ is a main advantage of hybrid systems based algorithm that makes them less prone to the so-called "oblivious detection" difficulty : the a priori jump probabilities captured in the regime dynamics make hybrid algorithms alert to detect regime changes whereas classical solutions are biased toward the no maneuver hypothesis after long quiescent periods.

Denoting by $\mathcal{R}_t$ and $\mathcal{N}_t$ the filtrations generated by the regime and observation (see below) process, and by $\mathcal{F}_t$ the joint σ-algebra ($\mathcal{F}_t =$

$\mathcal{R}_t \vee \mathcal{N}_{t-1}$), the regime dynamics can be written as

$$\phi_{t+1} = Q'\phi_t + \mu_t \tag{23}$$

with $\mu_t = \phi_{t+1} - E\{\phi_{t+1}|\mathcal{F}_t\}$ (note that μ_t is a $\mathcal{F}_t$-martingale).

The measurement process is rather peculiar in that the measured variable is an integer (typically the number of pixels related to the target extracted by some image processing). It is assumed that this integer attribute, noted n_t, can be linearly related to the actual regime :

$$n_t = H\phi_t + b_t \tag{24}$$

where $b_t = n_t - E\{n_t|\mathcal{F}_{t-1}\}$ is a $\mathcal{F}_{t-1}$-martingale the intensity of which is noted $E\{b_t b'_t|\mathcal{F}_{t-1}\} = B_t$. The H matrix contains the range normalization effect and, in some cases, such a linear relationship will have to be replaced by a non linear equation. This extension is not discussed here, see [26].

The innovation process associated with this measure is introduced as

$$\bar{n}_t = n_t - E\{n_t|\mathcal{N}_{t-1}\} \tag{25}$$

Because $\mathcal{N} \subset \mathcal{F}$ any $\mathcal{F}-$ martingale is also a $\mathcal{N}-$ martingale. From the martingale representation theorem (see e.g. [27, 28]) every $\mathcal{N}$ - martingale $\{\nu_t\}$ can be represented in terms of the innovation process $\bar{n}$ as

$$\nu_t = G_t \bar{n}_t \tag{26}$$

where the gain G_t is computed from the conditional covariances :

$$G_t = E\{\nu_t \bar{n}'_t|\mathcal{N}_{t-1}\} E\{\bar{n}_t \bar{n}'_t|\mathcal{N}_{t-1}\}^{-1} \tag{27}$$

The extrapolation of $\{\phi\}$ is naturally obtained as

$$\hat{\phi}_{t|t-1} = Q'\hat{\phi}_{t-1|t-1} \tag{28}$$

because of the martingale property of $\{\mu\}$. The gain computation is separated in two steps. First $E\{\bar{n}_t \bar{n}'_t|\mathcal{F}_{t-1}\}$ is computed :

$$E\{\bar{n}_t \bar{n}'_t|\mathcal{F}_{t-1}\} = E\{(n_t - H\hat{\phi}_{t|t-1})(n_t - H\hat{\phi}_{t|t-1})'|\mathcal{F}_{t-1}\} \tag{29}$$

but

$$n_t n'_t = H\phi_t \phi'_t H' + b_t b'_t + H\phi_t b'_t + b_t \phi'_t H' \tag{30}$$

Because ϕ is an indicator variable it is true that

$$\phi_t \phi_t' = diag(\phi_t) = diag(\phi_{t1}, \phi_{t2}, ..., \phi_{tM})$$

and finally

$$E\{\tilde{n}_t \tilde{n}_t' | \mathcal{F}_{t-1}\} = H diag(\hat{\phi}_t) H' + B_t - H\hat{\phi}_{t|t-1}\hat{\phi}_{t|t-1}' H' \qquad (31)$$

The gain numerator is $E\{\mu_{2t}\tilde{n}_t' | \mathcal{F}_{t-1}\}$:

$$
\begin{aligned}
E\{\mu_{2t}\tilde{n}_t' | \mathcal{F}_{t-1}\} &= E\{(\phi_t - \hat{\phi}_{t|t-1})(H\phi_t - H\hat{\phi}_{t|t-1})' | \mathcal{F}_{t-1}\} \\
&= E\{(\phi_t \phi_t' H' - \phi_t \hat{\phi}_{t|t-1})' H' | \mathcal{F}_{t-1}\} \qquad (32) \\
&= [diag(\hat{\phi}_{t|t-1}) - \hat{\phi}_{t|t-1}\hat{\phi}_{t|t-1}'] H'
\end{aligned}
$$

so that the gain can finally be expressed as a function of the regime estimate :

$$
\begin{aligned}
G_t &= [diag(\hat{\phi}_{t|t-1}) - \hat{\phi}_{t|t-1}\hat{\phi}_{t|t-1}'] H' \times \qquad (33) \\
&\quad [B_t + H diag(\hat{\phi}_{t|t-1}) H' - H\hat{\phi}_{t|t-1}\hat{\phi}_{t|t-1}' H']^{-1}
\end{aligned}
$$

To summarize, the regime estimate is propagated with :

$$\hat{\phi}_{t|t-1} = Q'\hat{\phi}_{t-1|t-1} \qquad (34)$$

$$\hat{\phi}_{t|t} = \hat{\phi}_{t|t-1} + G_t(n_t - H\hat{\phi}_{t|t-1}) \qquad (35)$$

with G_t as above.

A nice property of this filter is that it provides a measure of its own state of uncertainty at no extra cost. This is because the covariance matrix of the regime estimation error can be computed as

$$Cov(\tilde{\phi}_{t|t-1}) = diag(\hat{\phi}_{t|t-1}) - \hat{\phi}_{t|t-1}\hat{\phi}_{t|t-1}' \qquad (36)$$

This is very important in applications when this estimate is fed in a tracking loop where a high uncertainty should lead to some form of cautionnary feedback.

The above hybrid detector was used in an aerial target tracking scenario with $M = 2$ (two regimes) and $\phi_{t1} = 1$ in a no maneuver situation and $\phi_{t2} = 1$ when a maneuver occurs. Using the width attribute shown in Fig.12. the detector Eqs.(34)-(34)-(35) was simulated to obtain the $\hat{\phi}_{t2}$ behavior depicted in Fig. 13.

It appears that the maneuver can be reliably detected as soon as $t \sim 230$ based on comparing $\hat{\phi}_{t2}$ to a treshold or on a MAP regime decision (the true maneuver onset time is $t \sim 200$).

To give an idea of the encounter, Fig. 14. displays the CAD model of the target viewed by the sensor just prior to the maneuver onset and the corresponding infrared data. The same piece of information just after the maneuver is presented on Fig. 15.

As expected, it appears that the maneuver is signalled by the change of roll angle detected through the attributes extracted from the infrared image. The performance of the detector is apparent in that the aspect of the target on the infrared portion of Fig.15 is only slightly different from that on Fig.14. It is stressed that this encouraging sensitivity is obtained with a realistic simulation incorporating antenna error models and image creation disturbances.

At the time where the image specific filter detects the maneuver the position related measurement (here some target center-of-area) has not reflected the acceleration onset in any significant amount. This confirms the basic motivation that fast maneuver indicators are related to the attitude (and image derived) portion of the measurements while the position related information provides only slowlier evidence.

The next step of this research will be to couple the above maneuver detector to an IMM tracker where the image based regime estimate will be used to directly drive the IMM reconfiguration.

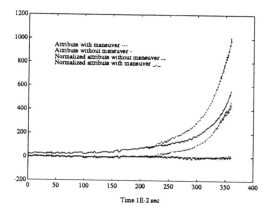

Figure 11: Number of pixels

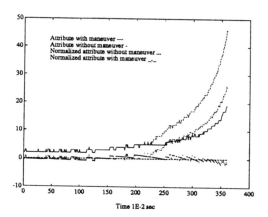

Figure 12: Width

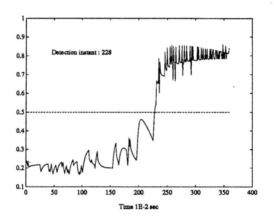

Figure 13: A posteriori maneuver probability ($\hat{\phi}_{t2}$

Figure 14: IR Image   CAD Image (before the maneuver)

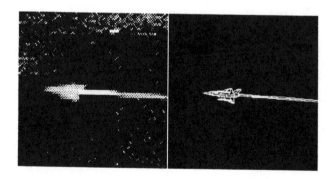

Figure 15: IR Image   CAD Image (at the detection instant)

ACKNOWLEDGEMENTS

This research was supported by DRET (DGA, Paris) under contract no.89/357 (M. Mariton), by NATO under collaborative research grant no. 890885 (M. Mariton and D.D. Sworder) and by the MICRO Program of the State of California under Project Number 89-021 (D.D. Sworder).

REFERENCES

References

[1] C.B. Chang and J.A. Tabaczynski, "Application of State Estimation to Target Tracking," *IEEE Trans. Automatic Control*, Vol.AC-26, 98-109 (1984).

[2] P.S. Maybeck, "Stochastic Models Estimation and Control," Academic Press, San Diego, 1979.

[3] R.G. Elliott, "Stochastic Calculus and Applications," Springer Verlag, New York, 1982.

[4] I. Rusnak and L. Meir, "Optimal Guidance for Accelerating Targets," *IEEE Transactions on Aerospace and Electronics Systems*, Vol.AES-26, 618-624 (1990).

[5] J.R. Cloutier, J.H. Evers and J.J. Feeler, "Assessment of Air to Ait Missile Guidance and Control Technology,", *IEEE Control Systems magazine*, 27-34 (1989).

[6] P.S. Maybeck and R.I. Suizu, "Adaptive tracker Fieldof View Variation Via Multiple Model Filtering," *IEEE Transactions on Aerospace and Electronics Systems*, Vol.AES-21, 529-539 (1985).

[7] H.A.P. Blom and Y. Bar-Shalom, "The Interacting Multiple Model Algorithm for Systems with martkovian Switching Coefficients", *IEEE Trans. Automatic Comtrol*, Vol.AC-33, 780-783 (1988).

[8] D.D. Sworder and G. Hutchins, "Maneuver Estimation Using Measuremnts of Orientation," *IEEE Transactions on Aerospace and Electronics Systems*, Vol.AES-6, 625-638 (1990).

[9] F. Dufour and M. Mariton, "Tracking a 3D Maneuvering Target with Passive Sensors," to appear *IEEE Transactions on Aerospace and Electronics Systems* (1991).

[10] H.A.P.Blom, "A sophisticated tracking algorithm for ATC surveillance data," in Proceedings of the International Radar Conference, Paris, France (1984).

[11] D.D. Sworder, "Control of Systems Subject to Sudden Changes in Character," *Proc. IEEE*, Vol.64, 1219-1225 (1976).

[12] M. Mariton, "Jump Linear Systems in Automatic Control," M. Dekker Inc., New York, 1990.

[13] Y. Bar-Shalom and T.E. Fortmann, "Tracking and Data Association, Academic Press, San Diego, 1988.

[14] A. Houles and Y. Bar-Shalom, "Multisensor Tracking of a Maneuvering Target in Clutter," *IEEE Transactions on Aerospace and Electronics Systems*, Vol.AES-25, 176-188 (1989).

[15] C.C. Lefas, "Using Roll Angle Measurements to Track Aircraft Maneuvers," *IEEE Transactions on Aerospace and Electronics Systems*, Vol.AES-20, 671-681 (1984).

[16] G. Hutchins and D.D. Sworder , "Image Fusion Algorithms for Tracking Maneuvering Targets," *AIAA Journal of Guidance Control and Dynamics*, (to appear 1991).

[17] D.D. Sworder and P.F. Singer, "The Role of Image Interpretation in Tracking and Guidance," *in* "Control and Dynamic Systems : Advances in Theory and Applications," Vol.38, (C.T. Leondes, ed.), 625-638, Academic Press, San Diego, 1988.

[18] E. Bekir, "Adaptive Kalman Filter for Tracking Maneuvering Targets,", *AIAA Journal of Guidance Control and Dynamics*, Vol.6, 414-416 (1983).

[19] D.D. Sworder and G. Hutchins, "Improved Tracking of an Agile Target," submitted to *AIAA Journal of Guidance Control and Dynamics* (1991).

[20] D.D. Sworder and G. Hutchins, "Image Enhanced Tracking," *IEEE Transactions on Aerospace and Electronics Systems*, Vol.AES-25, 701-709 (1990).

[21] D. Andrisani, F.P. Kuhl and D. Gleason, "A Non Linear Tracker using Attitude Measurements," *IEEE Trans. Aerospace and Electronic Systems*, Vol.AES-22, 533-539 (1986).

[22] H.H. Burke, "Circular Arc Aimed Munition (CAAM) : a Concept for Improving Gun Fire Prediction Systems," Proc. 21st Asilomar Conf. Signals Systems and Computers, Pacific Grove, 254-260 (1988).

[23] J.D. Kendrick, P.S. Maybeck and J.G. Reid, "Estimation of Aircraft Target Motion Using Orientation Measurements," *IEEE Transactions on Aerospace and Electronics Systems*, Vol. AES-17, no.2, 254-260 (1981).

[24] A.S. Willsky, "A Survey of Design Methods for Failure Detection in Dynamic Systems," *Automatica*, Vol.12, 601-611 (1976).

[25] M. Basseville and A. Benveniste, Eds., "Detection of Abrupt Changes in Signals and Systems," Lecture Notes in Control and Information Sciences, Vol.77, Springer Verlag, Berlin, 1986.

[26] D. Laneuville and M. Mariton, "Image Based Target Maneuver Detection," submitted to the IEEE 30th Conference Decision and Control (1991).

[27] C.S. Chou and P.A. Meyer, "Sur la Représentation des Martingales comme Intégrales Stochastiques dans les Processus Ponctuels," Séminaire de Probabilités, vol IX, Lecture Notes in Mathematics, Springer Verlag, Berlin, 226-236 (1975).

[28] A. Segall, "Stochastic Processes in Estimation Theory," *IEEE Trans. Information Theory*, Vol. IR-22, 275-286 (1976).

INDEX